Use R!

Series Editors:
Robert Gentleman Kurt Hornik Giovanni Parmigiani

Use R!

Albert: Bayesian Computation with R
Bivand/Pebesma/Gómez-Rubio: Applied Spatial Data Analysis with R
Cook/Swayne: Interactive and Dynamic Graphics for Data Analysis: With R and GGobi
Hahne/Huber/Gentleman/Falcon: Bioconductor Case Studies
Paradis: Analysis of Phylogenetics and Evolution with R
Pfaff: Analysis of Integrated and Cointegrated Time Series with R
Sarkar: Lattice: Multivariate Data Visualization with R
Spector: Data Manipulation with R

Roger S. Bivand • Edzer J. Pebesma
Virgilio Gómez-Rubio

Applied Spatial Data Analysis with R

Roger S. Bivand
Norwegian School of Economics
and Business Administration
Breiviksveien 40
5045 Bergen
Norway

Virgilio Gómez-Rubio
Department of Epidemiology
and Public Health
Imperial College London
St. Mary's Campus
Norfolk Place
London W2 1PG
United Kingdom

Edzer J. Pebesma
University of Utrecht
Department of Physical Geography
3508 TC Utrecht
Netherlands

Series Editors:

Robert Gentleman
Program in Computational Biology
Division of Public Health Sciences
Fred Hutchinson Cancer Research Center
1100 Fairview Ave. N, M2-B876
Seattle, Washington 98109-1024
USA

Kurt Hornik
Department für Statistik und Mathematik
Wirtschaftsuniversität Wien Augasse 2-6
A-1090 Wien
Austria

Giovanni Parmigiani
The Sidney Kimmel Comprehensive Cancer
Center at Johns Hopkins University
550 North Broadway
Baltimore, MD 21205-2011
USA

ISBN 978-0-387-78170-9 e-ISBN 978-0-387-78171-6
DOI 10.1007/978-0-387-78171-6

Library of Congress Control Number: 2008931196

Printed on acid-free paper

springer.com

Ewie

Voor Ellen, Ulla en Mandus

A mis padres, Victorina y Virgilio Benigno

Preface

We began writing this book in parallel with developing software for handling and analysing spatial data with R (R Development Core Team, 2008). Although the book is now complete, software development will continue, in the R community fashion, of rich and satisfying interaction with users around the world, of rapid releases to resolve problems, and of the usual joys and frustrations of getting things done. There is little doubt that without pressure from users, the development of R would not have reached its present scale, and the same applies to analysing spatial data analysis with R.

It would, however, not be sufficient to describe the development of the R project mainly in terms of narrowly defined utility. In addition to being a community project concerned with the development of world-class data analysis software implementations, it promotes specific choices with regard to how data analysis is carried out. R is open source not only because open source software development, including the dynamics of broad and inclusive user and developer communities, is arguably an attractive and successful development model.

R is also, or perhaps chiefly, open source because the analysis of empirical and simulated data in science should be reproducible. As working researchers, we are all too aware of the possibility of reaching inappropriate conclusions in good faith because of user error or misjudgement. When the results of research really matter, as in public health, in climate change, and in many other fields involving spatial data, good research practice dictates that someone else should be, at least in principle, able to check the results. Open source software means that the methods used can, if required, be audited, and journalling working sessions can ensure that we have a record of what we actually did, not what we thought we did. Further, using `Sweave`[1] – a tool that permits the embedding of R code for complete data analyses in documents – throughout this book has provided crucial support (Leisch, 2002; Leisch and Rossini, 2003).

[1] `http://www.statistik.lmu.de/~leisch/Sweave/`.

We acknowledge our debt to the members of R-core for their continuing commitment to the R project. In particular, the leadership and example of Professor Brian Ripley has been important to us, although our admitted 'muddling through' contrasts with his peerless attention to detail. His interested support at the Distributed Statistical Computing conference in Vienna in 2003 helped us to see that encouraging spatial data analysis in R was a project worth pursuing. Kurt Hornik's dedication to keep the Comprehensive R Archive Network running smoothly, providing package maintainers with superb, almost 24/7, service, and his dry humour when we blunder, have meant that the useR community is provided with contributed software in an unequalled fashion. We are also grateful to Martin Mächler for his help in setting up and hosting the R-Sig-Geo mailing list, without which we would have not had a channel for fostering the R spatial community.

We also owe a great debt to users participating in discussions on the mailing list, sometimes for specific suggestions, often for fruitful questions, and occasionally for perceptive bug reports or contributions. Other users contact us directly, again with valuable input that leads both to a better understanding on our part of their research realities and to the improvement of the software involved. Finally, participants at R spatial courses, workshops, and tutorials have been patient and constructive.

We are also indebted to colleagues who have contributed to improving the final manuscript by commenting on earlier drafts and pointing out better procedures to follow in some examples. In particular, we would like to mention Juanjo Abellán, Nicky Best, Peter J. Diggle, Paul Hiemstra, Rebeca Ramis, Paulo J. Ribeiro Jr., Barry Rowlingson, and Jon O. Skøien. We are also grateful to colleagues for agreeing to our use of their data sets. Support from Luc Anselin has been important over a long period, including a very fruitful CSISS workshop in Santa Barbara in 2002. Work by colleagues, such as the first book known to us on using R for spatial data analysis (Kopczewska, 2006), provided further incentives both to simplify the software and complete its description. Without John Kimmel's patient encouragement, it is unlikely that we would have finished this book.

Even though we have benefitted from the help and advice of so many people, there are bound to be things we have not yet grasped – so remaining mistakes and omissions remain our sole responsibility. We would be grateful for messages pointing out errors in this book; errata will be posted on the book website (`http://www.asdar-book.org`).

Bergen
Münster
London
April 2008

Roger S. Bivand
Edzer J. Pebesma
Virgilio Gómez-Rubio

Contents

Part II Analysing Spatial Data

1

Hello World: Introducing Spatial Data

1.1 Applied Spatial Data Analysis

Spatial data are everywhere. Besides those we collect ourselves ('is it raining?'), they confront us on television, in newspapers, on route planners, on computer screens, and on plain paper maps. Making a map that is suited to its purpose and does not distort the underlying data unnecessarily is not easy. Beyond creating and viewing maps, spatial data *analysis* is concerned with questions not directly answered by looking at the data themselves. These questions refer to hypothetical processes that generate the observed data. Statistical inference for such spatial processes is often challenging, but is necessary when we try to draw conclusions about questions that interest us.

Possible questions that may arise include the following:

- Does the spatial patterning of disease incidences give rise to the conclusion that they are clustered, and if so, are the clusters found related to factors such as age, relative poverty, or pollution sources?
- Given a number of observed soil samples, which part of a study area is polluted?
- Given scattered air quality measurements, how many people are exposed to high levels of black smoke or particulate matter (e.g. PM_{10}),[1] and where do they live?
- Do governments tend to compare their policies with those of their neighbours, or do they behave independently?

In this book we will be concerned with *applied* spatial data analysis, meaning that we will deal with data sets, explain the problems they confront us with, and show how we can attempt to reach a conclusion. This book will refer to the theoretical background of methods and models for data analysis, but emphasise hands-on, do-it-yourself examples using R; readers needing this background should consult the references. All data sets used in this book and all examples given are available, and interested readers will be able to reproduce them.

[1] Particulate matter smaller than about 10 µm.

In this chapter we discuss the following:

(i) Why we use R for analysing spatial data
(ii) The relation between R and geographical information systems (GIS)
(iii) What spatial data are, and the types of spatial data we distinguish
(iv) The challenges posed by their storage and display
(v) The analysis of observed spatial data in relation to processes thought to have generated them
(vi) Sources of information about the use of R for spatial data analysis and the structure of the book.

1.2 Why Do We Use R

1.2.1 ... In General?

The R system[2] (R Development Core Team, 2008) is a free software environment for statistical computing and graphics. It is an implementation of the S language for statistical computing and graphics (Becker et al., 1988). For data analysis, it can be highly efficient to use a special-purpose language like S, compared to using a general-purpose language.

For new R users without earlier scripting or programming experience, meeting a programming language may be unsettling, but the investment[3] will quickly pay off. The user soon discovers how analysis components – written or copied from examples — can easily be stored, replayed, modified for another data set, or extended. R can be extended easily with new dedicated components, and can be used to develop and exchange data sets and data analysis approaches. It is often much harder to achieve this with programs that require long series of mouse clicks to operate.

R provides many standard and innovative statistical analysis methods. New users may find access to both well-tried and trusted methods, and speculative and novel approaches, worrying. This can, however, be a major strength, because if required, innovations can be tested in a robust environment against legacy techniques. Many methods for analysing spatial data are less frequently used than the most common statistical techniques, and thus benefit proportionally more from the nearness to both the data and the methods that R permits. R uses well-known libraries for numerical analysis, and can easily be extended by or linked to code written in S, C, C++, Fortran, or Java. Links to various relational data base systems and geographical information systems exist, many well-known data formats can be read and/or written.

The level of voluntary support and the development speed of R are high, and experience has shown R to be environment suitable for developing professional, mission-critical software applications, both for the public and the

[2] `http://www.r-project.org`.

[3] A steep learning curve – the user learns a lot per unit time.

private sector. The S language can not only be used for low-level computation on numbers, vectors, or matrices but can also be easily extended with classes for new data types and analysis methods for these classes, such as methods for summarising, plotting, printing, performing tests, or model fitting (Chambers, 1998).

In addition to the core R software system, R is also a social movement, with many participants on a continuum from useRs just beginning to analyse data with R to developeRs contributing packages to the Comprehensive R Archive Network[4] (CRAN) for others to download and employ.

Just as R itself benefits from the open source development model, contributed package authors benefit from a world-class infrastructure, allowing their work to be published and revised with improbable speed and reliability, including the publication of source packages and binary packages for many popular platforms. Contributed add-on packages are very much part of the R community, and most core developers also write and maintain contributed packages. A contributed package contains R functions, optional sample data sets, and documentation including examples of how to use the functions.

1.2.2 ... for Spatial Data Analysis?

For over 10 years, R has had an increasing number of contributed packages for handling and analysing spatial data. All these packages used to make different assumptions about how spatial data were organised, and R itself had no capabilities for distinguishing coordinates from other numbers. In addition, methods for plotting spatial data and other tasks were scattered, made different assumptions on the organisation of the data, and were rudimentary. This was not unlike the situation for time series data at the time.

After some joint effort and wider discussion, a group[5] of R developers have written the R package **sp** to extend R with classes and methods for spatial data (Pebesma and Bivand, 2005). Classes specify a structure and define how spatial data are organised and stored. Methods are instances of functions specialised for a particular data class. For example, the summary method for all spatial data classes may tell the range spanned by the spatial coordinates, and show which coordinate reference system is used (such as degrees longitude/latitude, or the UTM zone). It may in addition show some more details for objects of a specific spatial class. A plot method may, for example create a map of the spatial data.

The **sp** package provides classes and methods for points, lines, polygons, and grids (Sect. 1.4, Chap. 2). Adopting a single set of classes for spatial data offers a number of important advantages:

---

[4] CRAN mirrors are linked from `http://www.r-project.org/`.

[5] Mostly the authors of this book with help from Barry Rowlingson and Paulo J. Ribeiro Jr.

(i) It is much easier to move data across spatial statistics packages. The classes are either supported directly by the packages, reading and writing data in the new spatial classes, or indirectly, for example by supplying data conversion between the **sp** classes and the package's classes in an interface package. This last option requires one-to-many links between the packages, which are easier to provide and maintain than many-to-many links.
(ii) The new classes come with a well-tested set of methods (functions) for plotting, printing, subsetting, and summarising spatial objects, or combining (overlaying) spatial data types.
(iii) Packages with interfaces to geographical information systems (GIS), for reading and writing GIS file formats, and for coordinate (re)projection code support the new classes.
(iv) The new methods include Lattice plots, conditioning plots, plot methods that combine points, lines, polygons, and grids with map elements (reference grids, scale bars, north arrows), degree symbols (as in 52°N) in axis labels, etc.

Chapter 2 introduces the classes and methods provided by **sp**, and discusses some of the implementation details. Further chapters will show the degree of integration of **sp** classes and methods and the packages used for statistical analysis of spatial data.

Figure 1.1 shows how the reception of **sp** classes has already influenced the landscape of contributed packages; interfacing other packages for handling and analysing spatial data is usually simple as we see in Part II. The shaded nodes of the dependency graph are packages (co)-written and/or maintained by the authors of this book, and will be used extensively in the following chapters.

1.3 R and GIS

1.3.1 What is GIS?

Storage and analysis of spatial data is traditionally done in Geographical Information Systems (GIS). According to the toolbox-based definition of Burrough and McDonnell (1998, p. 11), a GIS is '...a powerful set of tools for collecting, storing, retrieving at will, transforming, and displaying spatial data from the real world for a particular set of purposes'. Another definition mentioned in the same source refers to '...checking, manipulating, and analysing data, which are spatially referenced to the Earth'.

Its capacity to analyse and visualise data makes R a good choice for spatial data analysis. For some spatial analysis projects, using only R may be sufficient for the job. In many cases, however, R will be used in conjunction with GIS software and possibly a GIS data base as well. Chapter 4 will show how spatial data are imported from and exported to GIS file formats. As is often the case in applied data analysis, the real issue is not whether a given problem *can* be

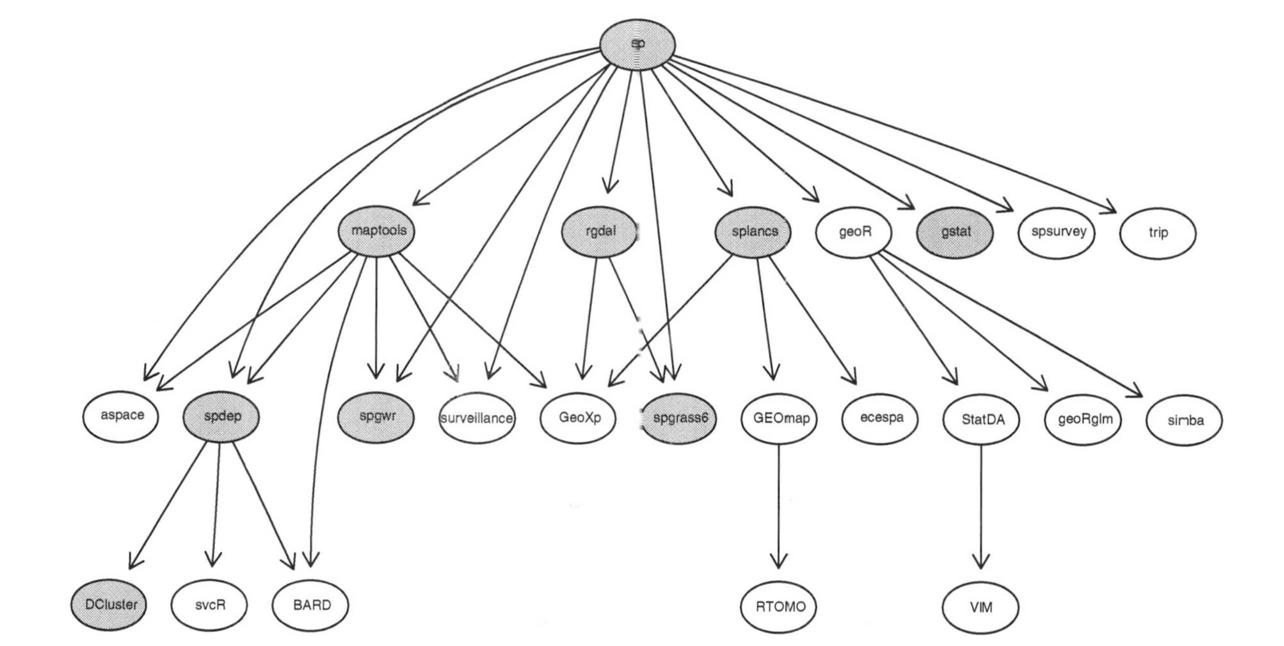

Fig. 1.1. Tree of R contributed packages on CRAN depending on or importing **sp** directly or indirectly; others suggest **sp** or use it without declaration in their package descriptions (status as of 2008-04-06)

solved using an environment such as R, but whether it can be solved *efficiently* with R. In some cases, combining different software components in a workflow may be the most robust solution, for example scripting in languages such as Python.

1.3.2 Service-Oriented Architectures

Today, much of the practice and research in geographical information systems is moving from toolbox-centred architectures (think of the 'classic' Arc/Info™ or ArcGIS™ applications) towards *service-centred* architectures (such as Google Earth™). In toolbox-centred architectures, the GIS application and data are situated on the user's computer or local area network. In service-centred architectures, the tools and data are situated on remote computers, typically accessed through Internet connections.

Reasons for this change are the increasing availability and bandwidth of the Internet, and also ownership and maintenance of data and/or analysis methods. For instance, data themselves may not be freely distributable, but certain derived products (such as visualisations or generalisations) may be. A service can be kept and maintained by the provider without end users having to bother about updating their installed software or data bases. The R system operates well under both toolbox-centred and service-centred architectures.

1.3.3 Further Reading on GIS

It seems appropriate to give some recommendations for further reading concerning GIS, not least because a more systematic treatment would not be appropriate here. Chrisman (2002) gives a concise and conceptually elegant introduction to GIS, with weight on using the data stored in the system; the domain focus is on land planning. A slightly older text by Burrough and McDonnell (1998) remains thorough, comprehensive, and perhaps a shade closer to the earth sciences in domain terms than Chrisman.

Two newer comprehensive introductions to GIS cover much of the same ground, but are published in colour. Heywood et al. (2006) contains less extra material than Longley et al. (2005), but both provide very adequate coverage of GIS as it is seen from within the GIS community today. To supplement these, Wise (2002) provides a lot of very distilled experience on the technicalities of handling geometries in computers in a compact form, often without dwelling on the computer science foundations; these foundations are given by Worboys and Duckham (2004). Neteler and Mitasova (2008) provide an excellent analytical introduction to GIS in their book, which also shows how to use the open source GRASS GIS, and how it can be interfaced with R.

It is harder to provide guidance with regard to service-centred architectures for GIS. The book by Shekar and Xiong (2008) work is a monumental, forward-looking collection with strong roots in computer and information science, and reflects the ongoing embedding of GIS technologies into database systems far more than the standard texts. Two hands-on alternatives show how service-centred architectures can be implemented at low cost by non-specialists, working, for example in environmental advocacy groups, or volunteer search and rescue teams (Mitchell, 2005; Erle et al., 2005); their approach is certainly not academic, but gets the job done quickly and effectively.

In books describing the handling of spatial data for data analysts (looking at GIS from the outside), Waller and Gotway (2004, pp. 38–67) cover most of the key topics, with a useful set of references to more detailed treatments; Banerjee et al. (2004, pp. 10–18) give a brief overview of cartography sufficient to get readers started in the right direction.

1.4 Types of Spatial Data

Spatial data have spatial reference: they have coordinate values and a system of reference for these coordinates. As a fairly simple example, consider the locations of volcano peaks on the Earth. We could list the coordinates for all known volcanoes as pairs of longitude/latitude decimal degree values with respect to the prime meridian at Greenwich and zero latitude at the equator. The World Geodetic System (WGS84) is a frequently used representation of the Earth.

Suppose we are interested in the volcanoes that have shown activity between 1980 and 2000, according to some agreed seismic registration system. This data set consists of points only. When we want to draw these points on a (flat) map, we are faced with the problem of projection: we have to translate from the spherical longitude/latitude system to a new, non-spherical coordinate system, which inevitably changes their relative positions. In Fig. 1.2, these data are projected using a Mollweide projection, and, for reference purposes, coast lines have been added. Chapter 4 deals with coordinate reference systems, and with transformations between them.

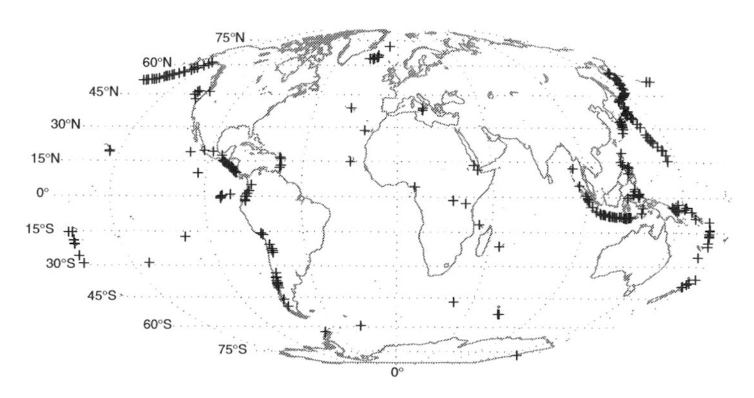

Fig. 1.2. Volcanoes of the world, with last known eruption 1964 or later (+); source: National Geophysical Data Center

If we also have the date and time of the last observed eruption at the volcano, this information is called an *attribute*: it is non-spatial in itself, but this attribute information is believed to exist for each spatial entity (volcano).

Without explicit attributes, points usually carry implicit attributes, for example all points in this map have the constant implicit attribute – they mark a 'volcano peak', in contrast to other points that do not. We represent the *purely spatial* information of entities by data models.The different types of data models that we distinguish here include the following:

Point, a single point location, such as a GPS reading or a geocoded address
Line, a set of ordered points, connected by straight line segments
Polygon, an area, marked by one or more enclosing lines, possibly containing holes
Grid, a collection of points or rectangular cells, organised in a regular lattice

The first three are vector data models and represent entities as exactly as possible, while the final data model is a raster data model, representing continuous surfaces by using a regular tessellation. All spatial data consist of positional information, answering the question 'where is it?'. In many applications these will be extended by attributes, answering the question 'what is where?'; Chrisman (2002, pp. 37–69) distinguishes a range of spatial and spatio-temporal queries of this kind. Examples for these four basic data models and of types with attributes will now follow.

The location (x, y coordinates) of a volcano may be sufficient to establish its position relative to other volcanoes on the Earth, but for describing a single volcano we can use more information. Let us, for example try to describe the topography of a volcano. Figure 1.3 shows a number of different ways to represent a continuous surface (such as topography) in a computer.

First, we can use a large number of points on a dense regular *grid* and store the attribute *altitude* for each point to approximate the surface. Grey tones are used to specify classes of these points on Fig. 1.3a.

Second, we can form contour *lines* connecting ordered points with equal altitude; these are overlayed on the same figure, and separately shown on Fig. 1.3b. Note that in this case, the contour lines were derived from the point values on the regular grid.

A *polygon* is formed when a set of line segments forms a closed object with no lines intersecting. On Fig. 1.3a, the contour lines for higher altitudes are closed and form polygons.

Lines and polygons may have *attributes*, for example the 140 contour line of Fig. 1.3a may have the label '140 m above sea level', or simply 140. Two closed contour lines have the attribute 160 m, but within the domain of this study area several non-closed contour lines have the attribute 110 m. The complete area inside the 140 m polygon (Fig. 1.3c) has the attribute 'more than 140 m above sea level', or $>$140. The area above the 160 m contour is represented by a polygon with a *hole* (Fig. 1.3d): its centre is part of the crater, which is below 160 m.

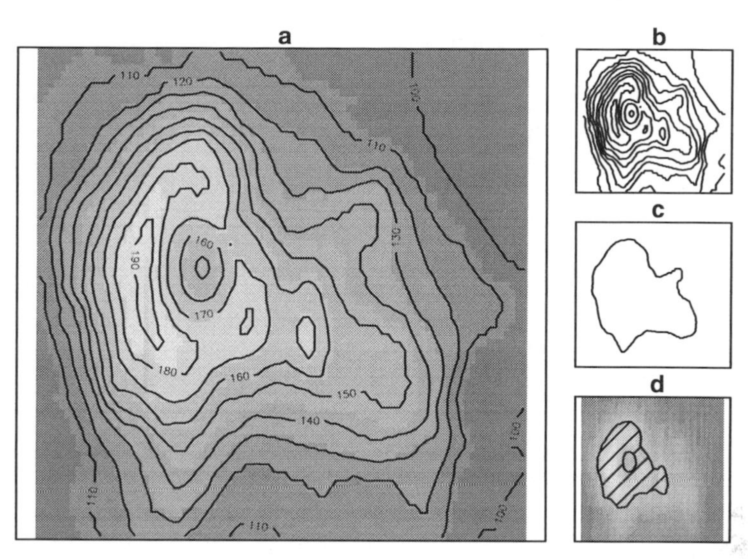

Fig. 1.3. Maunga Whau (Mt Eden) is one of about 50 volcanoes in the Auckland volcanic field. (**a**) Topographic information (altitude, m) for Maunga Whau on a $10 \times 10\,\mathrm{m}^2$ grid, (**b**) contour lines, (**c**) 140 m contour line: a closed polygon, (**d**) area above 160 m (*hashed*): a polygon with a hole

Polygons formed by contour lines of volcanoes usually have a more or less circular shape. In general, polygons can have arbitrary form, and may for certain cases even overlap. A special, but common case is when they represent the boundary of a single categorical variable, such as an administrative region. In that case, they cannot overlap and should divide up the entire study area: each point in the study area can and must be attributed to a single polygon, or lies on a boundary of one or more polygons.

A special form to represent spatial data is that of a grid: the values in each grid cell may represent an average over the area of the cell, or the value at the midpoint of the cell, or something more vague – think of image sensors. In the first case, we can see a grid as a special case of ordered points; in the second case, they are a collection of rectangular polygons. In any case, we can *derive* the position of each cell from the grid location, grid cell size, and the organisation of the grid cells. Grids are a common way to tessellate a plane. They are important because

- Devices such as digital cameras and remote sensing instruments register data on a regular grid
- Computer screens and projectors show data on a grid
- Many spatial or spatio-temporal models, such as climate models, discretise space by using a regular grid.

1.5 Storage and Display

As R is open source, we can find out the meaning of every single bit and byte manipulated by the software if we need to do so. Most users will, however, be happy to find that this is unlikely to be required, and is left to a small group of developers and experts. They will rely on the fact that many users have seen, tested, or used the code before.

When running an R session, data are usually read or imported using explicit commands, after which all data are *kept in memory*; users may choose to load a saved workspace or data objects. During an R session, the workspace can be saved to disk or chosen objects can be saved in a portable binary form for loading into the next session. When leaving an interactive R session, the question *Save workspace image?* may be answered positively to save results to disk; saving the session history is a very useful way of documenting what has been done, and is recommended as normal practice – consider choosing an informative file name.

Despite the fact that computers have greater memory capacity than they used to, R may not be suitable for the analysis of massive data sets, because data being analysed is held in memory. Massive data sets may, for example come from satellite imagery, or detailed global coast line information. It is in such cases necessary to have some idea about data size and memory management and requirements. Under such circumstances it is often still possible to use R as an analysis engine on part of the data sets. Smaller useful data sets can be obtained by selecting a certain region or by sub-sampling, aggregating or generalising the original data. Chapters 4 and 6 will give hints on how to do this.

Spatial data are usually displayed on maps, where the x- and y-axes show the coordinate values, with the aspect ratio chosen such that a unit in x equals a unit in y. Another property of maps is that elements are added for reference purposes, such as coast lines, rivers, administrative boundaries, or even satellite images.

Display of spatial data in R is a challenge on its own, and is dealt with in Chap. 3. For many users, the graphical display of statistical data is among the most compelling reasons to use R, as maps are traditionally amongst the strongest graphics we know.

The core R engine was not designed specifically for the display and analysis of maps, and the limited interactive facilities it offers have drawbacks in this area. Still, a large number of visualisations come naturally to R graphics, while they would take a substantial effort to accomplish in legacy GIS. For one thing, most GIS do not provide *conditioning plots*, where series of plots are organised in a regular lattice, share axes, and legends, and allow for systematic comparison across a large number of settings, scenarios, time, or other variables (e.g. Fig. 3.10). R provides on-screen graphics and has many graphics drivers, for example for vector graphics output to PostScript, Windows metafiles, PDF, and many bitmapped graphics formats. And, as mentioned, it works equally well as a front end or as a service providing back end for statistical analysis.

1.6 Applied Spatial Data Analysis

Statistical inference is concerned with drawing conclusions based on data and prior assumptions. The presence of a model of the data generating process may be more or less acknowledged in the analysis, but its reality will make itself felt sooner or later. The model may be manifest in the design of data collection, in the distributional assumptions employed, and in many other ways. A key insight is that observations in space cannot in general be assumed to be mutually independent, and that observations that are close to each other are likely to be similar (*ceteris paribus*). This spatial patterning – spatial autocorrelation – may be treated as useful information about unobserved influences, but it does challenge the application of methods of statistical inference that assume the mutual independence of observations.

Not infrequently, the prior assumptions are not made explicit, but are rather taken for granted as part of the research tradition of a particular scientific subdiscipline. Too little attention typically is paid to the assumptions, and too much to superficial differences; for example Venables and Ripley (2002, p. 428) comment on the difference between the covariance function and the semi-variogram in geostatistics, that '[m]uch heat and little light emerges from discussions of their comparison'.

To illustrate the kinds of debates that rage in disparate scientific communities analysing spatial data, we sketch two current issues: red herrings in geographical ecology and the interpretation of spatial autocorrelation in urban economics.

The red herring debate in geographical ecology was ignited by Lennon (2000), who claimed that substantive conclusions about the impact of environmental factors on, for example species richness had been undermined by not taking spatial autocorrelation into account. Diniz-Filho et al. (2003) replied challenging not only the interpretation of the problem in statistical terms, but pointing out that geographical ecology also involves the scale problem, that the influence of environmental factors is moderated by spatial scale.

They followed this up in a study in which the data were sub-sampled to attempt to isolate the scale problem. But they begin: 'It is important to note that we do not present a formal evaluation of this issue using statistical theory..., our goal is to illustrate heuristically that the often presumed bias due to spatial autocorrelation in OLS regression does not apply to real data sets' (Hawkins et al., 2007, p. 376).

The debate continues with verve in Beale et al. (2007) and Diniz-Filho et al. (2007). This is quite natural, as doubts about the impacts of environmental drivers on species richness raise questions about, for example, the effects of climate change. How to analyse spatial data is obviously of importance within geographical ecology. However, Diniz-Filho et al. (2007, p. 850) conclude that '[w]hen multiple assumptions are not being met, as in the case of virtually all geographical analyses, can a result from any single method (whether spatial or non-spatial) be claimed to be better? ... If different spatial methods themselves are unstable and generate conflicting results in real data, it makes no sense to claim that any particular method is always superior to any other'.

The urban economics debate is not as vigorous, but is of some practical interest, as it concerns the efficiency of services provided by local government. Revelli (2003) asks whether the spatial patterns observed in model residuals are a reaction to model misspecification, or do they signal the presence of substantive interaction between observations in space? In doing so, he reaches back to evocations of the same problem in the legacy literature of spatial statistics. As Cliff and Ord (1981, pp. 141–142) put it, 'two adjacent supermarkets will compete for trade, and yet their turnover will be a function of general factors such as the distribution of population and accessibility'. They stress that 'the presence of spatial autocorrelation may be attributable either to trends in the data or to interactions; ... [t]he choice of model must involve the scientific judgement of the investigator *and* careful testing of the assumptions'. When the fitted model is misspecified, it will be hard to draw meaningful conclusions, and the care advised by Cliff and Ord will be required.

One way of testing the assumptions is through changes in the policy context over time, where a behavioural model predicts changes in spatial autocorrelation – if the policy changes, the level of spatial interaction should change (Bivand and Szymanski, 1997; Revelli, 2003). Alternatives include using multiple levels in local government (Revelli, 2003), or different electoral settings, such as lame-duck administrations as controls (Bordignon et al., 2003). A recent careful study has used answers to a questionnaire survey to check whether interaction has occurred or not. It yields a clear finding that the observed spatial patterning in local government efficiency scores is related to the degree to which they compare their performance with that of other local government entities (Revelli and Tovmo, 2007).

This book will not provide explicit guidance on the choice of models, because the judgement of researchers in different scientific domains will vary. One aspect shared by both examples is that the participants stress the importance of familiarity with the core literature of spatial statistics. It turns out that many of the insights found there remain fundamental, despite the passage of time. Applied spatial data analysis seems to be an undertaking that, from time to time, requires the analyst to make use of this core literature.

Without attempting to be exhaustive in reviewing key books covering all the three acknowledged areas of spatial statistics – point processes, geostatistics, and areal data – we can make some choices. Bivand (2008, pp. 16–17) documents the enduring position of Ripley (1981)[6] and Cliff and Ord (1981) in terms of paper citations. Ripley (1988) supplements and extends the earlier work, and is worth careful attention. The comprehensive text by Cressie (1993) is referred to very widely; careful reading of the often very short passages of relevance to a research problem can be highly rewarding. Schabenberger and

[6] Reprinted in 2004.

Gotway (2005) cover much of the same material, incorporating advances made over the intervening period. Banerjee et al. (2004) show how the Bayesian approach to statistics can be used in applied spatial data analysis.

Beyond the core statistical literature, many disciplines have their own traditions, often collated in widely used textbooks. Public health and disease mapping are well provided for by Waller and Gotway (2004), as is ecology by Fortin and Dale (2005). O'Sullivan and Unwin (2003) cover similar topics from the point of view of geography and GIS. Like Banerjee et al. (2004), the disciplinary texts differ from the core literature not only in the way theoretical material is presented, but also in the availability of the data sets used in the books for downloading and analysis. Haining (2003) is another book providing some data sets, and an interesting bridge to the use of Bayesian approaches in the geographies of health and crime. Despite its age, Bailey and Gatrell (1995) remains a good text, with support for its data sets in R packages.

In an *R News* summary, Ripley (2001) said that one of the reasons for the relatively limited availability of spatial statistics functions in R at that time was the success of the S-PLUS™ spatial statistics module (Kaluzny et al., 1998). Many of the methods for data handling and analysis are now available in R complement and extend those in the S-PLUS™ module. We also feel that the new packaging system in S-PLUS™ constitutes an invitation, for instance to release packages like **sp** for S-PLUS™– during the development of the package, it was tested regularly under both compute engines. Although the names of functions and arguments for spatial data analysis differ between S-PLUS™ and R, users of the S-PLUS™ spatial statistics module should have no difficulty in 'finding their way around' our presentation.

To summarise the approach to applied spatial data analysis adopted here, we can say that – as with the definition of geography as 'what geographers do' – applied spatial data analysis can best be understood by observing what practitioners do and how they do it. Since practitioners may choose to conduct analyses in different ways, it becomes vital to keep attention on 'how they do it', which R facilitates, with its unrivalled closeness to both data and the implementation of methods. It is equally important to create and maintain bridges between communities of practitioners, be they innovative statisticians or dedicated field scientists, or (rarely) both in the same person. The R Spatial community attempts to offer such opportunities, without necessarily prescribing or proscribing particular methods, and this approach will be reflected in this book.

1.7 R Spatial Resources

There are a range of resources for analysing spatial data with R, one being this book. In using the book, it is worth bearing in mind the close relationships between the increase in the availability of software for spatial data analysis on CRAN and the activities of the informal community of users interested in spatial data analysis. Indeed, without contributions, advice, bug reports, and fruitful questions from users, very little would have been achieved. So before going on to present the structure of the book, we mention some of the more helpful online resources.

1.7.1 Online Resources

Since CRAN has grown to over 1,200 packages, finding resources is not simple. One opportunity is to use the collection of 'Task Views' available on CRAN itself. One of these covers spatial data analysis, and is kept more-or-less up to date. Other task views may also be relevant. These web pages are intended to be very concise, but because they are linked to the resources listed, including packages on CRAN, they can be considered as a kind of 'shop window'. By installing the **ctv** package and executing the command `install.views("Spatial")`, you will install almost all the contributed packages needed to reproduce the examples in this book (which may be downloaded from the book website).

The spatial task view is available on all CRAN mirrors, but may be accessed directly;[7] it provides a very concise summary of available contributed packages. It also specifically links two other resources, a mailing list dedicated to spatial data analysis with R and an R-Geo website. The R-sig-geo mailing list was started in 2003 after sessions on spatial statistics at the Distributed Statistical Computing conference organised in Vienna earlier the same year. By late 2007, the mailing list was being used by over 800 members, off-loading some of the spatial topic traffic from the main R-help mailing list. While R-help can see over 100 messages a day, R-sig-geo has moderate volume.

The archives of the mailing list are hosted in Zurich with the other R mailing list archives, and copies are held on Gmane and Nabble. This means that list traffic on an interesting thread can be accessed by general Internet search engines as well as the `RSiteSearch()` internal R search engine; a Google™ search on `R gstat kriging` picks up list traffic easily.

The second linked resource is the R-Geo website, generously hosted since its inception by Luc Anselin, and is currently hosted at the Spatial Analysis Laboratory (SAL) in the Department of Geography at the University of Illinois, Urbana-Champaign. Because the site uses a content management system, it may be updated at will, but does not duplicate the CRAN task view. When users report news or issues, including installation issues, with packages, this is the site where postings will be made.

1.7.2 Layout of the Book

This book is divided into two basic parts, the first presenting the shared R packages, functions, classes, and methods for handling spatial data. This part

[7] `http://CRAN.R-project.org/view=Spatial`.

is of interest to users who need to access and visualise spatial data, but who are not initially concerned with drawing conclusions from analysing spatial data per se. The second part showcases more specialised kinds of spatial data analysis, in which the relative position of observations in space may contribute to understanding the data generation process. This part is not an introduction to spatial statistics in itself, and should be read with relevant textbooks and papers referred to in the chapters.

Chapters 2 through 6 introduce spatial data handling in R. Readers needing to get to work quickly may choose to read Chap. 4 first, and return to other chapters later to see how things work. Those who prefer to see the naked structure first before using it will read the chapters in sequence, probably omitting technical subsections. The functions, classes, and methods are indexed, and so navigation from one section to another should be feasible.

Chapter 2 discusses in detail the classes for spatial data in R, as implemented in the **sp** package, and Chap. 3 discusses a number of ways of visualising for spatial data. Chapter 4 explains how coordinate reference systems work in the **sp** representation of spatial data in R, how they can be defined and how data can be transformed from one system to another, how spatial data can be imported into R or exported from R to GIS formats, and how R and the open source GRASS GIS are integrated. Chapter 5 covers methods for handling the classes defined in Chap. 2, especially for combining and integrating spatial data. Finally, Chap. 6 explains how the methods and classes introduced in Chap. 2 can be extended to suit one's own needs.

If we use the classification of Cressie (1993), we can introduce the applied spatial data analysis part of the book as follows: Chap. 7 covers the analysis of spatial point patterns, in which the relative position of points is compared with clustered, random, or regular generating processes. Chapter 8 presents the analysis of geostatistical data, with interpolation from values at observation points to prediction points. Chapters 9 and 10 deal with the statistical analysis of areal data, where the observed entities form a tessellation of the study area, and are often containers for data arising at other scales; Chap. 11 covers the special topic of disease mapping in R, and together they cover the analysis of lattice data, here termed areal data.

Data sets and code for reproducing the examples in this book are available from `http://www.asdar-book.org`; the website also includes coloured versions of the figures and other support material.

Part I

Handling Spatial Data in R

Handling Spatial Data

The key intuition underlying the development of the classes and methods in the **sp** package, and its closer dependent packages, is that users approaching R with experience of GIS will want to see 'layers', 'coverages', 'rasters', or 'geometries'. Seen from this point of view, **sp** classes should be reasonably familiar, appearing to be well-known data models. On the other hand, for statistician users of R, 'everything' is a `data.frame`, a rectangular table with rows of observations on columns of variables. To permit the two disparate groups of users to play together happily, classes have grown that look like GIS data models to GIS and other spatial data people, and look and behave like data frames from the point of view of applied statisticians and other data analysts.

This part of the book describes the classes and methods of the **sp** package, and in doing so also provides a practical guide to the internal structure of many GIS data models, as R permits the user to get as close as desired to the data. However, users will not often need to know more than that of Chap. 4 to read in their data and start work. Visualisation is covered in Chap. 3, and so a statistician receiving a well-organised set of data from a collaborator may even be able to start making maps in two lines of code, one to read the data and one to plot the variable of interest using lattice graphics. Note that coloured versions of figures may be found on the book website together with complete code examples, data sets, and other support material.

If life was always so convenient, this part of the book could be much shorter than it is. But combining spatial data from different sources often means that much more insight is needed into the data models involved. The data models themselves are described in Chap. 2, and methods for handling and combining them are covered in Chap. 5. Keeping track of which observation belongs to which geometry is also discussed here, seen from the GIS side as feature identifiers, and row names from the data frame side. In addition to data import and export, Chap. 4 also describes the use and transformation of coordinate reference systems for **sp** classes, and integration of the open source GRASS GIS and R. Finally, Chap. 6 explains how the methods and classes introduced in Chap. 2 can be extended to suit one's own needs.

2

Classes for Spatial Data in R

2.1 Introduction

Many disciplines have influenced the representation of spatial data, both in analogue and digital forms. Surveyors, navigators, and military and civil engineers refined the fundamental concepts of mathematical geography, established often centuries ago by some of the founders of science, for example by al-Khwārizmī. Digital representations came into being for practical reasons in computational geometry, in computer graphics and hardware-supported gaming, and in computer-assisted design and virtual reality. The use of spatial data as a business vehicle has been spurred in the early years of the present century by consumer broadband penetration and distributed server farms, with a prime example being Google Earth™.[1] There are often interactions between the graphics hardware required and the services offered, in particular for the fast rendering of scene views.

In addition, space and other airborne technologies have vastly increased the volumes and kinds of spatial data available. Remote sensing satellites continue to make great contributions to earth observation, with multi-spectral images supplementing visible wavelengths. The Shuttle Radar Topography Mission (SRTM) in February 2000 has provided elevation data for much of the earth. Other satellite-borne sensor technologies are now vital for timely storm warnings, amongst other things. These complement terrestrial networks monitoring, for example lightning strikes and the movement of precipitation systems by radar.

Surveying in the field has largely been replaced by aerial photogrammetry, mapping using air photographs usually exposed in pairs of stereo images. Legacy aerial photogrammetry worked with analogue images, and many research laboratories and mapping agencies have large archives of air photographs with coverage beginning from the 1930s. These images can be scanned to provide a digital representation at chosen resolutions. While

[1] `http://earth.google.com/`.

satellite imagery usually contains metadata giving the scene frame – the sensor direction in relation to the earth at scan time – air photographs need to be registered to known ground control points.

These ground control points were 'known' from terrestrial triangulation, but could be in error. The introduction of Global Positioning System (GPS) satellites has made it possible to correct the positions of existing networks of ground control points. The availability of GPS receivers has also made it possible for data capture in the field to include accurate positional information in a known coordinate reference system. This is conditioned by the requirement of direct line-of-sight to a sufficient number of satellites, not easy in mountain valleys or in city streets bounded by high buildings. Despite this limitation, around the world the introduction of earth observation satellites and revised ground control points have together caused breaks of series in published maps, to take advantage of the greater accuracy now available. This means that many older maps cannot be matched to freshly acquired position data without adjustment.

All of these sources of spatial data involve points, usually two real numbers representing position in a known coordinate reference system. It is possible to go beyond this simple basis by combining pairs of points to form line segments, combining line segments to form polylines, networks or polygons, or regular grid centres. Grids can be defined within a regular polygon, usually a rectangle, with given resolution – the size of the grid cells. All these definitions imply choices of what are known in geographical information systems (GIS) as data models, and these choices have most often been made for pragmatic reasons. All the choices also involve trade-offs between accuracy, feasibility, and cost.

Artificial objects are easiest to represent, like roads, bridges, buildings, or similar structures. They are crisply defined, and are not subject to natural change – unlike placing political borders along the centre lines or deepest channels of meandering rivers. Shorelines are most often natural and cannot be measured accurately without specifying measurement scale. Boundaries between areas of differing natural land cover are frequently indeterminate, with gradations from one land cover category to another. Say that we want to examine the spatial distribution of a species by land cover category; our data model of how to define the boundary between categories will affect the outcome, possibly strongly. Something of the same affects remote sensing, because the reported values of the observed pixels will hide sub-pixel variation.

It is unusual for spatial data to be defined in three dimensions, because of the close links between cartography and data models for spatial data. When there are multiple observations on the same attribute at varying heights or depths, they are most often treated as separate layers. GIS-based data models do not fit time series data well either, even though some environmental monitoring data series are observed in three dimensions and time. Some GIS software can handle voxels, the 3D equivalent of pixels – 2D raster cells – but the third dimension in spatial data is not handled satisfactorily, as is the case in computer-assisted design or medical imaging. On the other hand, many GIS packages do provide a 2.5D intermediate solution for viewing, by draping thematic layers, like land cover or a road network, over a digital elevation model. In this case, however, there is no 'depth' in the data model, as we can see when a road tunnel route is draped over the mountain it goes through.

2.2 Classes and Methods in R

In Chap. 1, we described R as a language and environment for data analysis. Although this is not the place to give an extended introduction to R,[2] it will be useful to highlight some of its features (see also Braun and Murdoch, 2007, for an up-to-date introduction). In this book, we will be quoting R commands in the text, showing which commands a user could give, and how the non-graphical output might be represented when printed to the console.

Of course, R can be used as a calculator to carry out simple tasks, where no values are assigned to variables, and where the results are shown without being saved, such as the area of a circle of radius 10:

```
> pi * 10^2
```

```
[1] 314.1593
```

Luckily, π is a built-in constant in R called `pi`, and so entering a rounded version is not needed. So this looks like a calculator, but appearances mislead. The first misleading impression is that the arithmetic is simply being 'done', while in fact it is being translated (parsed) into functions (operators) with arguments first, and then evaluated:

```
> "*"(pi, "^"(10, 2))
```

```
[1] 314.1593
```

When the operators or functions permit, vectors of values may be used as readily as scalar values (which are vectors of unit length) — here the ':' operator is used to generate an integer sequence of values:

```
> pi * (1:10)^2
```

```
 [1]   3.141593  12.566371  28.274334  50.265482  78.539816 113.097336
 [7] 153.938040 201.061930 254.469005 314.159265
```

The second misapprehension is that what is printed to the console is the 'result', when it is actually the outcome of applying the appropriate `print` method for the class of the 'result', with default arguments. If we store the value returned for the area of our circle in variable `x` using the assignment operator `<-`, we can print `x` with the default number of digits, or with more if

[2] Free documentation, including the very useful 'An Introduction to R' (Venables et al., 2008), may be downloaded from CRAN.

we so please. Just typing the variable name at the interactive prompt invokes the appropriate print method, but we can also pass it to the print method explicitly:

```
> x <- pi * 10^2
> x

[1] 314.1593

> print(x)

[1] 314.1593

> print(x, digits = 12)

[1] 314.159265359
```

We can say that the variable `x` contains an object of a particular class, in this case:

```
> class(x)

[1] "numeric"

> typeof(x)

[1] "double"
```

where `typeof` returns the storage mode of the object in variable `x`. It is the class of the object that determines the method that will be used to handle it; if there is no specific method for that class, it may be passed to a default method. These methods are also known as generic functions, often including at least `print`, `plot`, and `summary` methods. In the case of the `print` method, `numeric` is not provided for explicitly, and so the default method is used. The `plot` method, as its name suggests, will use the current graphics device to make a visual display of the object, dispatching to a specific method for the object class if provided. In comparison with the `print` method, the `summary` method provides a qualified view of the data, highlighting the key features of the object.

When the `S` language was first introduced, it did not use class/method mechanisms at all. They were introduced in Chambers and Hastie (1992) and `S` version 3, in a form that is known as `S3` classes or old-style classes. These classes were not formally defined, and 'just grew'; the vast majority of objects returned by model fitting functions belong to old-style classes. Using a non-spatial example from the standard data set `cars`, we can see that it is an object of class `data.frame`, stored in a `list`, which is a vector whose components can be arbitrary objects; `data.frame` has both names and summary methods:

```
> class(cars)

[1] "data.frame"
```

```
> typeof(cars)

[1] "list"

> names(cars)

[1] "speed" "dist"

> summary(cars)

     speed           dist
 Min.   : 4.0   Min.   :  2.00
 1st Qu.:12.0   1st Qu.: 26.00
 Median :15.0   Median : 36.00
 Mean   :15.4   Mean   : 42.98
 3rd Qu.:19.0   3rd Qu.: 56.00
 Max.   :25.0   Max.   :120.00
```

The data.frame contains two variables, one recording the speed of the observed cars in mph, the other the stopping distance measured in feet – the observations were made in the 1920s. When uncertain about the structure of something in our R workspace, revealed for example by using the ls function for listing the contents of the workspace, the str[3] method often gives a clear digest, including the size and class:

```
> str(cars)

'data.frame': 50 obs. of 2 variables:
$ speed:num 4 4 7 7 8 ...
$ dist :num 2 10 4 22 16 ...
```

Data frames are containers for data used everywhere in S since their full introduction in Chambers and Hastie (1992, pp. 45–94). Recent and shorter introductions to data frames are given by Crawley (2005, pp. 15–22), Crawley (2007, pp. 107–133), and Dalgaard (2002, pp. 18–19) and in the online documentation (Venables et al., 2008, pp. 27–29 in the R 2.6.2 release). Data frames view the data as a rectangle of rows of observations on columns of values of variables of interest. The representation of the values of the variables of interest can include integer and floating point numeric types, logical, character, and derived classes. One very useful derived class is the factor, which is represented as integers pointing to character levels, such as 'forest' or 'arable'. Printed, the values look like character values, but are not – when a data frame is created, all character variables included in it are converted to factor by default. Data frames also have unique row names, represented as an integer or character vector or as an internal mechanism to signal that

[3] str can take additional arguments to control its output.

the sequence from 1 to the number of rows in the data frame are used. The `row.names` function is used to access and assign data frame row names.

One of the fundamental abstractions used in R is the `formula` introduced in Chambers and Hastie (1992, pp. 13–44) – an online summary may be found in Venables et al. (2008, pp. 50–52 in the R 2.6.2 release). The abstraction is intended to make statistical modelling as natural and expressive as possible, permitting the analyst to focus on the substantive problem at hand. Because the `formula` abstraction is used in very many contexts, it is worth some attention. A `formula` is most often two-sided, with a response variable to the left of the $\sim$ (tilde) operator, and in this case a determining variable on the right:

```
> class(dist ~ speed)

[1] "formula"
```

These objects are typically used as the first argument to model fitting functions, such as `lm`, which is used to fit linear models. They will usually be accompanied by a `data` argument, indicating where the variables are to be found:

```
> lm(dist ~ speed, data = cars)

Call:
lm(formula = dist ~ speed, data = cars)

Coefficients:
(Intercept)        speed
    -17.579        3.932
```

This is a simple example, but very much more can be done with the `formula` abstraction. If we create a factor for the `speed` variable by cutting it at its quartiles, we can contrast how the `plot` method displays the relationship between two numerical variables and a numerical variable and a factor (shown in Fig. 2.1):

```
> cars$qspeed <- cut(cars$speed, breaks = quantile(cars$speed),
+     include.lowest = TRUE)
> is.factor(cars$qspeed)

[1] TRUE

> plot(dist ~ speed, data = cars)
> plot(dist ~ qspeed, data = cars)
```

Finally, let us see how the `formula` with the right-hand side factor is handled by `lm` – it is converted into 'dummy' variable form automatically:

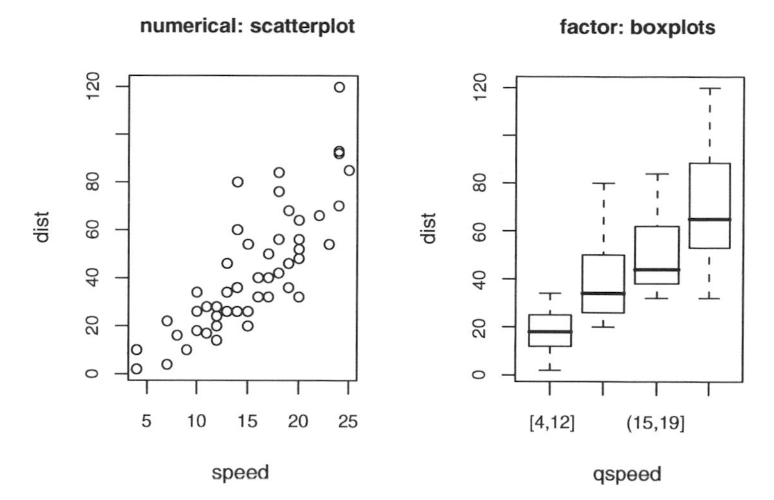

Fig. 2.1. Plot methods for a formula with numerical (*left panel*) and factor (*right panel*) right-hand side variables

```
> lm(dist ~ qspeed, data = cars)

Call:
lm(formula = dist ~ qspeed, data = cars)

Coefficients:
  (Intercept)  qspeed(12,15]  qspeed(15,19]  qspeed(19,25]
        18.20          21.98          31.97          51.13
```

Variables in the `formula` may also be transformed in different ways, for example using `log`. The `formula` is carried through into the object returned by model fitting functions to be used for prediction from new data provided in a `data.frame` with the same column names as the right-hand side variables, and the same level names if the variable is a factor.

New-style (`S4`) classes were introduced in the `S` language at release 4, and in Chambers (1998), and are described by Venables and Ripley (2000, pp. 75–121), and in subsequent documentation installed with R.[4] Old-style classes are most often simply lists with attributes; they are not defined formally. Although users usually do not change values inside old-style classes, there is nothing to stop them doing so, for example changing the representation of coordinates from floating point to integer numbers. This means that functions need to check, among other things, whether components of a class exist, and whether they are represented correctly, before they can be handled. The central advantage of new-style classes is that they have formal definitions

[4] There is little instructional material online, although this useR conference talk remains relevant: `http://www.ci.tuwien.ac.at/Conferences/useR-2004/Keynotes/Leisch.pdf`.

that specify the name and type of the components, called *slots*, that they contain. This simplifies the writing, maintenance, and use of the classes, because their format is known from the definition. For a further discussion of programming for classes and methods, see Sect. 6.1.

Because the classes provided by the **sp** package are new-style classes, we will be seeing how such classes work in practice below. In particular, we will be referring to the slots in class definitions; slots are specified in the definition as the representation of what the class contains. Many methods are written for the classes to be introduced in the remainder of this chapter, in particular coercion methods for changing the way an object is represented from one class to another. New-style classes can also check the validity of objects being created, for example to stop the user from filling slots with data that do not conform to the definition.

2.3 Spatial Objects

The foundation class is the `Spatial` class, with just two slots. The first is a bounding box, a matrix of numerical coordinates with column names `c('min', 'max')`, and at least two rows, with the first row eastings (x-axis) and the second northings (y-axis). Most often the bounding box is generated automatically from the data in subclasses of `Spatial`. The second is a `CRS` class object defining the coordinate reference system, and may be set to 'missing', represented by `NA` in R, by `CRS(as.character(NA))`, its default value. Operations on `Spatial*` objects should update or copy these values to the new `Spatial*` objects being created. We can use `getClass` to return the complete definition of a class, including its slot names and the types of their contents:

```
> library(sp)

> getClass("Spatial")

Slots:

Name:          bbox proj4string
Class:       matrix         CRS

Known Subclasses:
Class "SpatialPoints", directly
Class "SpatialLines", directly
Class "SpatialPolygons", directly
Class "SpatialPointsDataFrame", by class "SpatialPoints", distance 2
Class "SpatialPixels", by class "SpatialPoints", distance 2
Class "SpatialGrid", by class "SpatialPoints", distance 3
Class "SpatialPixelsDataFrame", by class "SpatialPoints", distance 3
Class "SpatialGridDataFrame", by class "SpatialPoints", distance 4
Class "SpatialLinesDataFrame", by class "SpatialLines", distance 2
Class "SpatialPolygonsDataFrame", by class "SpatialPolygons",
     distance 2
```

As we see, getClass also returns known subclasses, showing the classes that include the Spatial class in their definitions. This also shows where we are going in this chapter, moving from the foundation class to richer representations. But we should introduce the coordinate reference system (CRS) class very briefly; we will return to its description in Chap. 4.

```
> getClass("CRS")

Slots:


Name:   projargs
Class: character
```

The class has a character string as its only slot value, which may be a missing value. If it is not missing, it should be a PROJ.4-format string describing the projection (more details are given in Sect. 4.1.2). For geographical coordinates, the simplest such string is "+proj=longlat", using "longlat", which also shows that eastings always go before northings in **sp** classes. Let us build a simple Spatial object from a bounding box matrix, and a missing coordinate reference system:

```
> m <- matrix(c(0, 0, 1, 1), ncol = 2, dimnames = list(NULL,
+     c("min", "max")))
> crs <- CRS(projargs = as.character(NA))
> crs

CRS arguments: NA

> S <- Spatial(bbox = m, proj4string = crs)
> S

An object of class "Spatial"
Slot "bbox":
     min max
[1,]   0   1
[2,]   0   1

Slot "proj4string":
CRS arguments: NA
```

We could have used new methods to create the objects, but prefer to use helper functions with the same names as the classes that they instantiate. If the object is known not to be projected, a sanity check is carried out on the coordinate range (which here exceeds the feasible range for geographical coordinates):

```
> Spatial(matrix(c(350, 85, 370, 95), ncol = 2, dimnames = list(NULL,
+     c("min", "max"))), proj4string = CRS("+longlat"))

Error in validityMethod(object) : Geographical CRS given to
non-conformant data
```

2.4 SpatialPoints

The SpatialPoints class is the first subclass of Spatial, and a very important one. The extension of SpatialPoints to other subclasses means that explaining how this class works will yield benefits later on. In this section, we also look at methods for Spatial* objects, and at extending Spatial* objects to include attribute data, where each spatial entity, here a point, is linked to a row in a data frame. We take Spatial* objects to be subclasses of Spatial, and the best place to start is with SpatialPoints.

A two-dimensional point can be described by a pair of numbers (x, y), defined over a known region. To represent geographical phenomena, the maximum known region is the earth, and the pair of numbers measured in degrees are a geographical coordinate, showing where our point is on the globe. The pair of numbers define the location on the sphere exactly, but if we represent the globe more accurately by an ellipsoid model, such as the World Geodetic System 1984 – introduced after satellite measurements corrected our understanding of the shape of the earth – that position shifts slightly. Geographical coordinates can extend from latitude $90°$ to $-90°$ in the north–south direction, and from longitude $0°$ to $360°$ or equivalently from $-180°$ to $180°$ in the east–west direction. The Poles are fixed, but where the longitudes fall depends on the choice of prime meridian, most often Greenwich just east of London. This means that geographical coordinates define a point on the earth's surface unequivocally if we also know which ellipsoid model and prime meridian were used; the concept of datum, relating the ellipsoid to the distance from the centre of the earth, is introduced on p. 82.

Using the standard read.table function, we read in a data file with the positions of CRAN mirrors across the world. We extract the two columns with the longitude and latitude values into a matrix, and use str to view a digest:

```
> CRAN_df <- read.table("CRAN051001a.txt", header = TRUE)
> CRAN_mat <- cbind(CRAN_df$long, CRAN_df$lat)
> row.names(CRAN_mat) <- 1:nrow(CRAN_mat)
> str(CRAN_mat)

num [1:54, 1:2] 153 145 ...
- attr(*, "dimnames")=List of 2
..$ :chr [1:54] "1" "2" ...
..$ :NULL
```

The SpatialPoints class extends the Spatial class by adding a coords slot, into which a matrix of point coordinates can be inserted.

```
> getClass("SpatialPoints")

Slots:

Name:        coords        bbox proj4string
Class:       matrix      matrix         CRS
```

```
Extends: "Spatial"

Known Subclasses:
Class "SpatialPointsDataFrame", directly
Class "SpatialPixels", directly
Class "SpatialGrid", by class "SpatialPixels", distance 2
Class "SpatialPixelsDataFrame", by class "SpatialPixels", distance 2
Class "SpatialGridDataFrame", by class "SpatialGrid", distance 3
```

It has a `summary` method that shows the bounding box, whether the object is projected (here `FALSE`, because the string `"longlat"` is included in the projection description), and the number of rows of coordinates. Classes in **sp** are not atomic: there is no `SpatialPoint` class that is extended by `SpatialPoints`. This is because R objects are vectorised by nature, not atomic. A `SpatialPoints` object may, however, consist of a single point.

```
> llCRS <- CRS("+proj=longlat +ellps=WGS84")
> CRAN_sp <- SpatialPoints(CRAN_mat, proj4string = llCRS)
> summary(CRAN_sp)

Object of class SpatialPoints
Coordinates:
                  min      max
coords.x1 -122.95000 153.0333
coords.x2  -37.81667  57.0500
Is projected: FALSE
proj4string : [+proj=longlat +ellps=WGS84]
Number of points: 54
```

`SpatialPoints` objects may have more than two dimensions, but plot methods for the class use only the first two.

2.4.1 Methods

Methods are available to access the values of the slots of `Spatial` objects. The `bbox` method returns the bounding box of the object, and is used both for preparing plotting methods (see Chap. 3) and internally in handling data objects. The first row reports the west–east range and the second the south–north direction. If we want to take a subset of the points in a `SpatialPoints` object, the bounding box is reset, as we will see.

```
> bbox(CRAN_sp)

                  min      max
coords.x1 -122.95000 153.0333
coords.x2  -37.81667  57.0500
```

First, the other generic method for all `Spatial` objects, `proj4string`, will be introduced. The basic method reports the projection string contained as a

CRS object in the proj4string slot of the object, but it also has an assignment form, allowing the user to alter the current value, which can also be a CRS object containing a character NA value:

```
> proj4string(CRAN_sp)

[1] "+proj=longlat +ellps=WGS84"

> proj4string(CRAN_sp) <- CRS(as.character(NA))
> proj4string(CRAN_sp)

[1] NA

> proj4string(CRAN_sp) <- llCRS
```

Extracting the coordinates from a SpatialPoints object as a numeric matrix is as simple as using the coordinates method. Like all matrices, the indices can be used to choose subsets, for example CRAN mirrors located in Brazil in 2005:

```
> brazil <- which(CRAN_df$loc == "Brazil")
> brazil

[1] 4 5 6 7 8

> coordinates(CRAN_sp)[brazil, ]

      coords.x1 coords.x2
[1,] -49.26667 -25.41667
[2,] -42.86667 -20.75000
[3,] -43.20000 -22.90000
[4,] -47.63333 -22.71667
[5,] -46.63333 -23.53333
```

In addition, a SpatialPoints object can also be accessed by index, using the "[" operator, here on the coordinate values treated as an entity. The object returned is of the same class, and retains the projection information, but has a new bounding box:

```
> summary(CRAN_sp[brazil, ])

Object of class SpatialPoints
Coordinates:
                min       max
coords.x1 -49.26667 -42.86667
coords.x2 -25.41667 -20.75000
Is projected: FALSE
proj4string : [+proj=longlat +ellps=WGS84]
Number of points: 5
```

The "[" operator also works for negative indices, which remove those coordinates from the object, here by removing mirrors south of the Equator:

```
> south_of_equator <- which(coordinates(CRAN_sp)[, 2] <
+     0)
> summary(CRAN_sp[-south_of_equator, ])

Object of class SpatialPoints
Coordinates:
               min    max
coords.x1 -122.95 140.10
coords.x2   24.15  57.05
Is projected: FALSE
proj4string : [+proj=longlat +ellps=WGS84]
Number of points: 45
```

Because `summary` and `print` methods are so common in R, we used them here without special mention. They are provided for **sp** classes, with `summary` reporting the number of spatial entities, the projection information, and the bounding box, and `print` gives a view of the data in the object. As usual in S, the actual underlying data and the output of the `print` method may differ, for example in the number of digits shown.

An important group of methods for visualisation of `Spatial*` objects are presented in detail in Chap. 3; each such object class has a `plot` method. Other methods will also be introduced in Chap. 5 for combining (overlaying) different `Spatial*` objects, for sampling from `Spatial` objects, and for merging spatial data objects.

2.4.2 Data Frames for Spatial Point Data

We described data frames on p. 25, and we now show how our `SpatialPoints` object can be taught to behave like a `data.frame`. Here we use numbers in sequence to index the points and the rows of our data frame, because neither the place names nor the countries are unique.

```
> str(row.names(CRAN_df))

chr [1:54] "1" "2" ...
```

What we would like to do is to associate the correct rows of our data frame object with 'their' point coordinates – it often happens that data are collected from different sources, and the two need to be merged. The `SpatialPoints-DataFrame` class is the container for this kind of spatial point information, and can be constructed in a number of ways, for example from a data frame and a matrix of coordinates. If the matrix of point coordinates has row names and the `match.ID` argument is set to its default value of `TRUE`, then the matrix row names are checked against the row names of the data frame. If they match, but are not in the same order, the data frame rows are re-ordered to suit the points. If they do not match, no `SpatialPointsDataFrame` is constructed. Note that the new object takes two indices, the first for the spatial object, the second, if given, for the column. Giving a single index number, or range of numbers, or column name or names returns a new SpatialPointsDataFrame with the requested columns. Using other extraction operators, especially the $ operator, returns the data frame column referred to. These operators mimic the equivalent ones for other standard S classes as far as possible.

```
> CRAN_spdf1 <- SpatialPointsDataFrame(CRAN_mat, CRAN_df,
+     proj4string = llCRS, match.ID = TRUE)
> CRAN_spdf1[4, ]

          coordinates    place  north    east    loc      long
4 (-49.2667, -25.4167) Curitiba 25d25'S 49d16'W Brazil -49.26667
       lat
4 -25.41667

> str(CRAN_spdf1$loc)

Factor w/ 30 levels "Australia","Austria",..: 1 1 2 3 3 ...

> str(CRAN_spdf1[["loc"]])

Factor w/ 30 levels "Australia","Austria",..: 1 1 2 3 3 ...
```

If we re-order the data frame at random using sample, we still get the same result, because the data frame is re-ordered to match the row names of the points:

```
> s <- sample(nrow(CRAN_df))
> CRAN_spdf2 <- SpatialPointsDataFrame(CRAN_mat, CRAN_df[s,
+     ], proj4string = llCRS, match.ID = TRUE)
> all.equal(CRAN_spdf2, CRAN_spdf1)

[1] TRUE

> CRAN_spdf2[4, ]

          coordinates    place  north    east    loc      long
4 (-49.2667, -25.4167) Curitiba 25d25'S 49d16'W Brazil -49.26667
       lat
4 -25.41667
```

But if we have non-matching ID values, created by pasting pairs of letters together and sampling an appropriate number of them, the result is an error:

```
> CRAN_df1 <- CRAN_df
> row.names(CRAN_df1) <- sample(c(outer(letters, letters,
+     paste, sep = "")), nrow(CRAN_df1))

> CRAN_spdf3 <- SpatialPointsDataFrame(CRAN_mat, CRAN_df1,
+     proj4string = llCRS, match.ID = TRUE)

Error in SpatialPointsDataFrame(CRAN_mat, CRAN_df1,
proj4string = llCRS, : row.names of data and coords do not
match
```

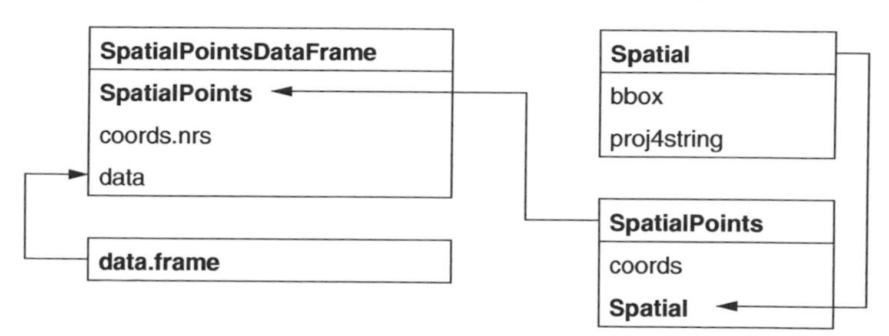

Fig. 2.2. Spatial points classes and their slots; arrows show subclass extensions

Let us examine the contents of objects of the `SpatialPointsDataFrame` class, shown in Fig. 2.2. Because the class extends `SpatialPoints`, it also inherits the information contained in the `Spatial` class object. The `data` slot is where the information from the data frame is kept, in a `data.frame` object.

```
> getClass("SpatialPointsDataFrame")

Slots:

Name:          data   coords.nrs       coords         bbox proj4string
Class:   data.frame      numeric       matrix       matrix         CRS

Extends:
Class "SpatialPoints", directly
Class "Spatial", by class "SpatialPoints", distance 2

Known Subclasses:
Class "SpatialPixelsDataFrame", directly, with explicit coerce
```

The `Spatial*DataFrame` classes have been designed to behave as far as possible like data frames, both with respect to standard methods such as `names`, and more demanding modelling functions like `model.frame` used in very many model fitting functions using `formula` and `data` arguments:

```
> names(CRAN_spdf1)

[1] "place" "north" "east"  "loc"   "long"  "lat"

> str(model.frame(lat ~ long, data = CRAN_spdf1), give.attr = FALSE)

'data.frame': 54 obs. of 2 variables:
$ lat :num -27.5 -37.8 ...
$ long:num 153 145 ...
```

Making our `SpatialPointsDataFrame` object from a matrix of coordinates and a data frame with or without ID checking is only one way to reach our goal, and others may be more convenient. We can construct the object by giving the `SpatialPointsDataFrame` function a `SpatialPoints` object as its first argument:

```
> CRAN_spdf4 <- SpatialPointsDataFrame(CRAN_sp, CRAN_df)
> all.equal(CRAN_spdf4, CRAN_spdf2)
```

```
[1] TRUE
```

We can also assign coordinates to a data frame – this approach modifies the original data frame. The coordinate assignment function can take a matrix of coordinates with the same number of rows as the data frame on the right-hand side, or an integer vector of column numbers for the coordinates, or equivalently a character vector of column names, assuming that the required columns already belong to the data frame.

```
> CRAN_df0 <- CRAN_df
> coordinates(CRAN_df0) <- CRAN_mat
> proj4string(CRAN_df0) <- llCRS
> all.equal(CRAN_df0, CRAN_spdf2)
```

```
[1] TRUE
```

```
> str(CRAN_df0, max.level = 2)
```

```
Formal class 'SpatialPointsDataFrame' [package "sp"] with 5 slots
..@ data :'data.frame': 54 obs. of 6 variables:
..@ coords.nrs :num(0)
..@ coords :num [1:54, 1:2] 153 145 ...
.. ..- attr(*, "dimnames")=List of 2
..@ bbox :num [1:2, 1:2] -123.0 -37.8 ...
.. ..- attr(*, "dimnames")=List of 2
..@ proj4string:Formal class 'CRS' [package "sp"] with 1 slots
```

Objects created in this way differ slightly from those we have seen before, because the `coords.nrs` slot is now used, and the coordinates are moved from the `data` slot to the `coords` slot, but the objects are otherwise the same:

```
> CRAN_df1 <- CRAN_df
> names(CRAN_df1)
```

```
[1] "place" "north" "east"  "loc"   "long"  "lat"
```

```
> coordinates(CRAN_df1) <- c("long", "lat")
> proj4string(CRAN_df1) <- llCRS
> str(CRAN_df1, max.level = 2)
```

```
Formal class 'SpatialPointsDataFrame' [package "sp"] with 5 slots
..@ data :'data.frame': 54 obs. of 4 variables:
..@ coords.nrs :int [1:2] 5 6
..@ coords :num [1:54, 1:2] 153 145 ...
.. ..- attr(*, "dimnames")=List of 2
..@ bbox :num [1:2, 1:2] -123.0 -37.8 ...
.. ..- attr(*, "dimnames")=List of 2
..@ proj4string:Formal class 'CRS' [package "sp"] with 1 slots
```

Transect and tracking data may also be represented as points, because the observation at each point contributes information that is associated with the point itself, rather than the line as a whole. Sequence numbers can be entered into the data frame to make it possible to trace the points in order, for example as part of a `SpatialLines` object as we see in the Sect. 2.5.

As an example, we use a data set[5] from satellite telemetry of a single loggerhead turtle crossing the Pacific from Mexico to Japan (Nichols et al., 2000).

```
> turtle_df <- read.csv("seamap105_mod.csv")
> summary(turtle_df)
```

```
       id              lat              lon
 Min.   :  1.00   Min.   :21.57   Min.   :-179.88
 1st Qu.: 99.25   1st Qu.:24.36   1st Qu.:-147.38
 Median :197.50   Median :25.64   Median :-119.64
 Mean   :197.50   Mean   :27.21   Mean   : -21.52
 3rd Qu.:295.75   3rd Qu.:27.41   3rd Qu.: 153.66
 Max.   :394.00   Max.   :39.84   Max.   : 179.93

                obs_date
 01/02/1997 04:16:53:  1
 01/02/1997 05:56:25:  1
 01/04/1997 17:41:54:  1
 01/05/1997 17:20:07:  1
 01/06/1997 04:31:13:  1
 01/06/1997 06:12:56:  1
 (Other)            :388
```

Before creating a `SpatialPointsDataFrame`, we will timestamp the observations, and re-order the input data frame by timestamp to make it easier to add months to Fig. 2.3, to show progress westwards across the Pacific:

```
> timestamp <- as.POSIXlt(strptime(as.character(turtle_df$obs_date),
+     "%m/%d/%Y %H:%M:%S"), "GMT")
> turtle_df1 <- data.frame(turtle_df, timestamp = timestamp)
> turtle_df1$lon <- ifelse(turtle_df1$lon < 0, turtle_df1$lon +
+     360, turtle_df1$lon)
> turtle_sp <- turtle_df1[order(turtle_df1$timestamp),
+     ]
> coordinates(turtle_sp) <- c("lon", "lat")
> proj4string(turtle_sp) <- CRS("+proj=longlat +ellps=WGS84")
```

The input data file is as downloaded, but without columns with identical values for all points, such as the number of the turtle (07667). We return to this data set in Chap. 6, examining the interesting contributed package **trip** by Michael Sumner, which proposes customised classes and methods for data of this kind.

[5] Data downloaded with permission from SEAMAP (Read et al., 2003), data set 105.

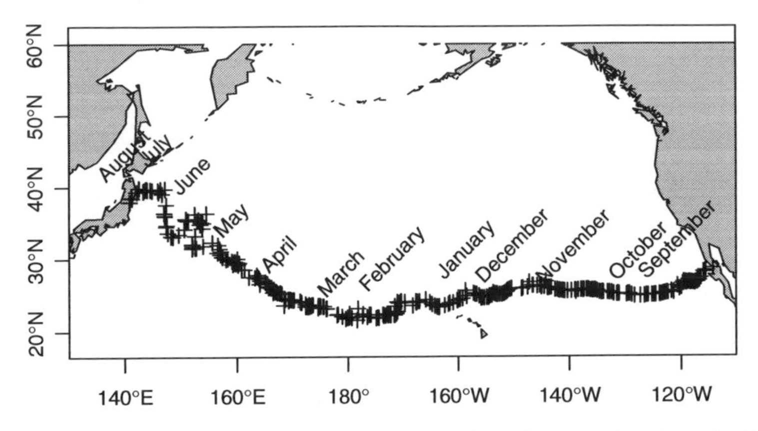

Fig. 2.3. Westward movements of a captive-raised adult loggerhead turtle (*Caretta caretta*) from 10 August 1996 to 12 August 1997

2.5 SpatialLines

Lines have been represented in S in a simple form as a sequence of points (see Becker et al. (1988), Murrell (2006, pp. 83–86)), based on lowering the graphic 'pen' at the first point and drawing to the successive points until an NA is met. Then the pen is raised and moved to the next non-NA value, where it is lowered, until the end of the set of points. While this is convenient for graphics output, it is less so for associating lines with data values, because the line is not subsetted into data objects in any other way than by NA values.

The approach adopted here is to start with a Line object that is a matrix of 2D coordinates, without NA values. A list of Line objects forms the Lines slot of a Lines object. An identifying character tag is also required, and will be used for constructing SpatialLines objects using the same approach as was used above for matching ID values for spatial points.

```
> getClass("Line")

Slots:

Name:  coords
Class: matrix

Known Subclasses: "Polygon"

> getClass("Lines")

Slots:

Name:       Lines          ID
Class:       list character
```

Neither Line nor Lines objects inherit from the Spatial class. It is the SpatialLines object that contains the bounding box and projection information for the list of Lines objects stored in its lines slot. This degree of complexity is required to be able to add observation values in a data frame, creating SpatialLinesDataFrame objects, and to use a range of extraction methods on these objects.

```
> getClass("SpatialLines")

Slots:

Name:        lines        bbox proj4string
Class:        list      matrix         CRS

Extends: "Spatial"

Known Subclasses: "SpatialLinesDataFrame"
```

Let us examine an example of an object of this class, created from lines retrieved from the **maps** package world database, and converted to a SpatialLines object using the map2SpatialLines function in **maptools**. We can see that the lines slot of the object is a list of 51 components, each of which must be a Lines object in a valid SpatialLines object.

```
> library(maps)
> japan <- map("world", "japan", plot = FALSE)
> p4s <- CRS("+proj=longlat +ellps=WGS84")
> SLjapan <- map2SpatialLines(japan, proj4string = p4s)
> str(SLjapan, max.level = 2)

Formal class 'SpatialLines' [package "sp"] with 3 slots
..@ lines :List of 51
..@ bbox :num [1:2, 1:2] 123.0 24.3 ...
.. ..- attr(*, "dimnames")=List of 2
..@ proj4string:Formal class 'CRS' [package "sp"] with 1 slots
```

SpatialLines and SpatialPolygons objects are very similar, as can be seen in Fig. 2.4 – the lists of component entities stack up in a hierarchical fashion. A very typical way of exploring the contents of these objects is to use lapply or sapply in combination with slot. The lapply and sapply functions apply their second argument, which is a function, to each of the elements of their first argument. The command used here can be read as follows: return the length of the Lines slot – how many Line objects it contains – of each Lines object in the list in the lines slot of SLjapan, simplifying the result to a numeric vector. If lapply was used, the result would have been a list. As we see, no Lines object contains more than one Line object:

```
> Lines_len <- sapply(slot(SLjapan, "lines"),function(x) length(slot(x,
+     "Lines")))
> table(Lines_len)
```

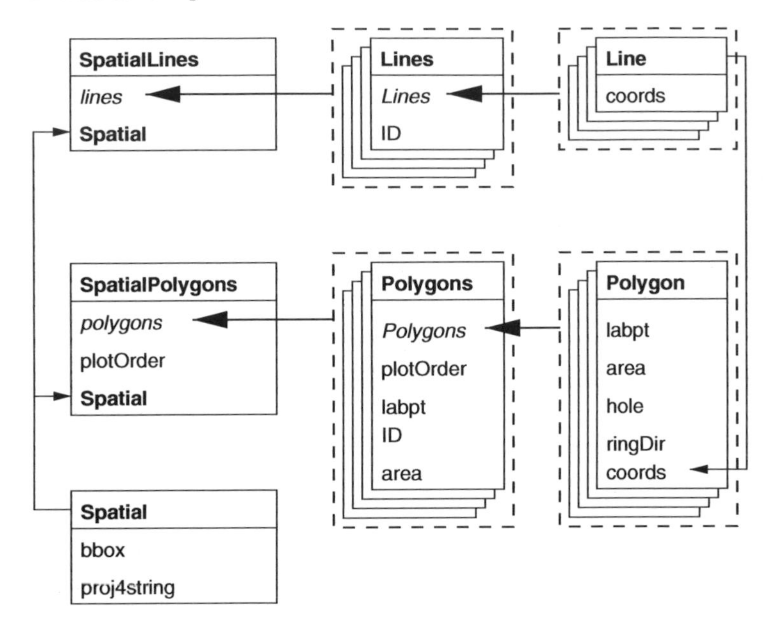

Fig. 2.4. SpatialLines and SpatialPolygons classes and slots; thin arrows show subclass extensions, thick arrows the inclusion of lists of objects

```
Lines_len
 1
51
```

We can use the `ContourLines2SLDF` function included in **maptools** in our next example, converting data returned by the base graphics function `contourLines` into a `SpatialLinesDataFrame` object; we used the volcano data set in Chap. 1, Fig. 1.3:

```
> volcano_sl <- ContourLines2SLDF(contourLines(volcano))
> t(slot(volcano_sl, "data"))
```

```
      C_1   C_2   C_3   C_4   C_5   C_6   C_7   C_8   C_9   C_10
level "100" "110" "120" "130" "140" "150" "160" "170" "180" "190"
```

We can see that there are ten separate contour level labels in the variable in the `data` slot, stored as a factor in the data frame in the object's `data` slot. As mentioned above, **sp** classes are new-style classes, and so the `slots` function can be used to look inside their slots.

To import data that we will be using shortly, we use another utility function in **maptools**, which reads shoreline data in 'Mapgen' format from the National Geophysical Data Center coastline extractor[6] into a `SpatialLines` object directly, here selected for the window shown as the object bounding box:

[6] `http://www.ngdc.noaa.gov/mgg/shorelines/shorelines.html`.

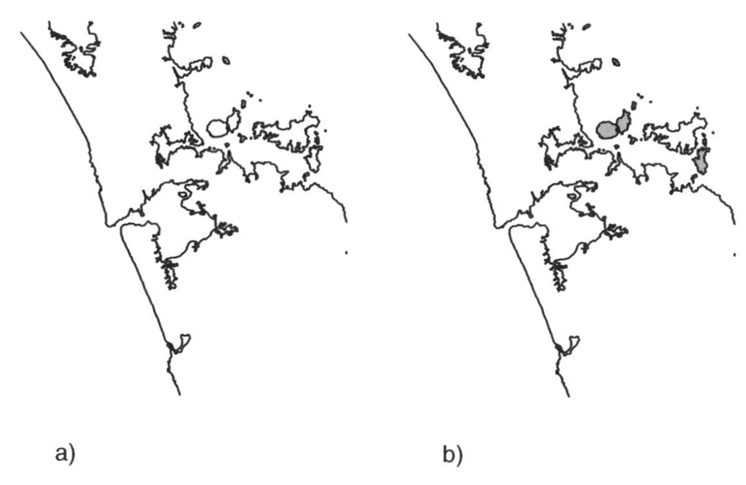

Fig. 2.5. Two maps of shorelines around Auckland: (**a**) line representation, (**b**) line representation over-plotted with islands converted to polygons and shaded. Note that Waiheke Island, the large island to the east, is not closed, and so not found as an island

```
> llCRS <- CRS("+proj=longlat +ellps=WGS84")
> auck_shore <- MapGen2SL("auckland_mapgen.dat", llCRS)
> summary(auck_shore)

Object of class SpatialLines
Coordinates:
     min   max
r1 174.2 175.3
r2 -37.5 -36.5
Is projected: FALSE
proj4string : [+proj=longlat +ellps=WGS84]
```

The shorelines are still just represented by lines, shown in Fig. 2.5, and so colour filling of apparent polygons formed by line rings is not possible. For this we need a class of polygon objects, discussed in Sect. 2.6. Lines, however, can be generalised by removing detail that is not required for analysis or visualisation – the **maps** and **RArcInfo** packages contain functions for line thinning. This operation can be performed successfully only on lines, because neighbouring polygons may have their shared boundary thinned differently. This leads to the creation of slivers, thin zones belonging to neither polygon or to both.

2.6 SpatialPolygons

The basic representation of a polygon in S is a closed line, a sequence of point coordinates where the first point is the same as the last point. A set of polygons is made of closed lines separated by NA points. Like lines, it is not easy to work with polygons represented this way. To have a data set to use for polygons, we first identify the lines imported above representing the shoreline around Auckland. Many are islands, and so have identical first and last coordinates.

```
> lns <- slot(auck_shore, "lines")
> table(sapply(lns, function(x) length(slot(x, "Lines"))))

 1
80

> islands_auck <- sapply(lns, function(x) {
+     crds <- slot(slot(x, "Lines")[[1]], "coords")
+     identical(crds[1, ], crds[nrow(crds), ])
+ })
> table(islands_auck)

islands_auck
FALSE  TRUE
   16    64
```

Since all the Lines in the auck_shore object contain only single Line objects, checking the equality of the first and last coordinates of the first Line object in each Lines object tells us which sets of coordinates can validly be made into polygons. The nesting of classes for polygons is the same as that for lines, but the successive objects have more slots.

```
> getClass("Polygon")

Slots:

Name:     labpt    area    hole ringDir  coords
Class: numeric numeric logical integer  matrix

Extends: "Line"
```

The Polygon class extends the Line class by adding slots needed for polygons and checking that the first and last coordinates are identical. The extra slots are a label point, taken as the centroid of the polygon, the area of the polygon in the metric of the coordinates, whether the polygon is declared as a hole or not – the default value is a logical NA, and the ring direction of the polygon (discussed later in Sect. 2.6.2). No check is made of whether lines cross or polygons have 'errors', in other words whether features are simple in the OpenGIS® (OpenGeoSpatial)[7] context; these are discussed briefly later on p. 122. GIS should do this, and we assume that data read into R can be trusted and contain only simple features.

```
> getClass("Polygons")
```

[7] http://www.opengeospatial.org/.

```
Slots:

Name:   Polygons plotOrder      labpt          ID       area
Class:       list   integer   numeric character    numeric
```

The Polygons class contains a list of valid Polygon objects, an identifying character string, a label point taken as the label point of the constituent polygon with the largest area, and two slots used as helpers in plotting using R graphics functions, given this representation of sets of polygons. These set the order in which the polygons belonging to this object should be plotted, and the gross area of the polygon, equal to the sum of all the constituent polygons. A Polygons object may, for example, represent an administrative district located on two sides of a river, or archipelago. Each of the parts should be seen separately, but data are only available for the larger entity.

```
> getClass("SpatialPolygons")

Slots:

Name:      polygons   plotOrder          bbox proj4string
Class:         list     integer        matrix         CRS

Extends: "Spatial"

Known Subclasses: "SpatialPolygonsDataFrame"
```

The top level representation of polygons is as a SpatialPolygons object, a set of Polygons objects with the additional slots of a Spatial object to contain the bounding box and projection information of the set as a whole. Like the Polygons object, it has a plot order slot, defined by default to plot its member polygons, stored in the polygons as a list of Polygons, in order of gross area, from largest to smallest. Choosing only the lines in the Auckland shoreline data set which are closed polygons, we can build a SpatialPolygons object.

```
> islands_sl <- auck_shore[islands_auck]
> list_of_Lines <- slot(islands_sl, "lines")
> islands_sp <- SpatialPolygons(lapply(list_of_Lines, function(x) {
+     Polygons(list(Polygon(slot(slot(x, "Lines")[[1]],
+         "coords"))), ID = slot(x, "ID"))
+ }), proj4string = CRS("+proj=longlat +ellps=WGS84"))
> summary(islands_sp)

Object of class SpatialPolygons
Coordinates:
         min       max
r1 174.30297 175.22791
r2 -37.43877 -36.50033
Is projected: FALSE
proj4string : [+proj=longlat +ellps=WGS84]
```

```
> slot(islands_sp, "plotOrder")

 [1] 45 54 37 28 38 27 12 11 59 53  5 25 26 46  7 55 17 34 30 16  6 43
[23] 14 40 32 19 61 42 15 50 21 18 62 23 22 29 24 44 13  2 36  9 63 58
[45] 56 64 52 39 51  1  8  3  4 20 47 35 41 48 60 31 49 57 10 33

> order(sapply(slot(islands_sp, "polygons"), function(x) slot(x,
+     "area")), decreasing = TRUE)

 [1] 45 54 37 28 38 27 12 11 59 53  5 25 26 46  7 55 17 34 30 16  6 43
[23] 14 40 32 19 61 42 15 50 21 18 62 23 22 29 24 44 13  2 36  9 63 58
[45] 56 64 52 39 51  1  8  3  4 20 47 35 41 48 60 31 49 57 10 33
```

As we saw with the construction of `SpatialLines` objects from raw coordinates, here we build a list of `Polygon` objects for each `Polygons` object, corresponding to a single identifying tag. A list of these `Polygons` objects is then passed to the `SpatialPolygons` function, with a coordinate reference system, to create the `SpatialPolygons` object. Again, like `SpatialLines` objects, `SpatialPolygons` objects are most often created by functions that import or manipulate such data objects, and seldom from scratch.

2.6.1 `SpatialPolygonsDataFrame` Objects

As with other spatial data objects, `SpatialPolygonsDataFrame` objects bring together the spatial representations of the polygons with data. The identifying tags of the `Polygons` in the `polygon` slot of a `SpatialPolygons` object are matched with the row names of the data frame to make sure that the correct data rows are associated with the correct spatial objects. The data frame is re-ordered by row to match the spatial objects if need be, provided all the objects can be matched to row names. If any differences are found, an error results. Both identifying tags and data frame row names are character strings, and so their sort order is also character, meaning that `"2"` follows `"11"` and `"111"`.[8]

As an example, we take a set of scores by US state of 1999 Scholastic Aptitude Test (SAT) used for spatial data analysis by Melanie Wall.[9] In the data source, there are also results for Alaska, Hawaii, and for the US as a whole. If we would like to associate the data with state boundary polygons provided in the **maps** package, it is convenient to convert the boundaries to a `SpatialPolygons` object – see also Chap. 4.

```
> library(maps)
> library(maptools)
> state.map <- map("state", plot = FALSE, fill = TRUE)
```

[8] Some **maptools** functions use Gregory R. Warnes' `mixedorder` sort from **gtools** to sort integer-like strings in integer order.

[9] `http://www.biostat.umn.edu/~melanie/Data/`, data here supplemented with variable names and state names as used in **maps**.

```
> IDs <- sapply(strsplit(state.map$names, ":"), function(x) x[1])
> state.sp <- map2SpatialPolygons(state.map, IDs = IDs,
+     proj4string = CRS("+proj=longlat +ellps=WGS84"))
```

Then we can use identifying tag matching to suit the rows of the data frame to the `SpatialPolygons`. Here, the rows of the data frame for which there are no matches will be dropped; all the `Polygons` objects are matched:

```
> sat <- read.table("state.sat.data_mod.txt", row.names = 5,
+     header = TRUE)
> str(sat)

'data.frame': 52 obs. of 4 variables:
$ oname :Factor w/ 52 levels "ala","alaska",..: 1 2 3 4 5 ...
$ vscore:int 561 516 524 563 497 ...
$ mscore:int 555 514 525 556 514 ...
$ pc :int 9 50 34 6 49 ...

> id <- match(row.names(sat), sapply(slot(state.sp, "polygons"),
+     function(x) slot(x, "ID")))
> row.names(sat)[is.na(id)]

[1] "alaska" "hawaii" "usa"

> state.spdf <- SpatialPolygonsDataFrame(state.sp, sat)
> str(slot(state.spdf, "data"))

'data.frame': 49 obs. of 4 variables:
$ oname :Factor w/ 52 levels "ala","alaska",..: 1 3 4 5 6 ...
$ vscore:int 561 524 563 497 536 ...
$ mscore:int 555 525 556 514 540 ...
$ pc :int 9 34 6 49 32 ...

> str(state.spdf, max.level = 2)

Formal class 'SpatialPolygonsDataFrame' [package "sp"] with 5 slots
..@ data :'data.frame': 49 obs. of 4 variables:
..@ polygons :List of 49
..@ plotOrder :int [1:49] 42 25 4 30 27 ...
..@ bbox :num [1:2, 1:2] -124.7 25.1 ...
.. ..- attr(*, "dimnames")=List of 2
..@ proj4string:Formal class 'CRS' [package "sp"] with 1 slots
```

If we modify the row name of `'arizona'` in the data frame to `'Arizona'`, there is no longer a match with a polygon identifying tag, and an error is signalled.

```
> rownames(sat)[3] <- "Arizona"
> SpatialPolygonsDataFrame(state.sp, sat)
```

```
Error in SpatialPolygonsDataFrame(state.sp, sat) : row.names
of data and Polygons IDs do not match
```

In subsequent analysis, Wall (2004) also drops District of Columbia. Rather than having to manipulate polygons and their data separately, when using a `SpatialPolygonsDataFrame` object, we can say:

```
> DC <- "district of columbia"
> not_dc <- !(row.names(slot(state.spdf, "data")) == DC)
> state.spdf1 <- state.spdf[not_dc, ]
> length(slot(state.spdf1, "polygons"))

[1] 48

> summary(state.spdf1)

Object of class SpatialPolygonsDataFrame
Coordinates:
          min       max
r1 -124.68134 -67.00742
r2   25.12993  49.38323
Is projected: FALSE
proj4string : [+proj=longlat +ellps=WGS84]
Data attributes:
     oname           vscore           mscore             pc
 ala     : 1   Min.   :479.0   Min.   :475.0   Min.   : 4.00
 ariz    : 1   1st Qu.:506.2   1st Qu.:505.2   1st Qu.: 9.00
 ark     : 1   Median :530.5   Median :532.0   Median :28.50
 calif   : 1   Mean   :534.6   Mean   :534.9   Mean   :35.58
 colo    : 1   3rd Qu.:563.0   3rd Qu.:558.5   3rd Qu.:63.50
 conn    : 1   Max.   :594.0   Max.   :605.0   Max.   :80.00
 (Other):42
```

2.6.2 Holes and Ring Direction

The hole and ring direction slots are included in `Polygon` objects as heuristics to address some of the difficulties arising from S not being a GIS. In a traditional vector GIS, and in the underlying structure of the data stored in **maps**, boundaries between polygons are stored only once as arcs between nodes (shared vertices between three or more polygons, possibly including the external space), and the polygons are constructed on the fly from lists of directed boundary arcs, including boundaries with the external space – void – not included in any polygon. This is known as the topological representation of polygons, and is appropriate for GIS software, but arguably not for other software using spatial data. It was mentioned above that it is the user's responsibility to provide line coordinates such that the coordinates represent the line object the user requires. If the user requires, for example, that a river channel does not cross itself, the user has to impose that limitation. Other users will not need such a limitation, as for example tracking data may very well involve an animal crossing its tracks.

The approach that has been chosen in **sp** is to use two markers commonly encountered in practice, marking polygons as holes with a logical (TRUE/FALSE) flag, the `hole` slot, and using ring direction – clockwise rings are taken as not being holes, anti-clockwise as being holes. This is needed because the non-topological representation of polygons has no easy way of knowing that a polygon represents an internal boundary of an enclosing polygon, a hole, or lake.

An approach that works when the relative status of polygons is known is to set the hole slot directly. This is done in reading GSHHS shoreline data, already used in Fig. 2.3 and described in Chap. 4. The data source includes a variable for each polygon, where the levels are land: 1, lake: 2, island in lake: 3, and lake on island in lake: 4. The following example takes a region of interest on the northern, Canadian shore of Lake Huron, including Manitoulin Island, and a number of lakes on the island, including Kongawong Lake.

```
> length(slot(manitoulin_sp, "polygons"))

[1] 1

> sapply(slot(slot(manitoulin_sp, "polygons")[[1]], "Polygons"),
+     function(x) slot(x, "hole"))

 [1] FALSE  TRUE FALSE  TRUE  TRUE FALSE  TRUE FALSE FALSE FALSE FALSE
[12] FALSE FALSE FALSE FALSE FALSE FALSE FALSE FALSE

> sapply(slot(slot(manitoulin_sp, "polygons")[[1]], "Polygons"),
+     function(x) slot(x, "ringDir"))

 [1]  1 -1  1 -1 -1  1 -1  1  1  1  1  1  1  1  1  1  1  1  1
```

In Fig. 2.6, there is only one `Polygons` object in the `polygons` slot of `manitoulin_sp`, representing the continental landmass, exposed along the northern edge, and containing the complete set of polygons. Within this is a large section covered by Lake Huron, which in turn is covered by islands and lakes on islands. Not having a full topological representation means that for plotting, we paint the land first, then paint the lake, then the islands, and finally the lakes on islands. Because the default plotting colour for holes is `'transparent'`, they can appear to be merged into the surrounding land – the same problem arises where the hole slot is wrongly assigned. The `plotOrder` slots in `Polygons` and `SpatialPolygons` objects attempt to get around this problem, but care is usually sensible if the spatial objects being handled are complicated.

2.7 SpatialGrid and SpatialPixel Objects

The point, line, and polygon objects we have considered until now have been handled one-by-one. Grids are regular objects requiring much less information to define their structure. Once the single point of origin is known, the extent

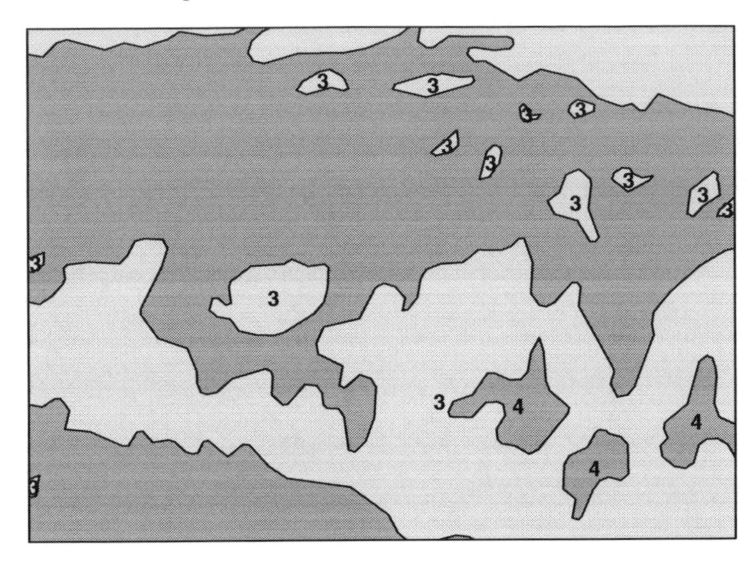

Fig. 2.6. The northern, Canadian shore of Lake Huron, including Manitoulin Island and lakes on the island; islands (*light grey*) and lakes on islands (*dark grey*) are marked with their GSHHS levels

of the grid can be given by the cell resolution and the numbers of rows and columns present in the full grid. This representation is typical for remote sensing and raster GIS, and is used widely for storing data in regular rectangular cells, such as digital elevation models, satellite imagery, and interpolated data from point measurements, as well as image processing.

```
> getClass("GridTopology")

Slots:

Name:  cellcentre.offset          cellsize         cells.dim
Class:           numeric           numeric           integer
```

As an example, we make a `GridTopology` object from the bounding box of the Manitoulin Island vector data set. If we choose a cell size of 0.01° in each direction, we can offset the south-west cell centre to make sure that at least the whole area is covered, and find a suitable number of cells in each dimension.

```
> bb <- bbox(manitoulin_sp)
> bb

     min   max
r1 277.0 278.0
r2  45.7  46.2

> cs <- c(0.01, 0.01)
> cc <- bb[, 1] + (cs/2)
> cd <- ceiling(diff(t(bb))/cs)
> manitoulin_grd <- GridTopology(cellcentre.offset = cc,
```

```
+     cellsize = cs, cells.dim = cd)
> manitoulin_grd

                       r1     r2
cellcentre.offset 277.005 45.705
cellsize            0.010  0.010
cells.dim         100.000 50.000
```

The object describes the grid completely, and can be used to construct a `SpatialGrid` object. A `SpatialGrid` object contains `GridTopology` and `Spatial` objects, together with two helper slots, `grid.index` and `coords`. These are set to zero and to the bounding box of the cell centres of the grid, respectively.

```
> getClass("SpatialGrid")

Slots:

Name:           grid   grid.index       coords         bbox
Class: GridTopology      integer       matrix       matrix

Name: proj4string
Class: CRS

Extends:
Class "SpatialPixels", directly, with explicit coerce
Class "SpatialPoints", by class "SpatialPixels", distance 2, with
     explicit coerce
Class "Spatial", by class "SpatialPixels", distance 3, with explicit
     coerce

Known Subclasses: "SpatialGridDataFrame"
```

Using the `GridTopology` object created above, and passing through the coordinate reference system of the original GSHHS data, the bounding box is created automatically, as we see from the summary of the object:

```
> p4s <- CRS(proj4string(manitoulin_sp))
> manitoulin_SG <- SpatialGrid(manitoulin_grd, proj4string = p4s)
> summary(manitoulin_SG)

Object of class SpatialGrid
Coordinates:
     min   max
r1 277.0 278.0
r2  45.7  46.2
Is projected: FALSE
proj4string : [+proj=longlat +datum=WGS84]
Number of points: 2
Grid attributes:
   cellcentre.offset cellsize cells.dim
r1           277.005     0.01       100
r2            45.705     0.01        50
```

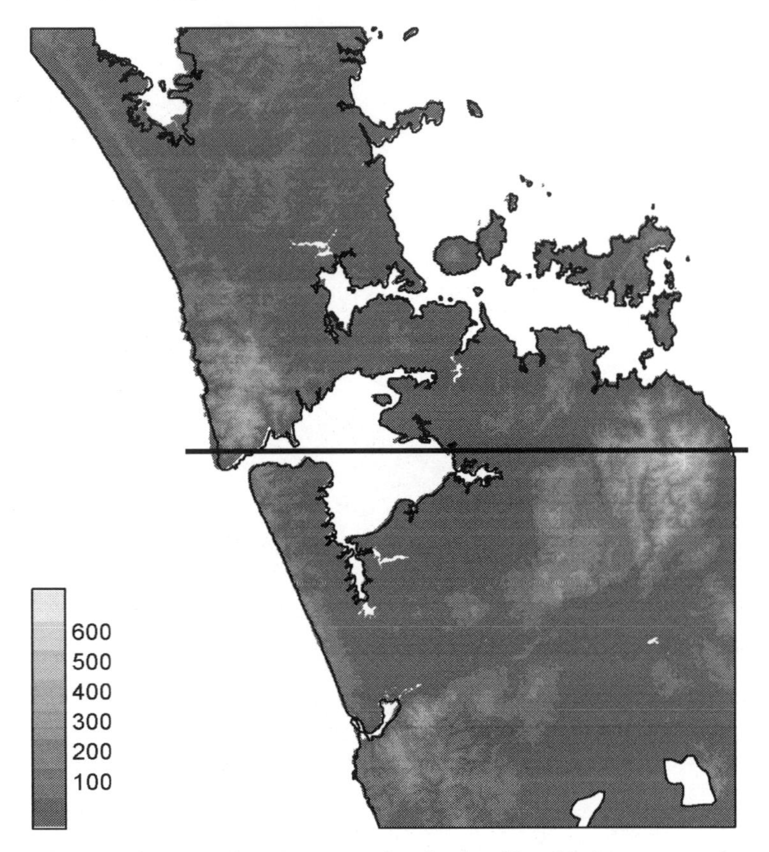

Fig. 2.7. SRTM elevation data in metres for the Auckland isthmus over-plotted with an excerpt from the GSHHS full resolution shoreline, including two lakes – there are detailed differences stemming from the very different technologies underlying the two data sources. A transect is marked for later use

As an example of using these classes with imported data, we use an excerpt from the Shuttle Radar Topography Mission (SRTM) flown in 2000, for the Auckland area[10] (Fig. 2.7). The data have been read from a Geotiff file into a `SpatialGridDataFrame` object – a `SpatialGrid` object extended with a `data` slot occupied by a `data.frame` object, filled with a single band of data representing elevation in metres.After checking the class of the data object, we examine in turn its slots. The `grid` slot contains the underlying `GridTopology` object, with the lower left cell centre coordinates, the pair of cell size resolution values, here both equal to 3 arcsec, and the numbers of columns and rows:

[10] Downloaded from the seamless data distribution system for 3 arcsec 'Finished' (90 m) data, `http://seamless.usgs.gov/`; the data can be downloaded as one degree square tiles, or cropped from a seamless raster database, as has been done here to avoid patching tiles together.

```
> class(auck_el1)

[1] "SpatialGridDataFrame"
attr(,"package")
[1] "sp"

> slot(auck_el1, "grid")

                             x             y
cellcentre.offset 1.742004e+02 -3.749958e+01
cellsize          8.333333e-04  8.333333e-04
cells.dim         1.320000e+03  1.200000e+03

> slot(auck_el1, "grid.index")

integer(0)

> slot(auck_el1, "coords")

            x         y
[1,] 174.2004 -37.49958
[2,] 175.2996 -36.50042

> slot(auck_el1, "bbox")

    min   max
x 174.2 175.3
y -37.5 -36.5

> object.size(auck_el1)

[1] 12674948

> object.size(slot(auck_el1, "data"))

[1] 12672392
```

The `grid.index` slot is empty, while the `coords` slot is as described earlier. It differs from the bounding box of the grid as a whole, contained in the `bbox` slot, by half a cell resolution value in each direction. The total size of the `SpatialGridDataFrame` object is just over 12 MB, almost all of which is made up of the `data` slot.

```
> is.na(auck_el1$band1) <- auck_el1$band1 <= 0 | auck_el1$band1 >
+     10000
> summary(auck_el1$band1)

     Min.   1st Qu.    Median      Mean   3rd Qu.      Max.      NA's
     1.00     23.00     53.00     78.05    106.00    686.00 791938.00
```

Almost half of the data are at or below sea level, and other values are spikes in the radar data, and should be set to `NA`. Once this is done, about half of the data are missing. In other cases, even larger proportions of raster grids are missing, suggesting that an alternative representation of the same data might

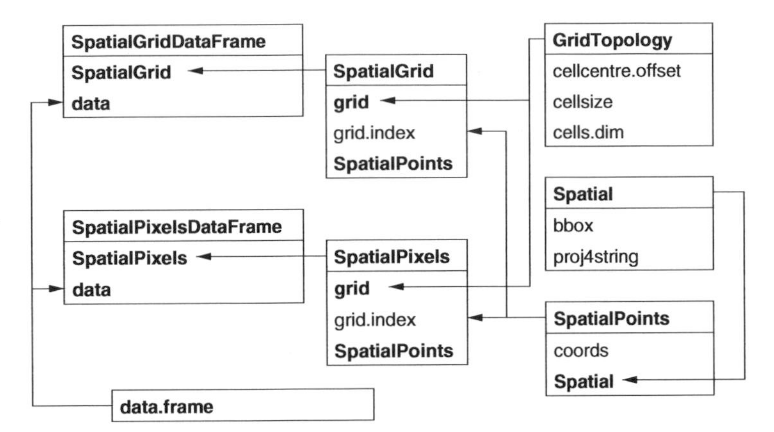

Fig. 2.8. `SpatialGrid` and `SpatialPixel` classes and their slots; arrows show subclass extensions

be attractive. One candidate from Terralib, discussed further in Chap. 4, is the cell representation of rasters, where the raster cells with data are represented by the coordinates of the cell centre, and by the sequence number of the cell among all the cells in the raster. In this representation, missing data are discarded, and savings in space and processing time can be large. It also permits cells to be stored, like points, in an external database. The class is here termed `SpatialPixels`, and has the same slots as `SpatialGrid` objects, but differently filled (Fig. 2.8). The `SpatialPixelsDataFrame` class is analogous.

```
> auck_el2 <- as(auck_el1, "SpatialPixelsDataFrame")
> object.size(auck_el2)

[1] 25349276

> object.size(slot(auck_el2, "grid.index"))

[1] 3168272

> object.size(slot(auck_el2, "coords"))

[1] 12673288

> sum(is.na(auck_el1$band1)) + nrow(slot(auck_el2, "coords"))

[1] 1584000

> prod(slot(slot(auck_el2, "grid"), "cells.dim"))

[1] 1584000
```

Returning to our example, we can coerce our `SpatialGridDataFrame` object to a `SpatialPixelsDataFrame` object. In this case, the proportion of missing to occupied cells is unfavourable, and when the `grid.index` and `coords` slots are populated with cell indices and coordinates, the output object is almost twice as large as its SpatialGridDataFrame equivalent. We can also see that the total number of cells – the product of the row and column dimensions – is equal to the number of coordinates in the output object plus the number of missing data values deleted by coercion. Had the number of attributes been 10, then the space saving relative to storing the full grid would have been 37%; with 100 attributes it would have been 48% for this particular case.

```
> auck_el_500 <- auck_el2[auck_el2$band1 > 500, ]

Warning messages:
1: grid has empty column/rows in dimension 1 in:
     points2grid(points, tolerance)
2: grid has empty column/rows in dimension 2 in:
     points2grid(points, tolerance)

> summary(auck_el_500)

Object of class SpatialPixelsDataFrame
Coordinates:
        min       max
x 175.18917 175.24333
y -37.10333 -37.01833
Is projected: FALSE
proj4string :
[+proj=longlat +ellps=WGS84 +datum=WGS84 +no_defs]
Number of points: 1114
Grid attributes:
  cellcentre.offset     cellsize cells.dim
x         175.18958 0.0008333333        65
y         -37.10292 0.0008333333       102
Data attributes:
   Min. 1st Qu.  Median    Mean 3rd Qu.    Max.
  501.0   523.0   552.0   559.4   591.0   686.0

> object.size(auck_el_500)

[1] 38940
```

Taking just the raster cells over 500 m, of which there are very few, less than 1% of the total, yields a much smaller object. In this case it has a smaller bounding box, and gaps between the pixels present, as the warning messages indicate.

We can also create a SpatialPixels object directly from a SpatialPoints object. As our example, we use the Meuse bank data set provided with **sp**. We can pass a SpatialPoints object to the SpatialPixels function, where the Spatial object components are copied across, and the points checked to see whether they lie on a regular grid. If they do, the function will return a SpatialPixels object:

```
> data(meuse.grid)
> mg_SP <- SpatialPoints(cbind(meuse.grid$x, meuse.grid$y))
> summary(mg_SP)
```

```
Object of class SpatialPoints
Coordinates:
             min    max
coords.x1 178460 181540
coords.x2 329620 333740
Is projected: NA
proj4string : [NA]
Number of points: 3103
```

```
> mg_SPix0 <- SpatialPixels(mg_SP)
> summary(mg_SPix0)
```

```
Object of class SpatialPixels
Coordinates:
             min    max
coords.x1 178440 181560
coords.x2 329600 333760
Is projected: NA
proj4string : [NA]
Number of points: 3103
Grid attributes:
          cellcentre.offset cellsize cells.dim
coords.x1            178460       40        78
coords.x2            329620       40       104
```

```
> prod(slot(slot(mg_SPix0, "grid"), "cells.dim"))
```

```
[1] 8112
```

As we can see from the product of the cell dimensions of the underlying grid, over half of the full grid is not present in the `SpatialPixels` representation, because many grid cells lie outside the study area. Alternatively, we can coerce a `SpatialPoints` object to a `SpatialPixels` object:

```
> mg_SPix1 <- as(mg_SP, "SpatialPixels")
> summary(mg_SPix1)
```

```
Object of class SpatialPixels
Coordinates:
             min    max
coords.x1 178440 181560
coords.x2 329600 333760
Is projected: NA
proj4string : [NA]
Number of points: 3103
Grid attributes:
          cellcentre.offset cellsize cells.dim
coords.x1            178460       40        78
coords.x2            329620       40       104
```

We have now described a coherent and consistent set of classes for spatial data. Other representations are also used by R packages, and we show further ways of converting between these representations and external formats in Chap. 4. Before treating data import and export, we discuss graphical methods for **sp** classes, to show that the effort of putting the data in formal classes may be justified by the relative ease with which we can make maps.

3

Visualising Spatial Data

A major pleasure in working with spatial data is their visualisation. Maps are amongst the most compelling graphics, because the space they map is the space we think we live in, and maps may show things we cannot see otherwise. Although one can work with all R plotting functions on the raw data, for example extracted from `Spatial` classes by methods like `coordinates` or `as.data.frame`, this chapter introduces the plotting methods for objects inheriting from class `Spatial` that are provided by package **sp**.

R has two plotting systems: the 'traditional' plotting system and the Trellis Graphics system, provided by package **lattice**, which is present in default R installations (Sarkar, 2008). The latter builds upon the 'grid' graphics model (Murrell, 2006). Traditional graphics are typically built incrementally: graphic elements are added in several consecutive function calls. Trellis graphics allow plotting of high-dimensional data by providing *conditioning plots*: organised lattices of plots with shared axes (Cleveland, 1993, 1994). This feature is particularly useful when multiple maps need to be compared, for example in case of a spatial time series, comparison across a number of species or variables, or comparison of different modelling scenarios or approaches. Trellis graphs are designed to avoid wasting space by repetition of identical information. The value of this feature, rarely found in other software, is hard to overestimate.

Waller and Gotway (2004, pp. 68–86) provide an introduction to statistical mapping, which may be deepened with reference to Slocum et al. (2005).

Package **sp** provides `plot` methods that build on the traditional R plotting system (`plot`, `image`, `lines`, `points`, etc.), as well as a 'new' generic method called `spplot` that uses the Trellis system (notably `xyplot` or `levelplot` from the **lattice** package) and can be used for conditioning plots. The `spplot` methods are introduced in a later sub-section, first we deal with the traditional plot system.

3.1 The Traditional Plot System

3.1.1 Plotting Points, Lines, Polygons, and Grids

In the following example session, we create points, lines, polygons, and grid object, from `data.frame` objects, retrieved from the **sp** package by function `data`, and plot them. The four plots obtained by the `plot` and `image` commands are shown in Fig. 3.1.

```
> library(sp)
> data(meuse)
```

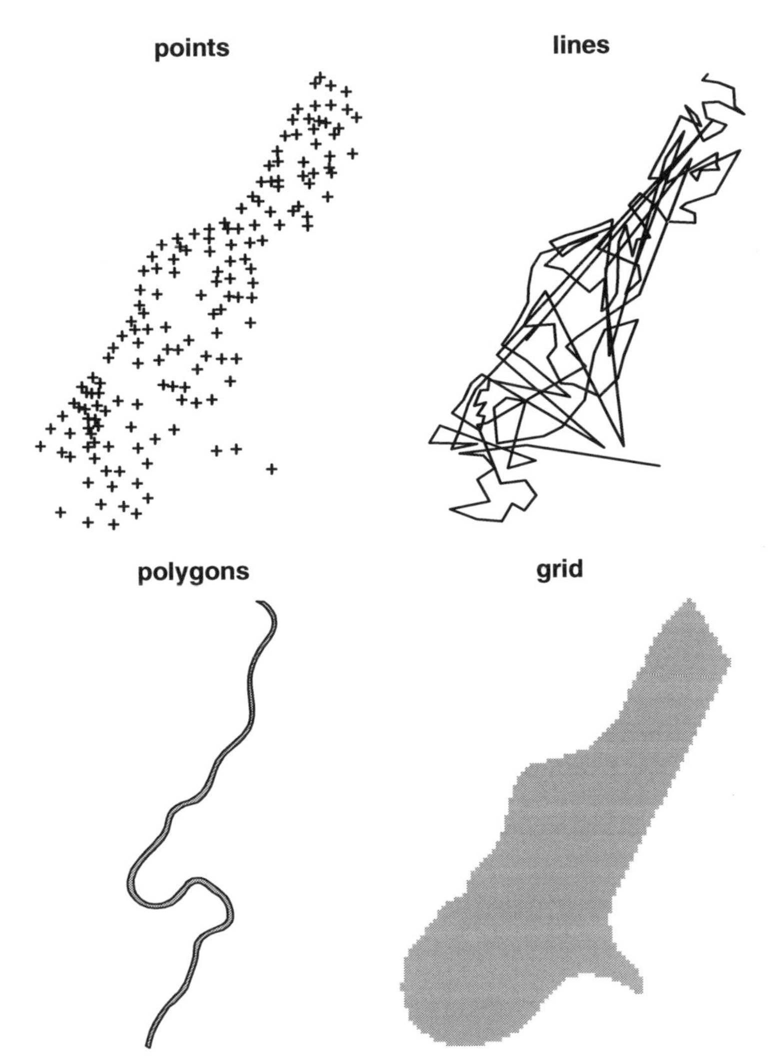

Fig. 3.1. The meuse data set: sample points, the sample path (*line*), the Meuse river (*ring*), and the gridded study area

```
> coordinates(meuse) <- c("x", "y")
> plot(meuse)
> title("points")
```

The SpatialPointsDataFrame object used is created from a data.frame provided with **sp**, and the plot method shows the points with the default symbol.

```
> cc <- coordinates(meuse)
> m.sl <- SpatialLines(list(Lines(list(Line(cc)))))
> plot(m.sl)
> title("lines")
```

A SpatialLines object is made by joining up the points in sequence, and plot draws the resulting zig-zags.

```
> data(meuse.riv)
> meuse.lst <- list(Polygons(list(Polygon(meuse.riv)),
+     "meuse.riv"))
> meuse.sr <- SpatialPolygons(meuse.lst)
> plot(meuse.sr, col = "grey")
> title("polygons")
```

We make a SpatialPolygons object from data provided with **sp** outlining the banks of the River Meuse.

```
> data(meuse.grid)
> coordinates(meuse.grid) <- c("x", "y")
> meuse.grid <- as(meuse.grid, "SpatialPixels")
> image(meuse.grid, col = "grey")
> title("grid")
```

Finally, we convert grid data for the same Meuse bank study area into a 'SpatialPixels' object and display it using the image method, with all cells set to 'grey'.

On each map, one unit in the x-direction equals one unit in the y-direction. This is the default when the coordinate reference system is not longlat or is unknown. For unprojected data in geographical coordinates (longitude/latitude), the default aspect ratio depends on the (mean) latitude of the area plotted. The default aspect can be adjusted by passing the asp argument.

A map becomes more readable when we combine several elements. We can display elements from those created above by using the add = TRUE argument in function calls:

```
> image(meuse.grid, col = "lightgrey")
> plot(meuse.sr, col = "grey", add = TRUE)
> plot(meuse, add = TRUE)
```

the result of which is shown in Fig. 3.2.

The over-plotting of polygons by points is the consequence of the order of plot commands. Up to now, the plots only show the geometry (topology, shapes) of the objects; we start plotting attributes (e.g. what has actually been measured at the sample points) in Sect. 3.1.5.

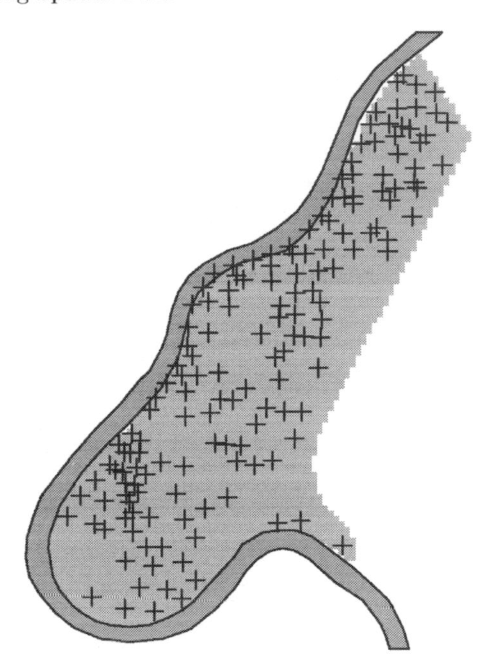

Fig. 3.2. Map elements combined into a single map

As an alternative to `plot(x,add=TRUE)`, one can use the commands `lines` for objects of class `SpatialLines` and `points` for `SpatialPoints`; text elements can be added by `text`.

3.1.2 Axes and Layout Elements

Maps often do not have axes, as the information carried in map axes can often be omitted. Especially, projected coordinates are usually long, hard to read and geographical reference is much easier when recognisable features such as administrative boundaries, rivers, coast lines, etc. are present. In the standard `plot` functions, the Boolean argument `axes` can be set to control axis plotting, and the function `axis` can be called to add axes, fine-tuning their appearance (tic placement, tic labels, and font size). The following commands result in Fig. 3.3:

```
> layout(matrix(c(1, 2), 1, 2))
> plot(meuse.sr, axes = TRUE)
> plot(meuse.sr, axes = FALSE)
> axis(1, at = c(178000 + 0:2 * 2000), cex.axis = 0.7)
> axis(2, at = c(326000 + 0:3 * 4000), cex.axis = 0.7)
> box()
```

Not plotting axes does not increase the amount of space R used for plotting the data.[1] R still reserves the necessary space for adding axes and titles later

[1] This is not true for Trellis plots; see Sect. 3.2.

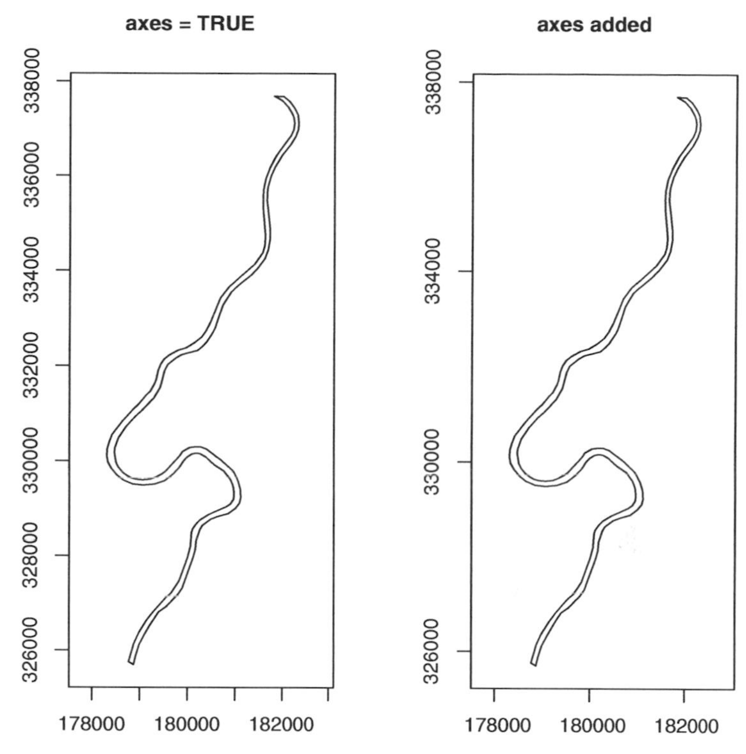

Fig. 3.3. Default axes (*left*) and custom axes (*right*) for the `meuse.riv` data

Table 3.1. Graphic arguments useful for controlling figure and plotting region

Argument	Meaning	Unit	Length
`fin`	Figure region	Inch	2
`pin`	Plotting region	Inch	2
`mai`	Plotting margins	Inch	4
`mar`	Plotting margins	Lines of text	4

see `?par` for more information

on. We can, however, explicitly instruct R not to reserve this space by using function `par`, which is intended to have side effects on the next plot on the current device. The `par`-settable arguments that are useful for controlling the physical size of the plot are listed in Table 3.1.

In Fig. 3.4, generated by

```
> oldpar = par(no.readonly = TRUE)
> layout(matrix(c(1, 2), 1, 2))
> plot(meuse, axes = TRUE, cex = 0.6)
> plot(meuse.sr, add = TRUE)
> title("Sample locations")
> par(mar = c(0, 0, 0, 0) + 0.1)
> plot(meuse, axes = FALSE, cex = 0.6)
```

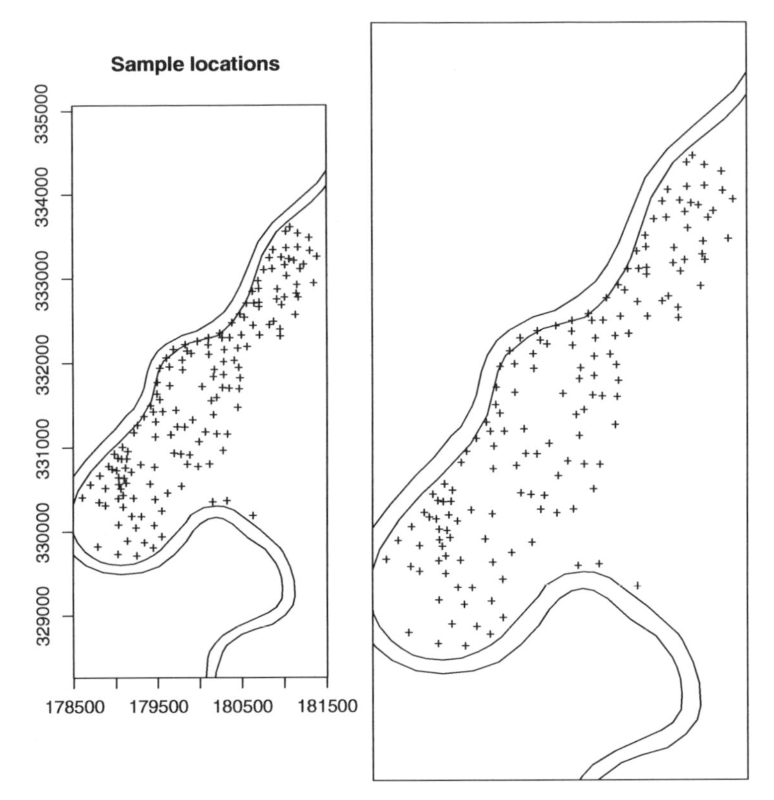

Fig. 3.4. Equal-area plots with (*left*) and without (*right*) the default space R reserves for axes and title(s)

```
> plot(meuse.sr, add = TRUE)
> box()
> par(oldpar)
```

the same data set is plotted twice within the same amount of space, at the left-hand side with R's default margins leaving space for axes, and on the right-hand side with maximised plotting space and no axes drawn.

Modifying the margins by setting `mar` in the `par` command, for example to `par(mar=c(3,3,2,1))` further optimises space usage when axes are drawn, leaving (little) space for a title. It should be noted that the margin sizes are absolute, expressed in units the height of a line of text, and so their effect on map scale decreases when the plotting region is enlarged.

The `plot` methods provided by package **sp** do not allow the printing of axis labels, such as 'Easting' and 'Northing', or 'x-coordinate' and 'y-coordinate'. The reason for this is technical, but mentioning axis names is usually obsolete once the graph is referred to as a map. The units of the coordinate reference system (such as metres) should be equal for both axes and do not need mentioning twice. Geographical coordinates are perhaps an exception, but this is made explicit by axis tic labels such as 52°N, or by adding a reference grid.

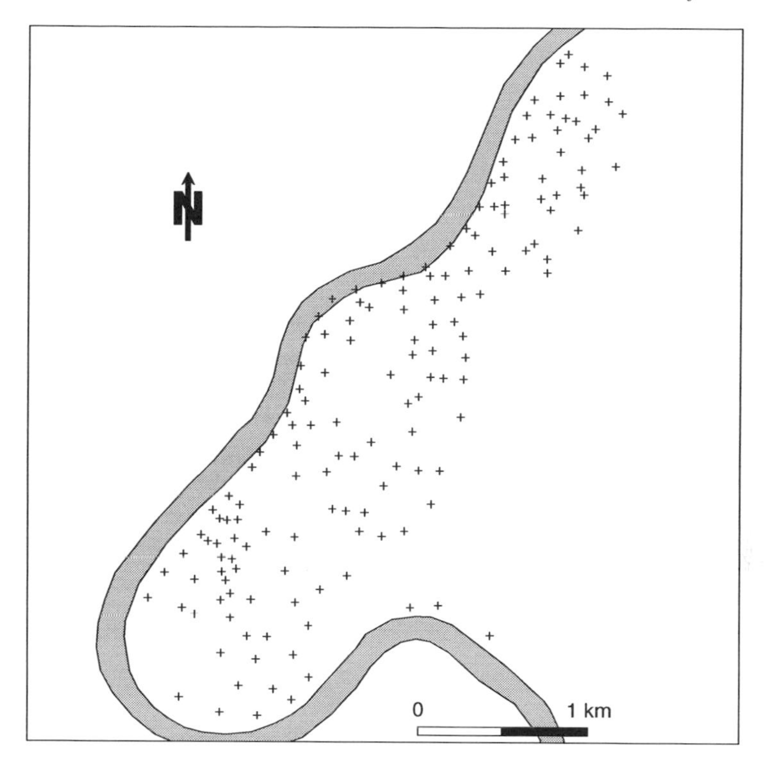

Fig. 3.5. Scale bar and north arrow as map elements

When we decide not to draw axes on a map, in addition to reference boundaries, we can provide the reader of a map with a guidance for distance and direction by plotting a scale bar and a north arrow, which can be placed interactively using `locator` followed by a few well-chosen clicks in the map (Fig. 3.5):

```
> plot(meuse)
> plot(meuse.sr, add = TRUE)
> plot(meuse)
> SpatialPolygonsRescale(layout.scale.bar(), offset = locator(1),
+   scale = 1000, fill = c("transparent", "black"), plot.grid = FALSE)
> text(locator(1), "0")
> text(locator(1), "1 km")
> SpatialPolygonsRescale(layout.north.arrow(), offset = locator(1),
+     scale = 400, plot.grid = FALSE)
```

When large numbers of maps for identical areas have to be produced with identical layout elements, it pays off to write a function that draws all layout elements. As an alternative, one may use conditioning plots; see the `spplot` method in Sect. 3.2.

3.1.3 Degrees in Axes Labels and Reference Grid

Unprojected data have coordinates in latitude and longitude degrees, with negative degrees referring to degrees west (of the prime meridian) and south (of the Equator). When unprojected spatial data are plotted using **sp** methods (`plot` or `spplot`), the axis label marks will give units in decimal degrees N/S/E/W, for example 50.5°N. An example is shown in Fig. 3.6.

When, for reference purposes, a grid needs to be added to a map, the function `gridlines` can be used to generate an object of class `SpatialLines`. By default it draws lines within the bounding box of the object at values where the default axes labels are drawn; other values can be specified. Grid lines may be latitude/longitude grids, and these are non-straight lines. This is accomplished by generating a grid for unprojected data, projecting it, and plotting it over the map shown. An example is given in Fig. 1.2. This is the code used to define and draw projected latitude/longitude grid lines and grid line labels for this figure, which uses the world map from package **maps**:

```
> library(maptools)
> library(maps)
> wrld <- map("world", interior = FALSE, xlim = c(-179,
+     179), ylim = c(-89, 89), plot = FALSE)
> wrld_p <- pruneMap(wrld, xlim = c(-179, 179))
> llCRS <- CRS("+proj=longlat +ellps=WGS84")
> wrld_sp <- map2SpatialLines(wrld_p, proj4string = llCRS)
> prj_new <- CRS("+proj=moll")
> library(rgdal)
> wrld_proj <- spTransform(wrld_sp, prj_new)
> wrld_grd <- gridlines(wrld_sp, easts = c(-179, seq(-150,
+     150, 50), 179.5), norths = seq(-75, 75, 15), ndiscr = 100)
> wrld_grd_proj <- spTransform(wrld_grd, prj_new)
> at_sp <- gridat(wrld_sp, easts = 0, norths = seq(-75,
+     75, 15), offset = 0.3)
```

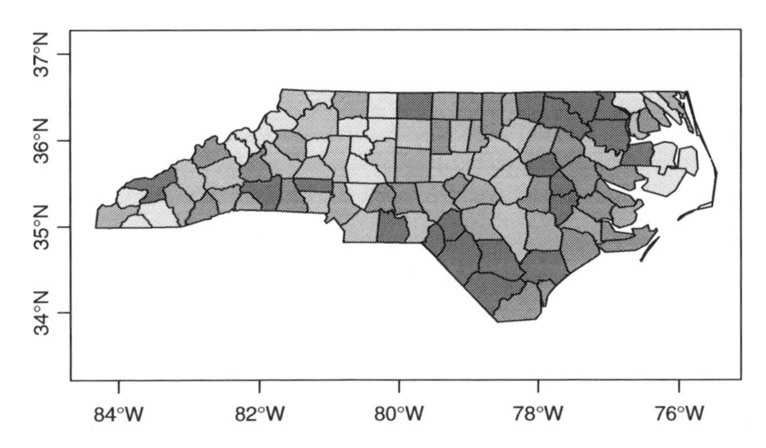

Fig. 3.6. Decimal degrees in axis labels: the North Carolina SIDS data

```
> at_proj <- spTransform(at_sp, prj_new)
> plot(wrld_proj, col = "grey60")
> plot(wrld_grd_proj, add = TRUE, lty = 3, col = "grey70")
> text(coordinates(at_proj),pos = at_proj$pos, offset = at_proj$offset,
+      labels = parse(text = as.character(at_proj$labels)),
+      cex = 0.6)
```

Here, function `gridat` returns an object to draw the labels for these 'gridded curves'.

3.1.4 Plot Size, Plotting Area, Map Scale, and Multiple Plots

R distinguishes between figure region, which is the size of the total figure including axes, title, etc., and plotting region, which is the area where the actual data are plotted. To control the total size of the figure, we can get and set the figure size in inches:

```
> par("pin")
> par(pin = c(4, 4))
```

If we want to enlarge the plotting window, we may have to close the current plotting device and re-open it specifying size, for example

```
> dev.off()
> X11(width = 10, height = 10)
```

on Unix machines; replace `X11` with `windows` on MS-Windows computers and with `quartz` on Mac OS X. When graphic output is written to files, we can use, for example

```
> postscript("file.ps", width = 10, height = 10)
```

The geographical (data) area that is shown on a plot is by default that of the data, extended with a 4% margin on each side. Because the plot size is fixed before plotting, only one of the axes will cover the entire plotting region, the other will be centred and have larger margins. We can control the data area plotted by passing `xlim` and `ylim` in a plot command, but by default they will still be extended with 4% on each side. To prevent this extension, we can set `par(xaxs="i")` and `par(yaxs="i")`. In the following example

```
> pin <- par("pin")
> dxy <- apply(bbox(meuse), 1, diff)
> ratio <- dxy[1]/dxy[2]
> par(pin = c(ratio * pin[2], pin[2]), xaxs = "i", yaxs = "i")
> plot(meuse, pch = 1)
> box()
```

we first set the aspect of the plotting region equal to that of the data points, and then we plot the points without allowing for the 4% extension of the range in all directions. The result (Fig. 3.7) is that in all four sides one plotting symbol is clipped by the plot border.

If we want to create more than one map in a single figure, as was done in Fig. 3.1, we can sub-divide the figure region into a number of sub-regions. We can split the figure into two rows and three columns either by

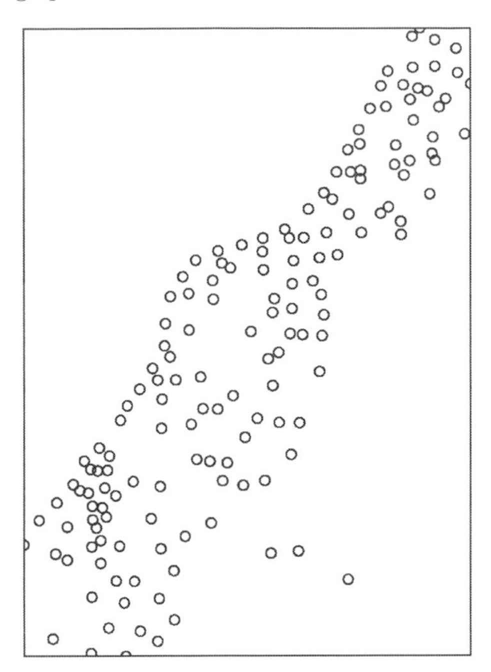

Fig. 3.7. Plotting region *exactly* equal to sample location ranges: border point symbols are clipped

```
> par(mfrow = c(2, 3))
```

or

```
> layout(matrix(1:6, 2, 3, byrow = TRUE))
```

Each time a plot command that would normally create a new plot is called (i.e. without `add = TRUE`), a plot is drawn in a new sub-area; this is done row-wise for this example, or column-wise when `byrow = FALSE`. Function `layout` also allows us to vary the height and width of the sub-areas.

Map scale is the ratio between the length of one unit on the map and one unit in the real world. It can only be controlled ahead of time when both the size of the plotting region, which is by default only a part of the figure size unless all margins are set to zero, and the plotting area are defined, or otherwise exactly known.

3.1.5 Plotting Attributes and Map Legends

Up to now we have only plotted the geometry or topology of the spatial objects. If in addition we want to show feature characteristics or *attributes* of the objects, we need to use type, size, or colour of the symbols, lines, or polygons. Grid cells are usually plotted as small adjacent squares, so their plotting is in some sense a special case of plotting polygons. Table 3.2 lists the

Table 3.2. Useful annotation arguments to be passed to `plot` or `image` methods

Class(es)	Argument	Meaning	Further help
SpatialLinesDataFrame	`col`	Colour	`?lines`
	`lwd`	Line width	`?lines`
	`lty`	Line type	`?lines`
SpatialPolygonsDataFrame	`border`	Border colour	`?polygon`
	`density`	Hashing density	`?polygon`
	`angle`	Hashing angle	`?polygon`
	`lty`	Line type	`?polygon`
	`pbg`	Hole colour	
SpatialPointsDataFrame	`pch`	Symbol	`?points`
	`col`	Colour	`?points`
	`bg`	Fill colour	`?points`
	`cex`	Symbol size	`?points`
SpatialPixelsDataFrame[a]	`zlim`	Attribute value limits	`?image.default`
and	`col`	Colours	`?image.default`
SpatialGridDataFrame	`breaks`	Break points	`?image.default`

[a]Use `image` to plot gridded data

graphic arguments that can be passed to the `plot` methods for the `Spatial` classes with attributes. When a specific colour, size, or symbol type refers to a specific numeric value or category label of an attribute, a map legend is needed to communicate this information. Example code for function `legend` is given below and shown in Fig. 3.8.

We provide `image` methods for objects of class `SpatialPixelsDataFrame` and `SpatialGridDataFrame`. As an example, we can plot interpolated (see Chap. 8) zinc concentration (`zinc.idw`) as a background image along with the data:

```
> grays = gray.colors(4, 0.55, 0.95)
> image(zn.idw, col = grays, breaks = log(c(100, 200, 400,
+     800, 1800)))
> plot(meuse.sr, add = TRUE)
> plot(meuse, pch = 1, cex = sqrt(meuse$zinc)/20, add = TRUE)
> legVals <- c(100, 200, 500, 1000, 2000)
> legend("left", legend = legVals, pch = 1, pt.cex = sqrt(legVals)/20,
+     bty = "n", title = "measured")
> legend("topleft", legend = c("100-200", "200-400", "400-800",
+     "800-1800"), fill = grays, bty = "n", title = "interpolated")
```

the result of which is shown in Fig. 3.8. This example shows how the `legend` command is used to place two legends, one for symbols and one for colours. In this example, rather light grey tones are used in order not to mask the black symbols drawn.

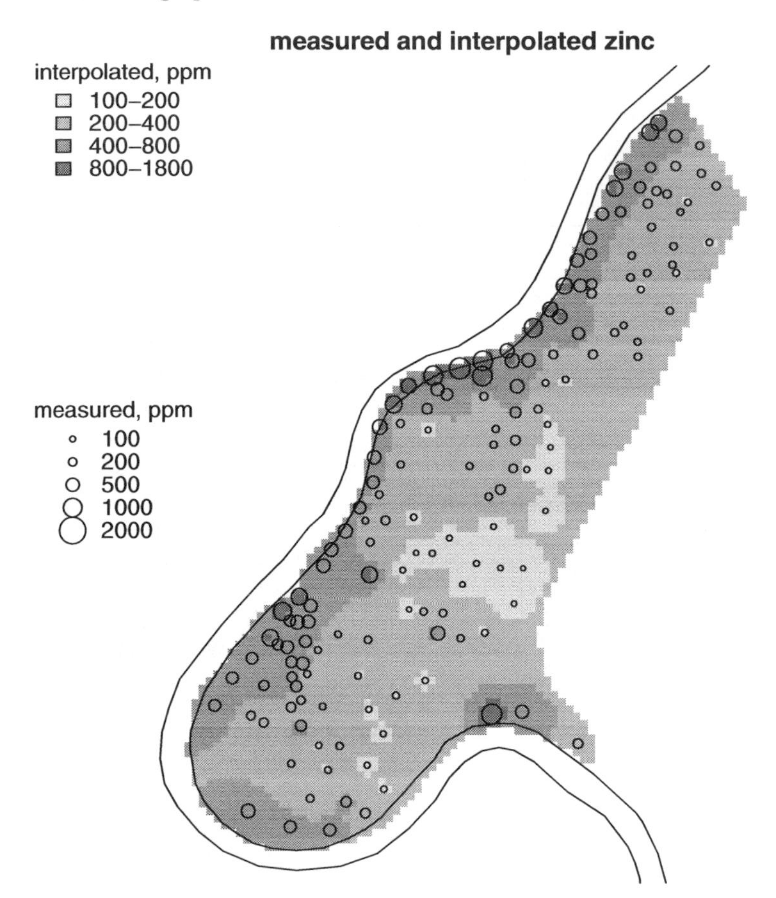

Fig. 3.8. Sample data points for zinc (ppm) plotted over an interpolated image, with symbol area proportional to measured concentration

3.2 Trellis/Lattice Plots with `spplot`

Apart from the traditional `plot` methods provided by package **sp**, a second method, called `spplot`, provides plotting of spatial data with attributes through the Trellis graphics system (Cleveland, 1993, 1994), which is for R provided (and extended) by package **lattice** (Sarkar, 2008). Trellis plots are a bit harder to deal with initially because plot annotation, the addition of information like `legend`, `lines`, `text`, etc., is handled differently and needs to be thought out first. The advantage they offer is that many maps can be composed into single (sets of) graphs, easily and efficiently.

3.2.1 A Straight Trellis Example

Consider the plotting of two interpolation scenarios for the zinc variable in the `meuse` data set, obtained on the direct scale and on the log scale. We can do this either by the `levelplot` function from package **lattice**, or by using `spplot`, which is for grids a simple wrapper around `levelplot`:

```
> library(lattice)
> levelplot(z ~ x + y | name, spmap.to.lev(zn[c("direct",
+     "log")]), asp = "iso")
> spplot(zn[c("direct", "log")])
```

The results are shown in Fig. 3.9. Function `levelplot` needs a `data.frame` as second argument with the grid values for both maps in a single column (`z`) and a factor (`name`) to distinguish between them. Helper function `spmap.to.lev` converts the `SpatialPixelsDataFrame` object to this format by replicating the

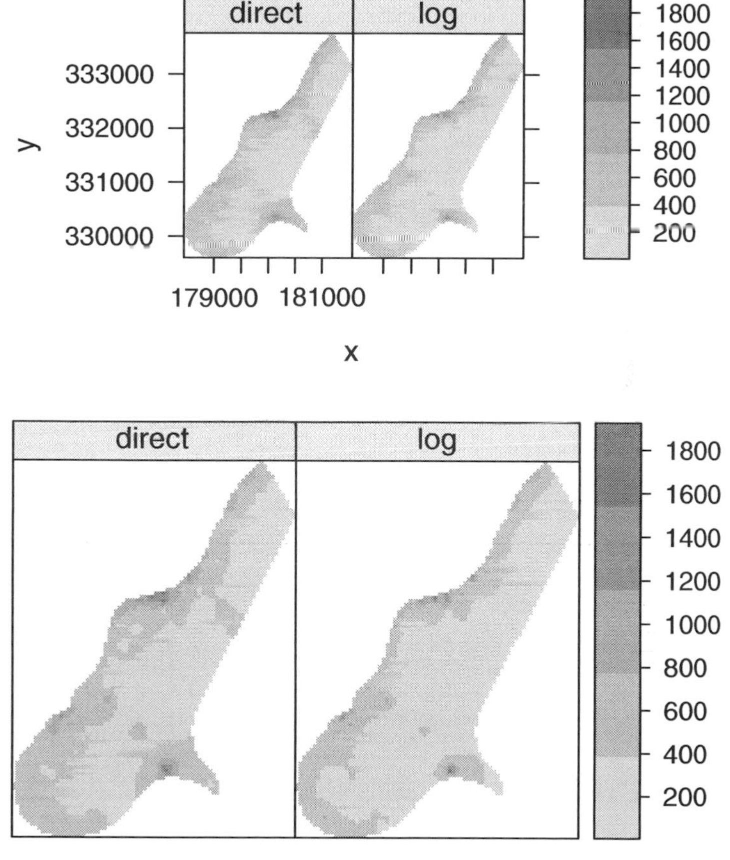

Fig. 3.9. Two interpolation scenarios for the `meuse` data set, plotted on the same total size. (*Top*) Example of `levelplot`, (*bottom*) example of the `spplot` wrapper, which turns off axes

coordinates, stacking the attribute variables, and adding a factor to distinguish the two maps. Function `spplot` plots each attribute passed in a single panel, which results in this case in two panels.

The `spplot` method does all this too, but hides many details. It provides a simple access to the functions provided by package **lattice** for plotting objects deriving from class `Spatial`, while retaining the flexibility offered by **lattice**. It also allows for adding geographic reference elements to maps.

Note that the plot shows four dimensions: the geographic space spanning x- and y-coordinates, the attribute values displayed in colour or grey tone, and the *panel* identifier, here the interpolation scenario but which may be used to denote, for example attribute variable or time.

3.2.2 Plotting Points, Lines, Polygons, and Grids

Function `spplot` plots spatial objects using colour (or grey tone) to denote attribute values. The first argument therefore has to be a spatial object with attributes.

Figure 3.10 shows a typical plot with four variables. If the goal is to compare the absolute levels in ppm across the four heavy metal variables, it makes

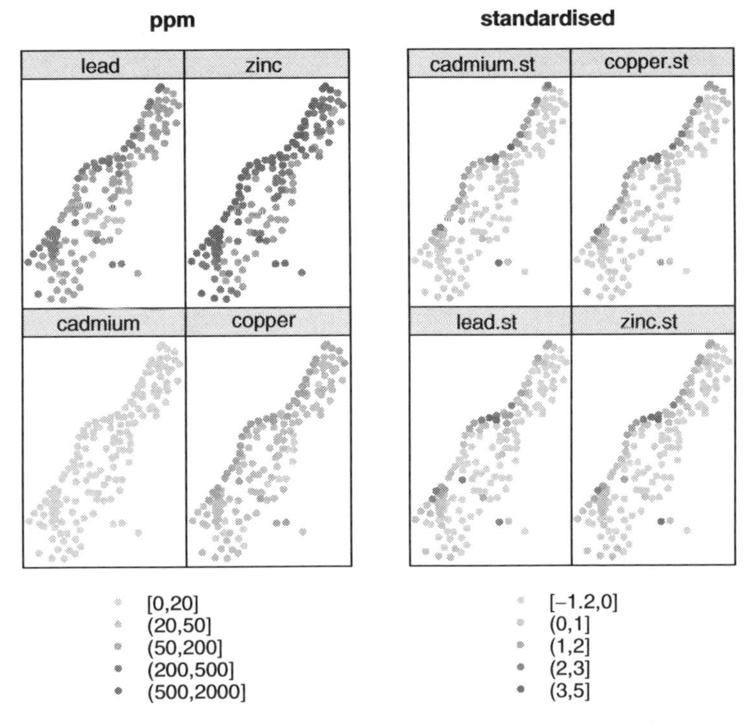

Fig. 3.10. Soil measurements for four heavy metals in the Meuse data set; (*left*) in ppm units, (*right*) each variable scaled to mean zero and unit standard variance

sense to plot them in a single figure with one legend. For such cases, the conditioning plots of `spplot` are ideal. Other cases in which multiple sub-maps are useful are, for example when different moments of time or different modelling scenarios are used to define the factor that splits the data over sub-plots (panels).

The first argument to `spplot` is a `Spatial*DataFrame` object with points, lines, polygons, or a grid. The second argument tells which attributes (column names or numbers) should be used; if omitted, all attributes are plotted. Further attributes control the plotting: colours, symbols, legend classes, size, axes, and geographical reference items to be added.

An example of a `SpatialLinesDataFrame` plot is shown in Fig. 3.11 (left). The R function `contourLines` is used to calculate the contourlines:

```
> library(maptools)
> data(meuse.grid)
> coordinates(meuse.grid) <- c("x", "y")
> meuse.grid <- as(meuse.grid, "SpatialPixelsDataFrame")
> im <- as.image.SpatialGridDataFrame(meuse.grid["dist"])
> cl <- ContourLines2SLDF(contourLines(im))
> spplot(cl)
```

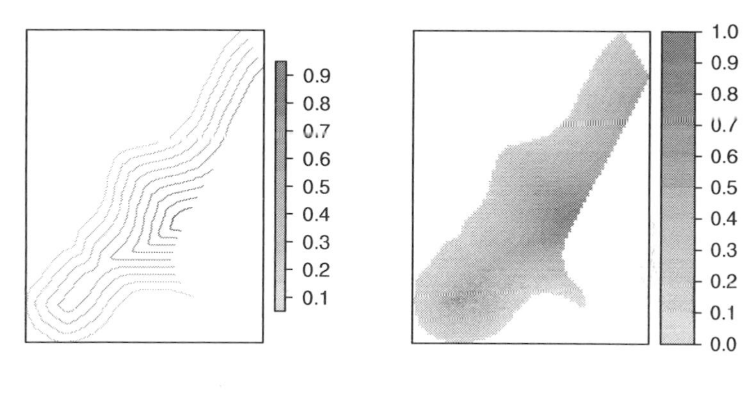

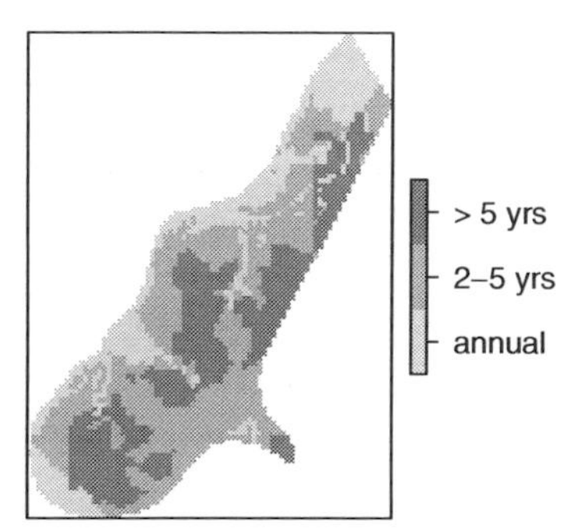

Fig. 3.11. (*Top left*) Contour lines for distance to river Meuse, levels represented by grey tones; (*top right*) grid plot of a numerical variable; (*bottom left*) plot of the factor variable flood frequency; note the different legend key

3.2.3 Adding Reference and Layout Elements to Plots

Method `spplot` takes a single argument, `sp.layout`, to annotate plots with lines, points, grids, polygons, text, or combinations of these. This argument contains either a single layout item or a list of layout items. A single layout item is a list object. Its first component is the name of the layout function to be called, followed by the object to be plotted and then optional arguments to adjust colour, symbol, size, etc. The layout functions provided are the following:

sp layout function	Object class	Useful arguments[a]
`sp.points`	`SpatialPoints`	`pch`, `cex`, `col`
`sp.polygons`	`SpatialPolygons`	`lty`, `lwd`, `col`
`sp.lines`	`SpatialLines`	`lty`, `lwd`, `col`
`sp.text`	text	(see `panel.text`)

[a]For help, see `?par`

An example of building an `sp.layout` structure is as follows:

```
> river <- list("sp.polygons", meuse.sr)
> north <- list("SpatialPolygonsRescale", layout.north.arrow(),
+     offset = c(178750, 332500), scale = 400)
> scale <- list("SpatialPolygonsRescale", layout.scale.bar(),
+     offset = c(180200, 329800), scale = 1000, fill = c("transparent",
+         "black"))
> txt1 <- list("sp.text", c(180200, 329950), "0")
> txt2 <- list("sp.text", c(181200, 329950), "1 km")
> pts <- list("sp.points", meuse, pch = 3, col = "black")
> meuse.layout <- list(river, north, scale, txt1, txt2,
+     pts)

> spplot(zn["log"], sp.layout = meuse.layout)
```

the result of which is shown in Fig. 3.12. Although the construction of this is more elaborate than annotating base plots, as was done for Fig. 3.5, this method seems better for the larger number of graphs as shown in Fig. 3.10.

A special layout element is `which` (integer), to control to which panel a layout item should be added. If `which` is present in the top-level list it applies to all layout items; in sub-lists with layout items it denotes the panel or set of panels in which the layout item should be drawn. Without `which`, layout items are drawn in each panel.

The order of items in the `sp.layout` argument matters; in principle objects are drawn in the order they appear. By default, when the object of `spplot` has points or lines, `sp.layout` items are drawn before the points to allow grids and polygons drawn as a background. For grids and polygons, `sp.layout` items are drawn afterwards (so the item will not be overdrawn by the grid and/or polygon). For grids, adding a list element `first = TRUE` ensures that the item is drawn *before* the grid is drawn (e.g. when filled polygons are

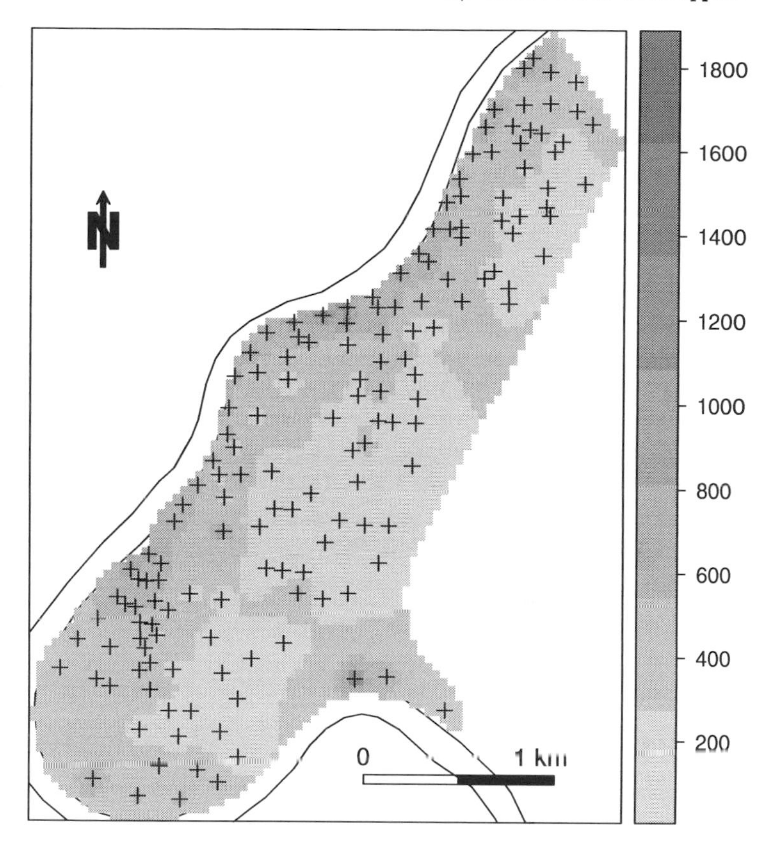

Fig. 3.12. Interpolated `spplot` image with layout elements

added). Transparency may help when combining layers; it is available for the PDF device and several other devices.

Function `sp.theme` returns a **lattice** theme that can be useful for plots made by `spplot`; use `trellis.par.set(sp.theme())` after a device is opened or changed to make this effective. Currently, this only sets the colours to `bpy.colors`.

3.2.4 Arranging Panel Layout

The default layout of `spplot` plots is computed by (i) the dimensions of the graphics device and the size of each panel and (ii) a row-wise ordering, starting top-left. The row-wise ordering can be started bottom-left if `as.table = FALSE` is passed to the `spplot` call. Note that `FALSE` is the default value for functions in **lattice**.

Besides `as.table`, panel layout can be modified with the `layout` and `skip` arguments. Argument `layout` is a numeric vector with the number of columns and the number of rows, for example `layout = c(3,4)` will result in three columns and four rows. Argument `skip` can be used to leave certain panels blank, in plotting order: `layout = c(3,3), skip = c(F,T,T,F,F,T,F,F,F)` will plot six panels in a lower triangular 3×3 panel matrix. Figure 8.10 gives an example of this. More information about `layout`, `skip`, and `as.table` can be found in the help for **lattice** function `xyplot`.

3.3 Interacting with Plots

The interaction R allows with plots in the traditional and **lattice** plot systems is rather limited, compared with stand-alone software written for interacting with data, or GIS. The main functionality is centred around which information is present at the location where a mouse is clicked.

3.3.1 Interacting with Base Graphics

Base graphics has two functions to interact with interactive (i.e. screen) graphic devices:

`locator` returns the locations of points clicked, in coordinates of the x- and y-axis

`identify` plots and returns the labels (by default: row number) of the items nearest to the location clicked, within a specified maximum distance (0.25 inch in plot units, by default).

Both functions wait for user input; left mouse clicks are registered; a right mouse click ends the input. An example session for `identify` may look like this:

```
> plot(meuse)
> meuse.id <- identify(coordinates(meuse))
```

and the result may look like the left side of Fig. 3.13. An example digitise session, followed by selection and re-plotting of points within the area digitised may be as follows:

```
> plot(meuse)
> region <- locator(type = "o")
> n <- length(region$x)
> p <- Polygon(cbind(region$x, region$y)[c(1:n, 1), ],
+     hole = FALSE)
> ps <- Polygons(list(p), ID = "region")
> sps <- SpatialPolygons(list(ps))
> plot(meuse[!is.na(overlay(meuse, sps)), ], pch = 16,
+     cex = 0.5, add = TRUE)
```

with results in the right-hand side of Fig. 3.13. Note that we 'manually' close the polygon by adding the first point to the set of points digitised.

To identify particular polygons, we can use `locator` and overlay the points with the polygon layer shown in Fig. 3.6:

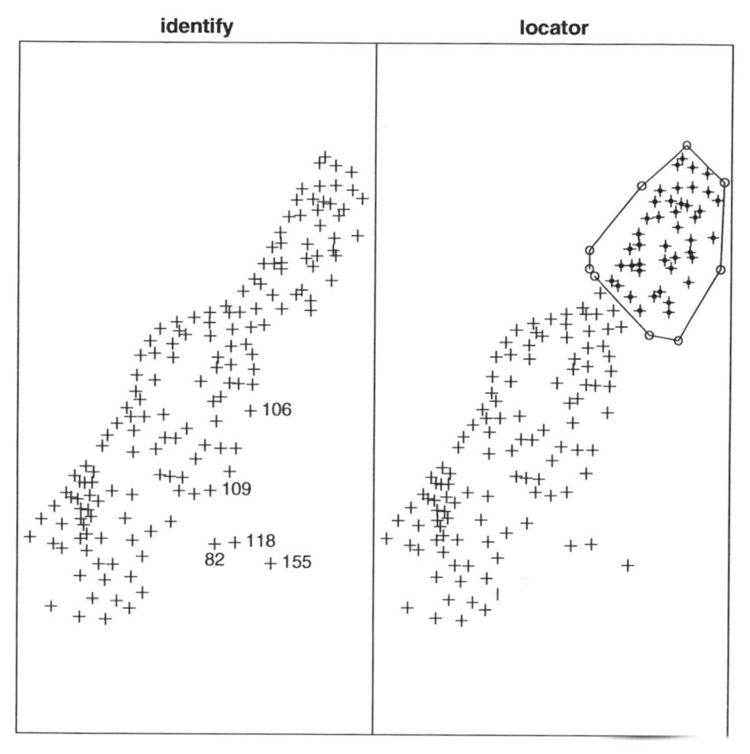

Fig. 3.13. Interaction with point plots. (*Left*) Individual identification of points; (*right*) digitising a region, highlighted points included in the region

```
> library(maptools)
> prj <- CRS("+proj=longlat +datum=NAD27")
> nc_shp <- system.file("shapes/sids.shp", package = "maptools")[1]
> nc <- readShapePoly(nc_shp, proj4string = prj)

> plot(nc)
> pt <- locator(type = "p")

> print(pt)

$x
[1] -78.69484


$y
[1] 35.8044

> overlay(nc, SpatialPoints(cbind(pt$x, pt$y), proj4string = prj))

    AREA PERIMETER CNTY_ CNTY_ID NAME  FIPS FIPSNO CRESS_ID BIR74
36 0.219      2.13  1938    1938 Wake 37183  37183       92 14484
   SID74 NWBIR74 BIR79 SID79 NWBIR79
36    16    4397 20857    31    6221
```

3.3.2 Interacting with spplot and Lattice Plots

In R, Trellis (**lattice**) plots have the same interaction functionality as base plots. However, the process is a bit more elaborate because multiple panels may be present. To select points with `spplot`, use

```
> ids <- spplot(meuse, "zinc", identify = TRUE)
```

This will show the points selected and return the selected points' row numbers.

In essence, and what the above function hides, we first select a panel, then identify within this panel, and finally unselect it, which is accomplished by the **lattice** functions

```
> library(lattice)
> trellis.focus("panel", column = 1, row = 1)
> ids <- panel.identify()
> trellis.unfocus()
```

Digitising can be done by the function `grid.locator` from package `grid`, which underlies the functionality in **lattice**. A single point is selected by

```
> library(grid)
> trellis.focus("panel", column = 1, row = 1)
> as.numeric(grid.locator())
> trellis.unfocus()
```

Package **sp** contains a simple function `spplot.locator` to return a digitised area, simulating the base plot `locator` behaviour. It returns a two-column matrix with spatial coordinates.

3.4 Colour Palettes and Class Intervals

3.4.1 Colour Palettes

R provides a number of colour palettes, and the functions providing them are self-descriptive: `rainbow`, `grey.colors`, `heat.colors`, `terrain.colors`, `topo.colors`, and `cm.colors` (`cm` for cyan-magenta) – `cm.colors` are the default palette in `spplot` and diverge from white. For quantitative data, shades in a single colour are usually preferred. These can be created by `colorRampPalette`, which creates a color interpolating function taking the required number of shades as argument, as in

```
> rw.colors <- colorRampPalette(c("red", "white"))
> image(meuse.grid["dist"], col = rw.colors(10))
```

Package **RColorBrewer** provides the palettes described (and printed) in Brewer et al. (2003) for continuous, diverging, and categorical variables. An interface for exploring how these palettes look on maps is found in the color-brewer applet.[2]

[2] See `http://www.colorbrewer.org/`.

It also has information on suitability of each of the palettes for colour-blind people, black-and-white photo-copying, projecting by LCD projectors, use on LCD or CRT screens, and for colour printing. Another, non-interactive, overview is obtained by

```
> library(RColorBrewer)
> example(brewer.pal)
```

Package **sp** provides the ramp `bpy.colors` (blue-pink-yellow), which has the advantage that it has many colors and that it prints well both on color and black-and-white printers.

3.4.2 Class Intervals

Although we can mimic continuous variation by choosing many (e.g. 100 or more) colours, matching map colours to individual colours in the legend is approximate. If we want to communicate changes connected to certain fixed levels, for example levels related to regulation, or if we for other reasons want differentiable or identifiable class intervals, we should limit the number of classes to, for example six or less.

Class intervals can be chosen in many ways, and some have been collected for convenience in the **classInt** package. The first problem is to assign class boundaries to values in a single dimension, for which many classification techniques may be used, including pretty, quantile, and natural breaks among others, or even simple fixed values. From there, the intervals can be used to generate colours from a colour palette as discussed earlier. Because there are potentially many alternative class memberships even for a given number of classes (by default from `nclass.Sturges`), choosing a communicative set matters.

We try just two styles, quantiles and Fisher-Jenks natural breaks for five classes (Slocum et al., 2005, pp. 85–86), among the many available – for further documentation see the help page of the `classIntervals` function. They yield quite different impressions, as we see:

```
> library(RColorBrewer)
> library(classInt)
> pal <- grey.colors(4, 0.95, 0.55, 2.2)
> q5 <- classIntervals(meuse$zinc, n = 5, style = "quantile")
> q5

style: quantile
  one of 14,891,626 possible partitions of this variable into 5 classes
  under 186.8 186.8 - 246.4 246.4 - 439.6 439.6 - 737.2     over 737.2
           31            31            31            31             31

> diff(q5$brks)

[1]   73.8   59.6  193.2  297.6 1101.8

> plot(q5, pal = pal)
```

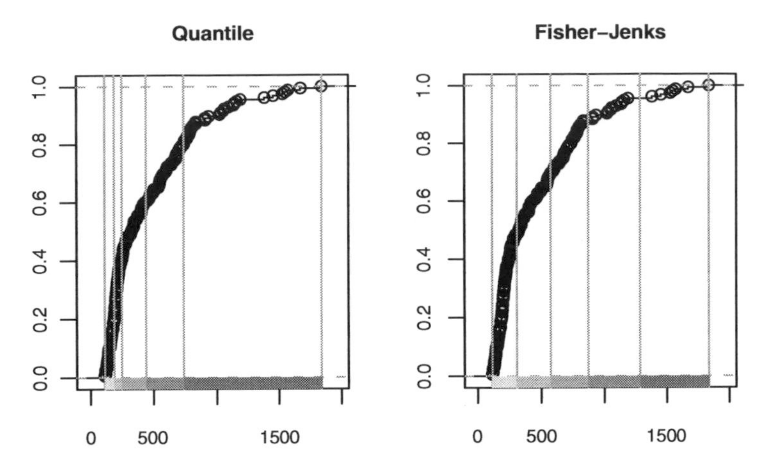

Fig. 3.14. Comparison of quantile and natural breaks methods for setting class intervals, Meuse bank zinc ppm

The empirical cumulative distribution function, used in the plot method for the `classIntervals` object returned, suggests that using quantiles is not necessarily a good idea. While of course the number of sites in each class is equal by definition, the observed values are far from uniformly distributed. Examining the widths of the classes using `diff` on the class breaks shows that many sites with moderate zinc values will be assigned to the darkest colour class. Figure 3.14 shows the plot of this class interval set compared with that for a five-class Fisher-Jenks classification. There are two implementations of this style, one named `'fisher'`, the other `'jenks'`. This 'natural breaks' set of class intervals is based on minimising the within-class variance, like many of the other styles available.

```
> fj5 <- classIntervals(meuse$zinc, n = 5, style = "fisher")
> fj5

style: fisher
  one of 14,891,626 possible partitions of this variable into 5 classes
   under 307.5  307.5 - 573.0  573.0 - 869.5 869.5 - 1286.5
            75             32             29              12
   over 1286.5
             7

> diff(fj5$brks)

[1] 194.5 265.5 296.5 417.0 552.5

> plot(fj5, pal = pal)
```

Once we are satisfied with the chosen class intervals and palette, we can go on to plot the data, using the `findColours` function to build a vector of colours and attributes, which can be used in constructing a legend:

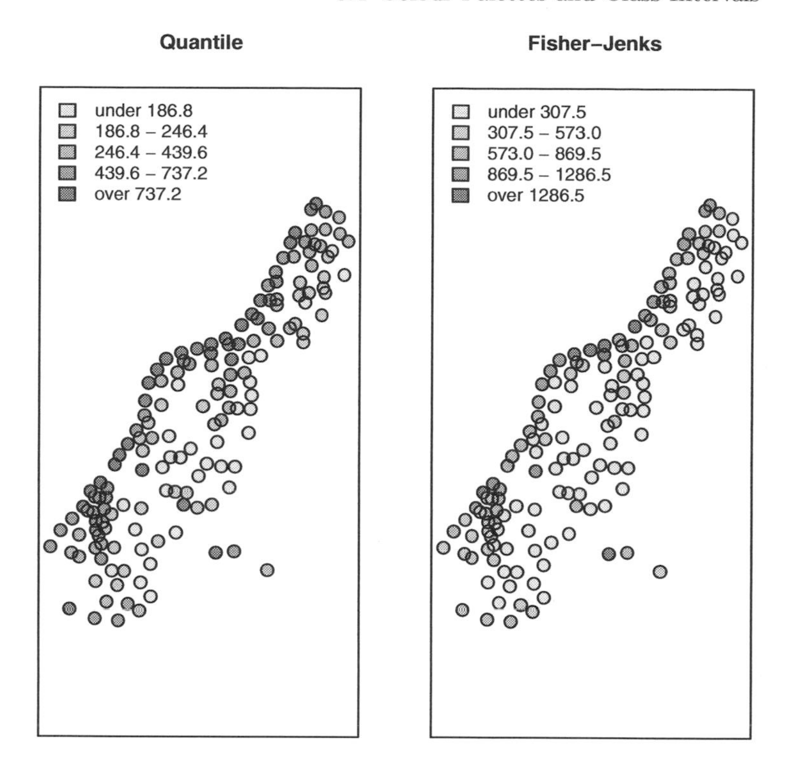

Fig. 3.15. Comparison of output maps made with quantile and natural breaks class intervals, Meuse bank zinc ppm

```
> q5Colours <- findColours(q5, pal)
> plot(meuse, col = q5Colours, pch = 19)
> legend("topleft", fill = attr(q5Colours, "palette"),
+     legend = names(attr(q5Colours, "table")), bty = "n")
```

The output for these two classifications is shown in Fig. 3.15, and does show that choice of representation matters. Using quantile-based class intervals, it appears that almost all the river bank sites are equally polluted, while the natural breaks intervals discriminate better.

For image, we can specify the `breaks` argument, as was done in Fig. 3.8. While the `classIntervals` function can be used with raster data, it may be prudent to search for class intervals using a sample of the input data, including the extremities to save time; this heuristic is used by many GIS. The default class interval style used by image is to divide the range into a number of classes of equal width (equivalent to the equal or pretty styles in `classIntervals`). With very skewed data, for example 2D density plots, this may give the impression of the data having disappeared, because almost all the cells will be in one extreme class, and only a few in other classes. Changing the class intervals will 'magically' reveal the data.

For the `spplot` methods for lines, polygons, and grids, we can pass the argument `pretty = TRUE`, which ensures that colour breaks coincide with legend values (see right-hand side of Fig. 3.11). To specify class intervals with `spplot`, for points data we can pass the `cuts` argument, and for lines, polygons, or grids we can pass the `at` argument. To also control the key tic marks and labels, we need to specify `colorkey` as well. For example, the middle plot of Fig. 3.11 was created by:

```
> cuts = (0:10)/10
> spplot(meuse.grid, "dist", colorkey = list(labels = list(at = cuts)),
+     at = cuts)
```

Having provided a framework for handling and visualising spatial data in R, we now move to demonstrate how user data may be imported into R, and the results of analysis exported.

4

Spatial Data Import and Export

Geographical information systems (GIS) and the types of spatial data they handle were introduced in Chap. 1. We now show how spatial data can be moved between **sp** objects in R and external formats, including the ones typically used by GIS. In this chapter, we first show how coordinate reference systems can be handled portably for import and export, going on to transfer vector and raster data, and finally consider ways of linking R and GIS more closely.

Before we begin, it is worth noting the importance of open source projects in making it possible to offer spatial data import and export functions in R. Many of these projects are now gathered in the Open Source Geospatial Foundation.[1] There are some projects which form the basis for the others, in particular the Geospatial Data Abstraction Library[2] (GDAL, pronounced Goödal, coordinated by Frank Warmerdam). Many of the projects also use the PROJ.4 Cartographic Projections library,[3] originally written by Gerald Evenden then of the United States Geological Survey, and modified and maintained by Frank Warmerdam. Without access to such libraries and their communities, it would not be possible to provide import or export facilities for spatial data in R. Many of the open source toolkits are also introduced in depth in Mitchell (2005). As we proceed, further links to relevant sources of information, such as mailing list archives, will be given.

In this chapter, we consider the representation of coordinate reference systems in a robust and portable way. Next, we show how spatial data may be read into R, and be written from R, using the most popular formats. The interface with GRASS GIS will be covered in detail, and finally the export of data for visualisation will be described.

First, we show how loading the package providing most of the interfaces to the software of these open source projects, **rgdal**, reports their status:

---

[1] `http://www.osgeo.org/`.

[2] `http://www.gdal.org/`.

[3] `http://proj.maptools.org/`.

```
> library(rgdal)

Geospatial Data Abstraction Library extensions to R successfully loaded
Loaded GDAL runtime: GDAL 1.5.1, released 2008/03/14
GDAL_DATA: /home/rsb/lib/R/rgdal/gdal
Loaded PROJ.4 runtime: Rel. 4.6.0, 21 Dec 2007
PROJ_LIB: /home/rsb/lib/R/rgdal/proj
```

We see that the release version numbers and dates of the external dynamically loaded libraries are reported. In addition, the values of the system environment variables `GDAL_DATA` and `PROJ_LIB` are set internally to support files bundled with **rgdal**, and are reset to their initial values when the package exits.[4]

4.1 Coordinate Reference Systems

Spatial data vary a great deal both in the ways in which their position attributes are recorded and in the adequacy of documentation of how position has been determined. This applies both to data acquired from secondary sources and to Global Positioning System input, or data capture from analogue maps by digitising. This also constitutes a specific difference from the analysis say of medical imagery, which in general requires only a local coordinate system; astronomy and the mapping of other planets also constitute a separate but linked field. Knowledge about the coordinate reference system is needed to establish the positional coordinates' units of measurement, obviously needed for calculating distances between observations and for describing the network topology of their relative positions. This knowledge is essential for integrating spatial data for the same study area, but coming from different sources. Waller and Gotway (2004, pp. 40–47) describe some of the key concepts and features to be dealt with here in an accessible fashion.

Coordinate reference systems (CRS) are at the heart of geodetics and cartography: how to represent a bumpy ellipsoid on the plane. We can speak of geographical CRS expressed in degrees and associated with an ellipse – a model of the shape of the earth, a prime meridian defining the origin in longitude, and a datum. The concept of a datum is arbitrary and anchors a specific geographical CRS to an origin point in three dimensions, including an assumed height above the assumed centre of the earth or above a standard measure of sea level. Since most of these quantities have only been subject to accurate measurement since the use of satellites for surveying became common, changes in ellipse and datum characteristics between legacy maps and newly collected data are common.

In contrast, projected CRS are expressed by a specific geometric model projecting to the plane and measures of length, as well as the underlying

[4] The report returned when loading **rgdal** may be suppressed by wrapping the call in `suppressPackageStartupMessages`.

ellipse, prime meridian, and datum. Most countries have multiple CRS, often for very good reasons. Surveyors in cities have needed to establish a local datum and a local triangulation network, and frequently these archaic systems continue to be used, forming the basis for property boundaries and other legal documents.

Cartography and surveying has seen the development of national triangulations and of stipulated national projections, or sub-national or zoned projections for larger countries. Typically, problems arise where these regimes meet. The choices of ellipse, prime meridian, and datum may differ, and the chosen projection and metric may also differ, or have different key parameters or origin offsets. On land, national borders tend to be described adequately with reference to the topography, but at sea, things change. It was because the coastal states around the North Sea basin had incompatible and not fully defined CRS that the European Petroleum Survey Group (EPSG; now Oil & Gas Producers (OGP) Surveying & Positioning Committee) began collecting a geodetic parameter data set[5] starting in 1986, based on earlier work in member companies.

4.1.1 Using the EPSG List

The EPSG list is under continuous development, with corrections being made to existing entries, and new entries being added as required. A copy of the list is provided in the **rgdal** package,[6] because it permits the conversion of a large number of CRS into the PROJ.4 style description used here. Since it allows for datum transformation as well as projection, the number of different coordinate reference systems is larger than that in the **mapproj** package. Datum transformation is based on transformation to the World Geodetic System of 1984 (WGS84), or inverse transformation from it to an alternative specified datum. WGS84 was introduced after measurements of earth from space had become very accurate, and forms a framework into which local and national systems may be fitted.

The **rgdal** package copy of the EPSG list can be read into a data frame and searched using `grep`, for example. We try to reproduce the example given by the Royal Netherlands Navy entitled 'From ED50 towards WGS84, or does your GPS receiver tell you the truth?'[7] A position has been read from a chart in the ED50 datum about a nautical mile west of the jetties of IJmuiden, but needs to be converted to the WGS84 datum for comparison with readings from a GPS satellite navigation instrument. We need to transform the chart coordinates in ED50 – ED50 is the European Datum 1950 – to coordinates in the WGS84 datum (the concept of a datum is described on p. 82). In this case to save space, the search string has been chosen to match exactly the row needed; entering just `ED50` gives 35 hits:

[5] `http://www.epsg.org/`.
[6] See installation note at chapter end, p. 111.
[7] `http://www.hydro.nl/articles/artikel2_en.htm`.

```
> EPSG <- make_EPSG()
> EPSG[grep("^# ED50$", EPSG$note), ]

    code   note                              prj4
149 4230 # ED50 +proj=longlat +ellps=intl +no_defs
```

The EPSG code is in the first column of the data frame and the PROJ.4 specification in the third column, with the known set of tags and values.

4.1.2 PROJ.4 CRS Specification

The PROJ.4 library uses a 'tag=value' representation of coordinate reference systems, with the tag and value pairs enclosed in a single character string. This is parsed into the required parameters within the library itself. The only values used autonomously in `CRS` class objects are whether the string is a character NA (missing) value for an unknown CRS, and whether it contains the string `longlat`, in which case the CRS contains geographical coordinates.[8] There are a number of different tags, always beginning with `+`, and separated from the value with `=`, using white space to divide the tag/value pairs from each other.[9] If we use the special tag `+init` with value `epsg:4230`, where `4230` is the EPSG code found above, the coordinate reference system will be populated from the tables supplied with the libraries (PROJ.4 and GDAL) and included in **rgdal**.

```
> CRS("+init=epsg:4230")

CRS arguments:
 +init=epsg:4230 +proj=longlat +ellps=intl +no_defs
```

The two tags that are known are `+proj` – projection, which takes the value `longlat` for geographical coordinates – and `+ellps` – ellipsoid, with value `intl` for the International Ellipsoid of 1909 (Hayford). There is, however, no `+towgs84` tag, and so without further investigation it will not be possible to make the datum transformation. Lots of information about CRS in general can be found in *Grids & Datums*,[10] a regular column in Photogrammetric Engineering & Remote Sensing. The February 2003 number covers the Netherlands and gives a three-parameter transformation – in some cases seven parameters are given to specify the shift between datums.[11] Adding these values gives a full specification:

---

[8] The value `latlong` is not used, although valid, because coordinates in **sp** class objects are ordered with eastings first followed by northings.

[9] In addition to the EPSG list, there are many examples at the PROJ.4 website, for example: `http://geotiff.maptools.org/proj_list/`.

[10] `http://www.asprs.org/resources/GRIDS/`.

[11] Searching the PROJ.4 mailing list can also provide useful hints: `http://news.gmane.org/gmane.comp.gis.proj-4.devel`.

```
> ED50 <- CRS("+init=epsg:4230 +towgs84=-87,-96,-120,0,0,0,0")
> ED50
```

```
CRS arguments:
 +init=epsg:4230 +towgs84=-87,-96,-120,0,0,0,0 +proj=longlat
+ellps=intl +no_defs
```

Datum transformation shifts coordinates between differently specified ellipsoids in all three dimensions, even if the data appear to be only 2D, because 2D data are assumed to be on the surface of the ellipsoid. It may seem unreasonable that the user is confronted with the complexities of coordinate reference system specification in this way. The EPSG list provides a good deal of help, but assumes that wrong help is worse than no help, and does not give transformation parameters where there is any ambiguity, and for the ED50 datum, parameter values do vary across Europe. Modern specifications are designed to avoid ambiguity, and so this issue will become less troublesome with time, although old maps are going to be a source of data for centuries to come.

4.1.3 Projection and Transformation

In the Dutch navy case, we do not need to project because the input and output coordinates are geographical:

```
> IJ.east <- as(char2dms("4d31'00\"E"), "numeric")
> IJ.north <- as(char2dms("52d28'00\"N"), "numeric")
> IJ.ED50 <- SpatialPoints(cbind(x = IJ.east, y = IJ.north),
+     ED50)
> res <- spTransform(IJ.ED50, CRS("+proj=longlat +datum=WGS84"))
> x <- as(dd2dms(coordinates(res)[1]), "character")
> y <- as(dd2dms(coordinates(res)[2], TRUE), "character")
> cat(x, y, "\n")
```

```
4d30'55.294"E 52d27'57.195"N
```

```
> spDistsN1(coordinates(IJ.ED50), coordinates(res), longlat = TRUE) *
+     1000
```

```
[1] 124.0994
```

```
> library(maptools)
> gzAzimuth(coordinates(IJ.ED50), coordinates(res))
```

```
[1] -134.3674
```

Using correctly specified coordinate reference systems, we can reproduce the example successfully, with a 124 m shift between a point plotted in the inappropriate WGS84 datum and the correct ED50 datum for the chart:

> 'For example: one who has read his position 52d28′00″N/ 4d31′00″E (ED50) from an ED50-chart, right in front of the jetties of IJmuiden, has to adjust this co-ordinate about 125 m to the Southwest .... The corresponding co-ordinate in WGS84 is 52d27′57″N/ 4d30′55″E.'

The work is done by the `spTransform` method, taking any `Spatial*` object, and returning an object with coordinates transformed to the target `CRS`. There is no way of warping regular grid objects, because for arbitrary transformations, the new positions will not form a regular grid. The solution in this case is to convert the object to point locations, transform them to the new CRS, and interpolate to a suitably specified grid in the new CRS.

Two helper functions are also used here to calculate the difference between the points in ED50 and WGS84: `spDistsN1` and `gzAzimuth`. Function `spDistsN1` measures distances between a matrix of points and a single point, and uses Great Circle distances on the WGS84 ellipsoid if the `longlat` argument is `TRUE`. It returns values in kilometres, and so we multiply by 1,000 here to obtain metres. `gzAzimuth` gives azimuths calculated on the sphere between a matrix of points and a single point, which must be geographical coordinates, with north zero, and negative azimuths west of north.

So far in this section we have used an example with geographical coordinates. There are many different projections to the plane, often chosen to give an acceptable representation of the area being displayed. There exist no all-purpose projections, all involve distortion when far from the centre of the specified frame, and often the choice of projection is made by a public mapping agency.

```
> EPSG[grep("Atlas", EPSG$note), 1:2]

    code                                note
578 2163 # US National Atlas Equal Area

> CRS("+init=epsg:2163")

+init=epsg:2163 +proj=laea +lat_0=45 +lon_0=-100 +x_0=0 +y_0=0
    +a=6370997 +b=6370997 +units=m +no_defs
```

For example, the US National Atlas has chosen a particular CRS for its view of the continental US, with a particular set of tags and values to suit. The projection chosen has the value `laea`, which, like many other values used to represent CRS in PROJ.4 and elsewhere, is rather cryptic. Provision is made to access descriptions within the PROJ.4 library to make it easier to interpret the values in the CRS. The `projInfo` function can return several kinds of information in tabular form, and those tables can be examined to shed a little more light on the tag values.

```
> proj <- projInfo("proj")
> proj[proj$name == "laea", ]

   name                  description
47 laea Lambert Azimuthal Equal Area
```

```
> ellps <- projInfo("ellps")
> ellps[grep("a=6370997", ellps$major), ]
```

```
     name        major           ell                  description
42 sphere a=6370997.0 b=6370997.0 Normal Sphere (r=6370997)
```

It turns out that this CRS is in the Lambert Azimuthal Equal Area projection, using the sphere rather than a more complex ellipsoid, with its centre at 100° west and 45° north. This choice is well-suited to the needs of the Atlas, a compromise between coverage, visual communication, and positional accuracy.

All this detail may seem unnecessary, until the analysis we need to complete turns out to depend on data in different coordinate reference systems. At that point, spending time establishing as clearly as possible the CRS for our data will turn out to have been a wise investment. The same consideration applies to importing and exporting data – if their CRS specifications are known, transferring positional data correctly becomes much easier. Fortunately, for any study region the number of different CRS used in archived maps is not large, growing only when the study region takes in several jurisdictions. Even better, all modern data sources are much more standardised (most use the WGS84 datum), and certainly much better at documenting their CRS specifications.

4.1.4 Degrees, Minutes, and Seconds

In common use, the sign of the coordinate values may be removed and the value given a suffix of E or N for positive values of longitude or latitude and W or S for negative values. In addition, values are often recorded traditionally not as decimal degrees, but as degrees, minutes, and decimal seconds, or some truncation of this. These representations raise exactly the same questions as for time series, although time can be mapped onto the infinite real line, while geographical coordinates are cyclical – move 360° and you return to your point of departure. For practical purposes, geographical coordinates should be converted to decimal degree form; this example uses the Netherlands point that we have already met:

```
> IJ.dms.E <- "4d31'00\"E"
> IJ.dms.N <- "52d28'00\"N"
```

We convert these character strings to class `'DMS'` objects, using function `char2dms`:

```
> IJ_east <- char2dms(IJ.dms.E)
> IJ_north <- char2dms(IJ.dms.N)
> IJ_east
```

```
[1] 4d31'E
```

```
> IJ_north
```

```
[1] 52d28'N
```

```
> getSlots("DMS")
```

```
        WS       deg       min       sec        NS
"logical" "numeric" "numeric" "numeric" "logical"
```

The DMS class has slots to store representations of geographical coordinates, however, they might arise, but the char2dms() function expects the character input format to be as placed, permitting the degree, minute, and second symbols to be given as arguments. We get decimal degrees by coercing from class 'DMS' to class 'numeric' with the as() function:

```
> c(as(IJ_east, "numeric"), as(IJ_north, "numeric"))
```

```
[1]  4.516667 52.466667
```

4.2 Vector File Formats

Spatial vector data are points, lines, polygons, and fit the equivalent **sp** classes. There are a number of commonly used file formats, most of them proprietary, and some newer ones which are adequately documented. GIS are also more and more handing off data storage to database management systems, and some database systems now support spatial data formats. Vector formats can also be converted outside R to formats for which import is feasible.

GIS vector data can be either topological or simple. Legacy GIS were topological, desktop GIS were simple (sometimes known as spaghetti). The **sp** vector classes are simple, meaning that for each polygon all coordinates are stored without checking that boundaries have corresponding points. A topological representation in principal stores each point only once, and builds arcs (lines between nodes) from points, polygons from arcs – the GRASS 6 open source GIS has such a topological representation of vector features. Only the **RArcInfo** package tries to keep some traces of topology in importing legacy ESRI™ ArcInfo™ binary vector coverage data (or e00 format data) – **maps** uses topology because that was how things were done when the underlying code was written. The import of ArcGIS™ coverages is described fully in Gómez-Rubio and López-Quílez (2005); conversion of imported features into **sp** classes is handled by the pal2SpatialPolygons function in **maptools**.

It is often attractive to make use of the spatial databases in the **maps** package. They can be converted to **sp** class objects using functions such as map2SpatialPolygons in the **maptools** package. An alternative source of coastlines is the Rgshhs function in **maptools**, interfacing binary databases of varying resolution distributed by the 'Global Self-consistent, Hierarchical, High-resolution Shoreline Database' project.[12]

[12] http://www.soest.hawaii.edu/wessel/gshhs/gshhs.html.

The best resolution databases are rather large, and so **maptools** ships only with the coarse resolution one; users can install and use higher resolution databases locally. Figures 2.3 and 2.7, among others in earlier chapters, have been made using these sources.

A format that is commonly used for exchanging vector data is the shapefile. This file format has been specified by ESRI™, the publisher of ArcView™ and ArcGIS™, which introduced it initially to support desktop mapping using ArcView™.[13] This format uses at least three files to represent the data, a file of geometries with an `*.shp` extension, an index file to the geometries `*.shx`, and a legacy `*.dbf` DBF III file for storing attribute data. Note that there is no standard mechanism for specifying missing attribute values in this format. If a `*.prj` file is present, it will contain an ESRI™ well-known text CRS specification. The shapefile format is not fully compatible with the OpenGIS® Simple Features Specification (see p. 122 for a discussion of this specification). Its incompatibility is, however, the same as that of the `SpatialPolygons` class, using a collection of polygons, both islands and holes, to represent a single observation in terms of attribute data.

4.2.1 Using OGR Drivers in rgdal

Using the OGR vector functions of the Geospatial Data Abstraction Library, interfaced in **rgdal**,[14] lets us read spatial vector data for which drivers are available. A driver is a software component plugged-in on demand – here the OGR library tries to read the data using all the formats that it knows, using the appropriate driver if available. OGR also supports the handling of coordinate reference systems directly, so that if the imported data have a specification, it will be read.

The availability of OGR drivers differs from platform to platform, and can be listed using the `ogrDrivers` function. The function also lists whether the driver supports the creation of output files. Because the drivers often depend on external software, the choices available will depend on the local computer installation. It is frequently convenient to convert from one external file format to another using utility programs such as `ogr2ogr` in binary FWTools releases, which typically include a wide range of drivers.[15]

The `readOGR` function takes at least two arguments – they are the data source name (`dsn`) and the layer (`layer`), and may take different forms for different drivers. It is worth reading the relevant web pages[16] for the format being imported. For ESRI™ shapefiles, `dsn` is usually the name of the directory containing the three (or more) files to be imported (given as `"."` if the working

[13] The format is fully described in this white paper: `http://shapelib.maptools.org/dl/shapefile.pdf`.

[14] See installation note at chapter end.

[15] `http://fwtools.maptools.org`.

[16] `http://ogr.maptools.org/ogr_formats.html`.

directory), and `layer` is the name of the shapefile without the `".shp"` extension. Additional examples are given on the function help page for file formats, but it is worth noting that the same function can also be used where the data source name is a database connection, and the layer is a table, for example using PostGIS in a PostgreSQL database.

We can use the classic Scottish lip cancer data set by district downloaded from the additional materials page for Chap. 9 in Waller and Gotway (2004).[17] There are three files making up the shapefile for Scottish district boundaries at the time the data were collected – the original study and extra data in a separate text file are taken from Clayton and Kaldor (1987). The shapefile appears to be in geographical coordinates, but no `*.prj` file is present, so after importing from the working directory, we assign a suitable coordinate reference system.

```
> scot_LL <- readOGR(".", "scot")

OGR data source with driver: ESRI Shapefile
Source: ".", layer: "scot"
with  56  rows and  2  columns

> proj4string(scot_LL) <- CRS("+proj=longlat ellps=WGS84")
> scot_LL$ID

 [1] 12 13 19  2 17 16 21 50 15 25 26 29 43 39 40 52 42 51 34 54 36 46
[23] 41 53 49 38 44 30 45 48 47 35 28  4 20 33 31 24 55 18 56 14 32 27
[45] 10 22  6  8  9  3  5 11  1  7 23 37
```

The Clayton and Kaldor data are for the same districts, but with the rows ordered differently, so that before combining the data with the imported polygons, they need to be matched first (matching methods are discussed in Sect. 5.5.2):

```
> scot_dat <- read.table("scotland.dat", skip = 1)
> names(scot_dat) <- c("District", "Observed", "Expected",
+     "PcAFF", "Latitude", "Longitude")
> scot_dat$District

 [1]  1  2  3  4  5  6  7  8  9 10 11 12 13 14 15 16 17 18 19 20 21 22
[23] 23 24 25 26 27 28 29 30 31 32 33 34 35 36 37 38 39 40 41 42 43 44
[45] 45 46 47 48 49 50 51 52 53 54 55 56

> library(maptools)
> scot_dat1 <- scot_dat[match(scot_LL$ID, scot_dat$District),
+     ]
> row.names(scot_dat1) <- sapply(slot(scot_LL, "polygons"),
+     function(x) slot(x, "ID"))
> scot_LLa <- spCbind(scot_LL, scot_dat1)
> all.equal(scot_LLa$ID, scot_LLa$District)
```

[17] http://www.sph.emory.edu/~lwaller/WGindex.htm.

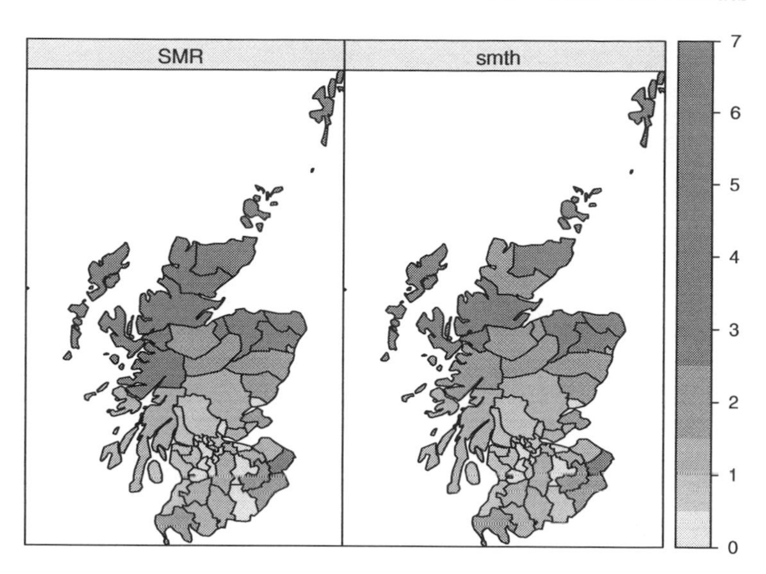

Fig. 4.1. Comparison of relative risk and EB smoothed relative risk for Scottish lip cancer

```
[1] TRUE

> names(scot_LLa)

[1] "NAME"      "ID"        "District"  "Observed"  "Expected"
[6] "PcAFF"     "Latitude"  "Longitude"
```

Figure 4.1 compares the relative risk by district with the Empirical Bayes smooth values – we return to the actual techniques involved in Chap. 11, here the variables are being added to indicate how results may be exported from R below. The relative risk does not take into account the possible uncertainty associated with unusual incidence rates in counties with relatively small populations at risk, while Empirical Bayes smoothing shrinks such values towards the rate for all the counties taken together.

```
> library(spdep)
> O <- scot_LLa$Observed
> E <- scot_LLa$Expected
> scot_LLa$SMR <- probmap(O, E)$relRisk/100
> library(DCluster)
> scot_LLa$smth <- empbaysmooth(O, E)$smthrr
```

Finally, we project the district boundaries to the British National Grid as described by Waller and Gotway (2004):

```
> scot_BNG <- spTransform(scot_LLa, CRS("+init=epsg:27700"))
```

We export these data in two forms, first as Keyhole Markup Language (KML) overlays for Google Earth™. The underlying coordinate reference system for Google Earth™ is geographical, in the WGS84 datum, so we can export the district boundaries as we imported them, using the `writeOGR` function, choosing only a single variable. This function, like `readOGR`, uses drivers to handle different data formats, with `driver="KML"` in this case. Next we take the district centroids and export them as a `SpatialPointsDataFrame`, with the district name, the observed and expected values, and the two rates:

```
> writeOGR(scot_LLa["ID"], dsn = "scot_district.kml",layer = "borders",
+     driver = "KML")
> llCRS <- CRS("+proj=longlat ellps=WGS84")
> scot_SP_LL <- SpatialPointsDataFrame(coordinates(scot_LLa),
+     proj4string = llCRS, data = as(scot_LLa, "data.frame")[c("NAME",
+         "Observed", "Expected", "SMR", "smth")])
> writeOGR(scot_SP_LL, dsn = "scot_rates.kml", layer = "rates",
+     driver = "KML")
```

The output format for Google Earth™ is fairly simple, and it will probably become possible in time to make the export objects more intelligent, but just being able to distribute spatial output permitting an online recipient simply to display results by opening a file does offer considerable opportunities, as Fig. 4.2 illustrates. We could also have written an image overlay, as we see later in Sect. 4.3.2.

Fig. 4.2. Scottish district boundaries and centroid pointers shown in Google Earth™

We can of course export to a shapefile, using `driver="ESRI Shapefile"`, or to other file formats for which output drivers are implemented:

```
> drv <- "ESRI Shapefile"
> writeOGR(scot_BNG, dsn = ".", layer = "scot_BNG", driver = drv)

> list.files(pattern = "^scot_BNG")

[1] "scot_BNG.dbf" "scot_BNG.prj" "scot_BNG.shp" "scot_BNG.shx"
[5] "scot_BNG.txt"
```

The output now contains a `*.prj` file with the fully specified coordinate reference system for the British National Grid, to which we projected the data object.

4.2.2 Other Import/Export Functions

If the **rgdal** package is not available, there are two other packages that can be used for reading and writing shapefiles. The **shapefiles** package is written without external libraries, using file connections. It can be very useful when a shapefile is malformed, because it gives access to the raw numbers. The **maptools** package contains a local copy of the library used in OGR for reading shapefiles (the DBF reader is in the **foreign** package), and provides a low-level import `read.shape` function, a helper function `getinfo.shape` to identify whether the shapefile contains points, lines, or polygons.

```
> getinfo.shape("scot_BNG.shp")

Shapefile type: Polygon, (5), # of Shapes: 56
```

There are three functions to read these kinds of data: `readShapePoints`, `readShapeLines`, and `readShapePoly`. They are matched by equivalent exporting functions: `writePolyShape`, `writeLinesShape`, `writePointsShape`, using local copies of shapelib functions otherwise available in **rgdal** in the OGR framework. The **RArcInfo** package also provides local access to OGR functionality, for reading ArcGIS™ binary vector coverages, but with the addition of a utility function for converting `e00` format files into binary coverages; full details are given in Gómez-Rubio and López-Quílez (2005).

4.3 Raster File Formats

There are very many raster and image formats; some allow only one band of data, others assume that data bands are Red-Green-Blue (RGB), while yet others are flexible and self-documenting. The simplest formats are just rectangular blocks of uncompressed data, like a matrix, but sometimes with row indexing reversed. Others are compressed, with multiple bands, and may be interleaved so that subscenes can be retrieved without unpacking the whole image. There are now a number of R packages that support image import and export, such as the **rimage** and **biOps** packages and the **EBImage** package in the Bioconductor project. The requirements for spatial raster data handling include respecting the coordinate reference system of the image, so that specific solutions are needed. There is, however, no direct support for the transformation or 'warping' of raster data from one coordinate reference system to another.

4.3.1 Using GDAL Drivers in rgdal

Many drivers are available in **rgdal** in the `readGDAL` function, which – like `readOGR` – finds a usable driver if available and proceeds from there. Using arguments to `readGDAL`, subregions or bands may be selected, and the data may be decimated, which helps handle large rasters. The simplest approach is just to read all the data into the R workspace – here we will the same excerpt from the Shuttle Radar Topography Mission (SRTM) flown in 2000, for the Auckland area as in Chap. 2.

```
> auck_el1 <- readGDAL("70042108.tif")

70042108.tif has GDAL driver GTiff
and has 1200 rows and 1320 columns

> summary(auck_el1)

Object of class SpatialGridDataFrame
Coordinates:
    min   max
x 174.2 175.3
y -37.5 -36.5
Is projected: FALSE
proj4string :
[+proj=longlat +ellps=WGS84 +datum=WGS84 +no_defs
+towgs84=0,0,0]
Number of points: 2
Grid attributes:
  cellcentre.offset     cellsize cells.dim
x         174.20042 0.0008333333      1320
y         -37.49958 0.0008333333      1200
Data attributes:
      Min.    1st Qu.     Median       Mean    3rd Qu.       Max.
-3.403e+38  0.000e+00  1.000e+00 -1.869e+34  5.300e+01  6.860e+02

> is.na(auck_el1$band1) <- auck_el1$band1 <= 0 | auck_el1$band1 >
+     10000
```

The `readGDAL` function is actually a wrapper for substantially more powerful R bindings for GDAL written by Timothy Keitt. The bindings allow us to handle very large data sets by choosing sub-scenes and re-sampling, using the `offset`, `region.dim`, and `output.dim` arguments. The bindings work by opening a data set known by GDAL using a `GDALDriver` class object, but only reading the required parts into the workspace.

```
> x <- GDAL.open("70042108.tif")
> xx <- getDriver(x)
> xx

An object of class "GDALDriver"
Slot "handle":
<pointer: 0x83945f0>

> getDriverLongName(xx)

[1] "GeoTIFF"

> x

An object of class "GDALReadOnlyDataset"
Slot "handle":
<pointer: 0x83d4708>

> dim(x)

[1] 1200 1320

> GDAL.close(x)
```

Here, `x` is a derivative of a `GDALDataset` object, and is the GDAL data set handle; the data are not in the R workspace, but all their features are there to be read on demand. An open GDAL handle can be read into a SpatialGridDataFrame, so that `readGDAL` may be done in pieces if needed. Information about the file to be accessed may also be shown without the file being read, using the GDAL bindings packaged in the utility function `GDALinfo`:

```
> GDALinfo("70042108.tif")

rows         1200
columns      1320
bands        1
ll.x         174.2
ll.y         -36.5
res.x        0.0008333333
res.y        0.0008333333
oblique.x    0
oblique.y    0
driver       GTiff
projection   +proj=longlat +ellps=WGS84 +datum=WGS84 +no_defs
file         70042108.tif
```

We use the Meuse grid data set to see how data may be written out using GDAL, adding a set of `towgs84` values from the GDAL bug tracker.[18] The `writeGDAL` function can be used directly for drivers that support file creation. For other file formats, which can be made as copies of a prototype, we need to create an intermediate GDAL data set using `create2GDAL`, and then use functions operating on the GDAL data set handle to complete. First we simply output inverse distance weighted interpolated values of Meuse Bank logarithms of zinc ppm as a GeoTiff file.

```
> library(gstat)
> log_zinc <- krige(log(zinc) ~ 1, meuse, meuse.grid)["var1.pred"]

[inverse distance weighted interpolation]

> proj4string(log_zinc) <- CRS(proj4string(meuse.grid))
> summary(log_zinc)

Object of class SpatialPixelsDataFrame
Coordinates:
     min    max
x 178440 181560
y 329600 333760
Is projected: TRUE
proj4string :
[+init=epsg:28992
+towgs84=565.237,50.0087,465.658,-0.406857,0.350733,-1.87035,4.0812
+proj=sterea +lat_0=52.15616055555555 +lon_0=5.38763888888889
+k=0.9999079 +x_0=155000 +y_0=463000 +ellps=bessel +units=m
+no_defs]
Number of points: 3103
Grid attributes:
  cellcentre.offset cellsize cells.dim
x            178460       40        78
y            329620       40       104
Data attributes:
   Min. 1st Qu.  Median    Mean 3rd Qu.    Max.
  4.791   5.484   5.694   5.777   6.041   7.482

> writeGDAL(log_zinc, fname = "log_zinc.tif", driver = "GTiff",
+     type = "Float32", options = "INTERLEAVE=PIXEL")

> GDALinfo("log_zinc.tif")

rows        104
columns     78
bands       1
ll.x        178440
ll.y        333760
```

[18] http://trac.osgeo.org/gdal/ticket/1987; newer versions of PROJ4/GDAL may include the correct `+towgs84` parameter values.

```
res.x        40
res.y        40
oblique.x    0
oblique.y    0
driver       GTiff
projection   +proj=sterea +lat_0=52.15616055555555
+lon_0=5.38763888888889 +k=0.9999079 +x_0=155000
+y_0=463000 +ellps=bessel +units=m +no_defs
file         log_zinc.tif
```

The output file can for example be read into ENVI™ directly, or into ArcGIS™ via the 'Calculate statistics' tool in the Raster section of the Toolbox, and displayed by adjusting the symbology classification.

4.3.2 Writing a Google Earth™ Image Overlay

Our next attempt to export a raster will be more ambitious; in fact we can use this technique to export anything that can be plotted on a PNG graphics device. We export a coloured raster of interpolated log zinc ppm values to a PNG file with an alpha channel for viewing in Google Earth™. Since the target software requires geographical coordinates, a number of steps will be needed. First we make a polygon to bound the study area and project it to geographical coordinates:

```
> library(maptools)
> grd <- as(meuse.grid, "SpatialPolygons")
> proj4string(grd) <- CRS(proj4string(meuse))
> grd.union <- unionSpatialPolygons(grd, rep("x", length(slot(grd,
+     "polygons"))))
> ll <- CRS("+proj=longlat +datum=WGS84")
> grd.union.ll <- spTransform(grd.union, ll)
```

Next we construct a suitable grid in geographical coordinates, as our target object for export, using the `GE_SpatialGrid` wrapper function. This grid is also the container for the output PNG graphics file, so `GE_SpatialGrid` also returns auxiliary values that will be used in setting up the `png` graphics device within R. We use the `overlay` method to set grid cells outside the river bank area to NA, and then discard them by coercion to a SpatialPixelsDataFrame:

```
> llGRD <- GE_SpatialGrid(grd.union.ll)
> llGRD_in <- overlay(llGRD$SG, grd.union.ll)
> llSGDF <- SpatialGridDataFrame(grid = slot(llGRD$SG,
+     "grid"), proj4string = CRS(proj4string(llGRD$SG)),
+     data = data.frame(in0 = llGRD_in))
> llSPix <- as(llSGDF, "SpatialPixelsDataFrame")
```

We use `idw` from the **gstat** package to make an inverse distance weighted interpolation of zinc ppm values from the soil samples available, also, as here, when the points are in geographical coordinates; interpolation will be fully presented in Chap. 8:

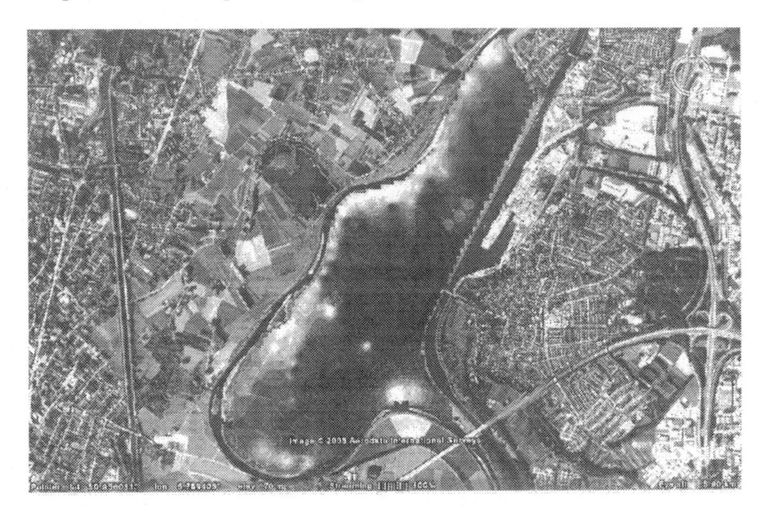

Fig. 4.3. Interpolated log zinc ppm for the Meuse Bank data set shown in Google Earth™

```
> meuse_ll <- spTransform(meuse, CRS("+proj=longlat +datum=WGS84"))
> llSPix$pred <- idw(log(zinc) ~ 1, meuse_ll, llSPix)$var1.pred
```

```
[inverse distance weighted interpolation]
```

Since we have used `GE_SpatialGrid` to set up the size of an R `png` graphics device, we can now use it as usual, here with `image`. In practice, any base graphics methods and functions can be used to create an image overlay. Finally, after closing the graphics device, we use `kmlOverlay` to write a `*.kml` file giving the location of the overlay and which will load the image at that position when opened in Google Earth™, as shown in Fig. 4.3:

```
> png(file = "zinc_IDW.png", width = llGRD$width,height = llGRD$height,
+     bg = "transparent")
> par(mar = c(0, 0, 0, 0), xaxs = "i", yaxs = "i")
> image(llSPix, "pred", col = bpy.colors(20))
> dev.off()
> kmlOverlay(llGRD, "zinc_IDW.kml", "zinc_IDW.png")
```

4.3.3 Other Import/Export Functions

There is a simple `readAsciiGrid` function in **maptools** that reads ESRI™ Arc ASCII grids into SpatialGridDataFrame objects; it does not handle CRS and has a single band. The companion `writeAsciiGrid` is for writing Arc ASCII grids. It is also possible to use connections to read and write arbitrary binary files, provided that the content is not compressed. Functions in the R image analysis packages referred to above may also be used to read and write a number of image formats. If the grid registration slots in objects of classes defined in the **pixmap** package are entered manually, these objects may also be used to hold raster data.

4.4 Grass

GRASS[19] is a major open source GIS, originally developed as the Geographic Resources Analysis Support System by the U.S. Army Construction Engineering Research Laboratories (CERL, 1982–1995), and subsequently taken over by its user community. GRASS has traditional strengths in raster data handling, but two advances (floating point rasters and support for missing values) were not completed when development by CERL was stopped. These were added for many modules in the GRASS 5.0 release; from GRASS 6.0, new vector support has been added. GRASS is a very large but very simple system – it is run as a collection of separate programs built using shared libraries of core functions. There is then no GRASS 'program', just a script setting environment variables needed by the component programs. GRASS does interact with the OSGeo stack of applications, and GRASS functionality, including the R interface, is available in the Quantum GIS desktop GIS application.

An R package to interface with GRASS has been available on CRAN – **GRASS** – since the release of GRASS 5.0. It provided a compiled interface to raster and sites data, but not vector data, and included a frozen copy of the core GRASS GIS C library, modified to suit the fact that its functions were being used in an interactive, longer-running program like R. The **GRASS** package is no longer being developed, but continues to work for users of GRASS 5. The GRASS 5 interface is documented in Neteler and Mitasova (2004, pp. 333–354) and Bivand (2000).

The current GRASS releases, from GRASS 6.0, with GRASS 6.2 released in October 2006 and 6.3 due in 2008, have a different interface, using the **sp** classes presented in Chap. 2. Neteler and Mitasova (2008) describe GRASS 6 fully, and present this interface on pp. 353–364. The **spgrass6** package depends on other packages for moving data between GRASS and R, chiefly using **rgdal**, because GRASS also uses GDAL and OGR as its main import/export mechanisms. The interface works by exchanging temporary files in formats that both GRASS and **rgdal** know. This kind of loose coupling is less of a burden than it was before, with smaller, slower machines. This is why the GRASS 5 interface was tight-coupled, with R functions reading from and writing to the GRASS database directly. Using GRASS plug-in drivers in GDAL/OGR is another possibility for reading GRASS data directly into R through **rgdal**, without needing **spgrass6**; **spgrass6** can use these plug-in drivers if present for reading GRASS data.

GRASS uses the concept of a working region or window, specifying both the viewing rectangle and – for raster data – the resolution. The data in the GRASS database can be from a larger or smaller region and can have a different resolution, and are re-sampled to match the working region for analysis. This current window should determine the way in which raster data are retrieved and transferred.

[19] `http://grass.osgeo.org/`.

GRASS also uses the concepts of a location, with a fixed and uniform coordinate reference system, and of mapsets within the location. The location is typically chosen at the start of a work session, and with the location, the user will have read access to possibly several mapsets, and write access to some, probably fewer, to avoid overwriting the work of other users of the location.

Intermediate temporary files are the chosen solution for interaction between GRASS and R in **spgrass6**, using shapefiles for vector data and single band BIL (Band Interleaved by Line) binary files for raster data. Note that missing values are defined and supported for GRASS raster data, but that missing values for vector data are not uniformly defined or supported.

Support for GRASS under Cygwin is provided and known to function satisfactorily. Native Windows GRASS has been tested only with Quantum GIS, but is expected to become more widely available as GRASS 6.3 has now been released, using MSYS rather than Cygwin for Unix shell emulation (a Windows **spgrass6** binary is on CRAN). Mac OSX is treated as Unix, and **spgrass6** is installed from source (like **rgdal**). The **spgrass6** package should be installed with the packages it depends upon, **sp** and **rgdal**.

R is started from within a GRASS session from the command line, and the **spgrass6** loaded with its dependencies:

```
> system("g.version", intern = TRUE)

[1] "GRASS 6.3.cvs (2007) "

> library(spgrass6)
> gmeta6()

gisdbase    /home/rsb/topics/grassdata
location    spearfish60
mapset      rsb
rows        477
columns     634
north       4928010
south       4913700
west        589980
east        609000
nsres       30
ewres       30
projection  +proj=utm +zone=13 +a=6378206.4 +rf=294.9786982 +no_defs
+nadgrids=/home/rsb/topics/grass63/grass-6.3.cvs/etc/nad/conus
+to_meter=1.0
```

The examples used here are taken from the 'Spearfish' sample data location (South Dakota, USA, 103.86W, 44.49N), perhaps the most typical for GRASS demonstrations. The `gmeta6` function is simply a way of summarising the current settings of the GRASS location and region within which we are working. Data moved from GRASS over the interface will be given category labels if present. The interface does not support the transfer of factor level labels from R to GRASS, nor does it set colours or quantisation rules. The `readRAST6` command here reads elevation values into a `SpatialGridDataFrame` object, treating the values returned as floating point and the geology categorical layer into a factor:

```
> spear <- readRAST6(c("elevation.dem", "geology"), cat = c(FALSE,
+     TRUE))
> summary(spear)

Object of class SpatialGridDataFrame
Coordinates:
              min     max
coords.x1  589980  609000
coords.x2 4913700 4928010
Is projected: TRUE
proj4string :
[+proj=utm +zone=13 +a=6378206.4 +rf=294.9786982 +no_defs
+nadgrids=/home/rsb/topics/grass63/grass-6.3.cvs/etc/nad/conus
+to_meter=1.0]
Number of points: 2
Grid attributes:
  cellcentre.offset cellsize cells.dim
1            589995       30       634
2           4913715       30       477
Data attributes:
 elevation.dem        geology
 Min.   : 1066   sandstone:74959
 1st Qu.: 1200   limestone:61355
 Median : 1316   shale    :46423
 Mean   : 1354   sand     :36561
 3rd Qu.: 1488   igneous  :36534
 Max.   : 1840   (Other)  :37636
 NA's   :10101   NA's     : 8950
```

When the `cat` argument is set to `TRUE`, the GRASS category labels are imported and used as factor levels; checking back, we can see that they agree:

```
> table(spear$geology)

metamorphic  transition     igneous   sandstone   limestone
      11693         142       36534       74959       61355
      shale sandy shale    claysand        sand
      46423       11266       14535       36561

> system("r.stats --q -cl geology", intern = TRUE)

 [1] "1 metamorphic 11693" "2 transition 142"    "3 igneous 36534"
 [4] "4 sandstone 74959"   "5 limestone 61355"   "6 shale 46423"
 [7] "7 sandy shale 11266" "8 claysand 14535"    "9 sand 36561"
[10] "* no data 8950"
```

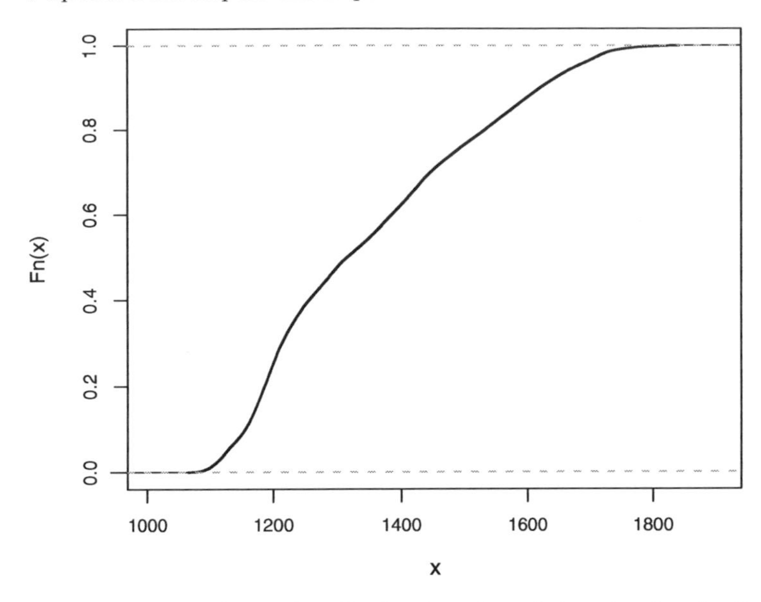

Fig. 4.4. Empirical cumulative distribution function of elevation for the Spearfish location

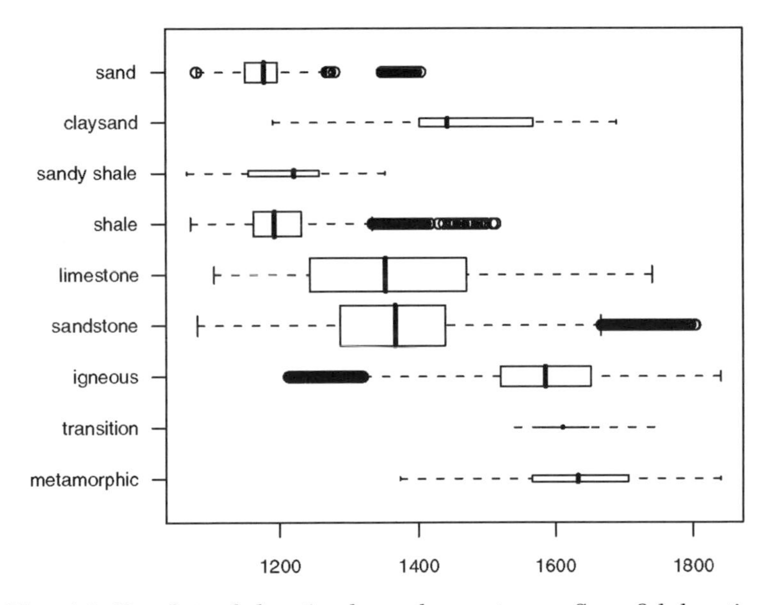

Fig. 4.5. Boxplots of elevation by geology category, Spearfish location

Figure 4.4 shows an empirical cumulative distribution plot of the elevation values, giving readings of the proportion of the study area under chosen elevations. In turn Fig. 4.5 shows a simple boxplot of elevation by geology category, with widths proportional to the share of the geology category in the total area.

We have used the `readRAST6` function to read from GRASS rasters into R; the `writeRAST6` function allows a single named column of a `SpatialGridDataFrame` object to be exported to GRASS.

The **spgrass6** package also provides functions to move vector features and associated attribute data to R and back again; unlike raster data, there is no standard mechanism for handling missing values. The `readVECT6` function is used for importing vector data into R, and `writeVECT6` for exporting to GRASS. The first data set to be imported from GRASS contains the point locations of sites where insects have been monitored, the second is a set of stream channel centre-lines:

```
> bugsDF <- readVECT6("bugsites")
```

```
> vInfo("streams")
```

```
    points      lines boundaries  centroids      areas    islands
         0        104         12          4          4          4
     faces    kernels
         0          0
```

```
> streams <- readVECT6("streams", type = "line,boundary",
+     remove.duplicates = FALSE)
```

The `remove.duplicates` argument is set to TRUE when there are only, for example lines or areas, and the number present is greater than the data count (the number of rows in the attribute data table). The `type` argument is used to override type detection when multiple types are non-zero, as here, where we choose lines and boundaries, but the function guesses areas, returning just filled water bodies.

Because different mechanisms are used for passing information concerning the GRASS location coordinate reference system for raster and vector data, the PROJ.4 strings often differ slightly, even though the actual CRS is the same. We can see that the representation for the point locations of beetle sites does differ here; the vector representation is more in accord with standard PROJ.4 notation than that for the raster layers, even though they are the same. In the summary of the `spear` object above, the ellipsoid was represented by `+a` and `+rf` tags instead of the `+ellps` tag using the `clrk66` value:

```
> summary(bugsDF)
```

```
Object of class SpatialPointsDataFrame
Coordinates:
              min     max
coords.x1  590232  608471
coords.x2 4914096 4920512
Is projected: TRUE
proj4string :
[+proj=utm +zone=13 +ellps=clrk66 +datum=NAD27 +units=m
+no_defs +nadgrids=@conus,@alaska,@ntv2_0.gsb,@ntv1_can.dat]
```

```
Number of points: 90
Data attributes:
      cat                   str1
 Min.   : 1.00   Beetle site:90
 1st Qu.:23.25
 Median :45.50
 Mean   :45.50
 3rd Qu.:67.75
 Max.   :90.00
```

This necessitates manual assignment from one representation to the other in some occasions, and is due to GRASS using non-standard but equivalent extensions to PROJ.4.

There are a number of helper functions in the **spgrass6** package, one `gmeta2grd` to generate a GridTopology object from the current GRASS region settings. This is typically used for interpolation from point data to a raster grid, and may be masked by coercion from a SpatialGrid to a SpatialPixels object having set cells outside the study area to NA. A second utility function for vector data uses the fact that GRASS 6 uses a topological vector data model. The `vect2neigh` function returns a data frame with the left and right neighbours of arcs on polygon boundaries, together with the length of the arcs. This can be used to modify the weighting of polygon contiguities based on the length of shared boundaries. Like GRASS, GDAL/OGR, PROJ.4, and other OSGeo projects, the functions offered by **spgrass6** are changing, and current help pages should be consulted to check correct usage.

4.4.1 Broad Street Cholera Data

Even though we know that John Snow already had a working hypothesis about cholera epidemics, his data remain interesting, especially if we use a GIS to find the street distances from mortality dwellings to the Broad Street pump in Soho in central London. Brody et al. (2000) point out that John Snow did not use maps to 'find' the Broad Street pump, the polluted water source behind the 1854 cholera epidemic, because he associated cholera with water contaminated with sewage, based on earlier experience. The accepted opinion of the time was that cholera was most probably caused by a 'concentrated noxious atmospheric influence', and maps could just as easily have been interpreted in support of such a point source.

The specific difference between the two approaches is that the atmospheric cause would examine straight-line aerial distances between the homes of the deceased and an unknown point source, while a contaminated water source would rather look at the walking distance along the street network to a pump or pumps. The basic data to be used here were made available by Jim Detwiler, who had collated them for David O'Sullivan for use on the cover of O'Sullivan and Unwin (2003), based on earlier work by Waldo Tobler and others. The files were a shapefile with counts of deaths at front doors of houses and a georeferenced copy of the Snow map as an image; the files were registered in the British National Grid CRS. The steps taken in GRASS were to set up a suitable location in the CRS, to import the image file, the file of mortalities, and the file of pump locations.

To measure street distances, the building contours were first digitised as a vector layer, cleaned, converted to raster leaving the buildings outside the street mask, buffered out 4 m to include all the front door points within the street mask, and finally distances measured from each raster cell in the buffered street network to the Broad Street pump and to the nearest other pump. These operations in summary were as follows:

```
v.digit -n map=vsnow4 bgcmd="d.rast map=snow"
v.to.rast input=vsnow4 output=rfsnow use=val value=1
r.buffer input=rfsnow output=buff2 distances=4
r.cost -v input=buff2 output=snowcost_not_broad \
  start_points=vpump_not_broad
r.cost -v input=buff2 output=snowcost_broad start_points=vpump_broad
```

The main operation here is `r.cost`, which uses the value of 2.5 m stored in each cell of `buff2`, which has a resolution of 2.5 m, to cumulate distances from the start points in the output rasters. The operation is carried out for the other pumps and for the Broad Street pump. This is equivalent to finding the line of equal distances shown on the extracts from John Snow's map shown in Brody et al. (2000, p. 65). It is possible that there are passages through buildings not captured by digitising, so the distances are only as accurate as can now be reconstructed.

```
> buildings <- readVECT6("vsnow4")
> sohoSG <- readRAST6(c("snowcost_broad", "snowcost_not_broad"))
```

For visualisation, we import the building outlines, and the two distance rasters. Next we import the death coordinates and counts, and overlay the deaths on the distances, to extract the distances for each house with mortalities – these are added to the `deaths` object, together with a logical variable indicating whether the Broad Street pump was closer (for this distance measure) or not:

```
> deaths <- readVECT6("deaths3")
> o <- overlay(sohoSG, deaths)
> deaths <- spCbind(deaths, as(o, "data.frame"))
> deaths$b_nearer <- deaths$snowcost_broad < deaths$snowcost_not_broad

> by(deaths$Num_Cases, deaths$b_nearer, sum)

INDICES: FALSE
[1] 221
-----------------------------------------------------
INDICES: TRUE
[1] 357
```

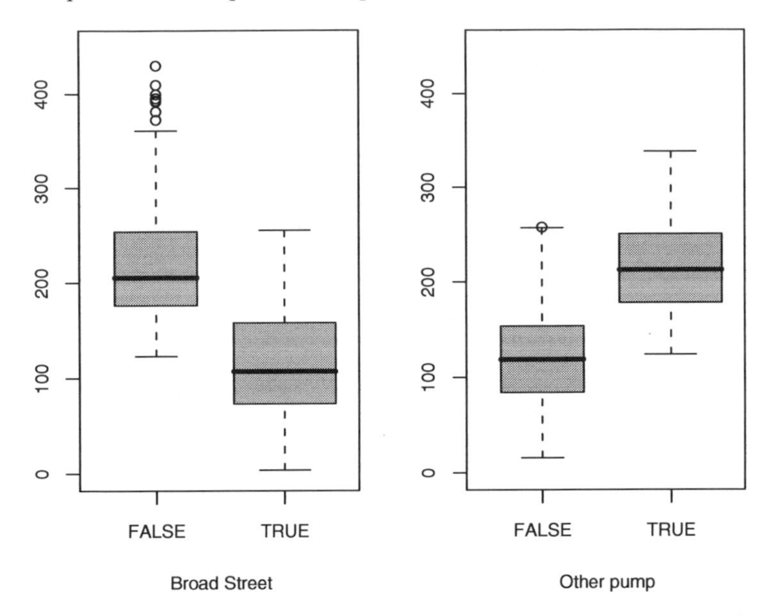

Fig. 4.6. Comparison of walking distances from homes of fatalities to the Broad Street pump or another pump by whether the Broad Street pump was closer or not

There are not only more mortalities in houses closer to the Broad Street pump, but the distributions of distances are such that their inter-quartile ranges do not overlap. This can be seen in Fig. 4.6, from which a remaining question is why some of the cases appear to have used the Broad Street pump in spite of having a shorter distance to an alternative. Finally, we import the locations of the pumps to assemble a view of the situation, shown in Fig. 4.7. The grey scaled streets indicate the distance of each 2.5 m raster cell from the Broad Street pump along the street network. The buildings are overlaid on the raster, followed by proportional symbols for the number of mortalities per affected house, coded for whether they are closer to the Broad Street pump or not, and finally the pumps themselves.

```
> nb_pump <- readVECT6("vpump_not_broad")
> b_pump <- readVECT6("vpump_broad")
```

4.5 Other Import/Export Interfaces

The classes for spatial data introduced in **sp** have made it easier to implement and maintain the import and export functions described earlier in this chapter. In addition, they have created opportunities for writing other interfaces, because the structure of the objects in R is better documented. In this section, a number of such interfaces will be presented, with others to come in

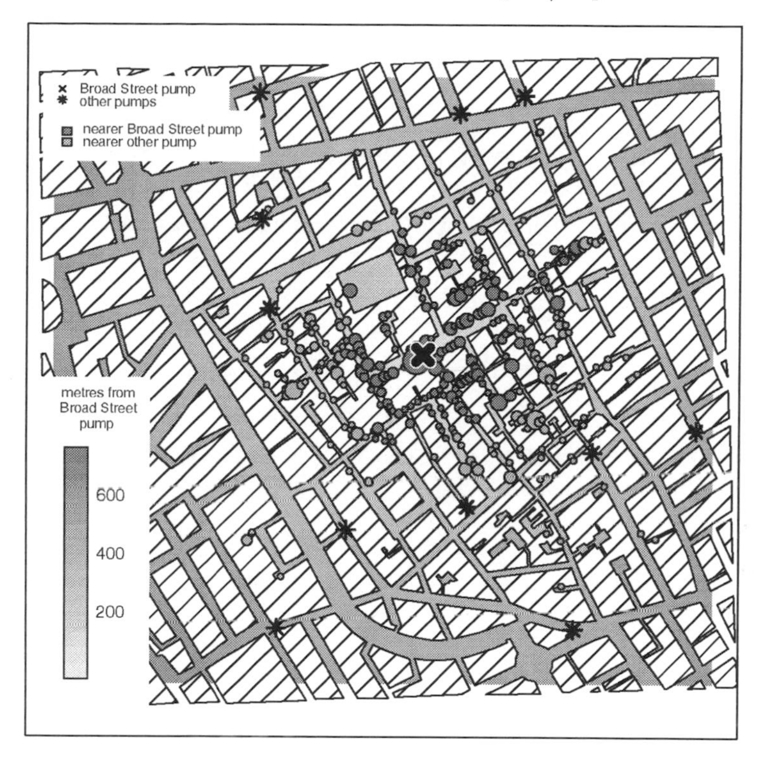

Fig. 4.7. The 1854 London cholera outbreak near Golden Square

the future, hosted in **maptools** or other packages. Before going on to discuss interfaces with external applications, conversion wrappers for R packages will be mentioned.

The **maptools** package contains interface functions to convert selected **sp** class objects to classes used in the **spatstat** for point pattern analysis – these are written as coercion methods to **spatstat** `ppp` and `owin` classes. **maptools** also contains the `SpatialLines2PolySet` and `SpatialPolygons2PolySet` functions to convert **sp** class objects to `PolySet` class objects as defined in the **PBSmapping** package, and a pair of matching functions in the other direction. This package provides a number of GIS procedures needed in fisheries research (PBS is the name of the Pacific Biological Station in Nanaimo, British Columbia, Canada).

There are also interface functions in the **adehabitat** package for conversion between **sp** class objects and **adehabitat** `kasc` and `asc` gridded objects, **adehabitat** `area` polygon objects, and **adehabitat** `traj` and `ltraj` trajectory objects. The package itself is documented in Calenge (2006), and includes many tools for the analysis of space and habitat use by animals.

4.5.1 Analysis and Visualisation Applications

While many kinds of data analysis can be carried out within the R environment, it is often very useful to be able to write out files for use in other applications or for sharing with collaborators not using R. These functions live in **maptools** and will be extended as required. The `sp2tmap` function converts a `SpatialPolygons` object for use with the Stata™ tmap contributed command,[20] by creating a data frame with the required columns. The data frame returned by the function is exported using `write.dta` from the **foreign** package, which should also be used to export the attribute data with the polygon tagging key. The `sp2WB` function exports a `SpatialPolygons` object as a text file in S-PLUS™ map format to be imported by WinBUGS.

The **GeoXp** package provides some possibilities for interactive statistical data visualisation within R, including mapping. The R graphics facilities are perhaps better suited to non-interactive use, however, especially as it is easy to write data out to Mondrian.[21] Mondrian provides fully linked multiple plots, and although the screen can become quite 'busy', users find it easy to explore their data in this environment. The function `sp2Mondrian` in **maptools** writes out two files, one with the data, the other with the spatial objects from a `SpatialPolygonsDataFrame` object for Mondrian to read; the polygon format before Mondrian 1.0 used a single file and may still be used, controlled by an additional argument.

```
> sp2Mondrian(scot_BNG, "scot_BNG.txt")
```

The example we use here is the Scottish lip cancer `SpatialPolygons-DataFrame` object in the British National Grid projection (Mondrian assumes a planar map). A screen shot of two Mondrian plots is shown in Fig. 4.8, with a map view and a parallel boxplot, where a group of districts has been selected on the map, and is shown as a subset on the linked display.

4.5.2 TerraLib and aRT

The **aRT** package[22] provides an advanced modular interface to TerraLib.[23] TerraLib is a GIS classes and functions library intended for the development of multiple GIS tools. Its main aim is to enable the development of a new generation of GIS applications, based on the technological advances on spatial databases. TerraLib defines the way that spatial data are stored in a database system, and can use MySQL, PostgreSQL, Oracle, or Access as a back-end. The library itself can undertake a wide range of GIS operations on the data stored in the database, as well as storing and retrieving the data as spatial objects from the database system.

[20] `http://www.stata.com/search.cgi?query=tmap`.
[21] `http://rosuda.org/Mondrian/`.
[22] `http://leg.ufpr.br/aRT/`.
[23] `http://www.terralib.org/`.

Fig. 4.8. Screen shot of two linked Mondrian plots: a map of the Empirical Bayes smooth and a parallel boxplot for four variables, with the selected districts on the map (three northern mainland counties, Outer Hebrides, Orkney, and Shetland) split out as overlay boxplots

The **aRT** package interfaces **sp** classes with TerraLib classes, permitting data to flow between R, used as a front-end system interacting with the user, through TerraLib and the back-end database system. One of the main objectives of **aRT** is to do spatial queries and operations in R. Because these operations are written to work efficiently in TerraLib, a wide range of overlay and buffering operations can be carried out, without them being implemented in R itself. Operations on the geometries, such as whether they touch, how far apart they are, whether they contain holes, polygon unions, and many others, can be handed off to TerraLib.

A further innovation is the provision of a wrapper for the R compute engine, allowing R with **aRT** to be configured with TerraLib between the back-end database system and a front-end application interacting with the user. This application, for example TerraView, can provide access through menus to spatial data analysis functionality coded in R using **aRT**.[24] All of this software is released under open source licences, and offers considerable opportunities for building non-proprietary customised systems for medium and larger organisations able to commit resources to C++ programming. Organisations running larger database systems are likely to have such resources anyway, so **aRT** and TerraLib provide a real alternative for fresh spatial data handling projects.

[24] Andrade Neto and Ribeiro Jr. (2005).

4.5.3 Other GIS and Web Mapping Systems

The **Rpad** package[25] is a general-purpose interactive, web-based analysis program, and Rpad pages are interactive workbook-type sheets based on R. Some of the examples[26] use spatial data directly, and it is quite practical to handle smaller or moderate-sized point data without needing bulky client-side applets. Integration with R results is automatic, without the need for much extra software on the server side.

An interface package – **RSAGA** – has been provided for SAGA GIS;[27] like the GRASS 6 interface, it uses `system` to pass commands to external software.

For raster or polygon data, it may be sensible to use MapServer[28] or alternative software. MapServer integrates and renders spatial data for the web, and because it uses the same open source geospatial toolset as the packages described in this chapter (PROJ.4, GDAL/OGR), it is possible to add R as a compute engine to websites running MapServer with little difficulty. It is also possible to set up MapServer as a Web Feature Server, which serves the actual map data encapsulated in XML wrappers instead of rendering the map for display on a web browser; Mitchell (2005) contains a good deal of information on providing web mapping facilities.

In the discussion above, integration between R and GIS has principally taken the form of file transfer. It is possible to use other mechanisms, similar in nature to the embedding of R in TerraView using **aRT**. One example is given by Tait et al. (2004), using the R `StatConnector` (D)COM mechanism to use R as a back-end from ArcGIS™. The specific context is the need to provide epidemiologists using ArcGIS™ for animal disease control and detection with point pattern analysis tools, using a GIS interface. The prototype was a system using **splancs** running in R to calculate results from data passed from ArcGIS™, with output passed back to ArcGIS™ for display. A practical difficulty of embedding both R and **splancs** on multiple workstations is that of software installation and maintenance.

A second example is ArcRstats,[29] for producing multivariate habitat prediction rasters using ArcGIS™ and R for interfacing classification and regression trees, generalised linear models, and generalised additive models. It is implemented using the Python interface introduced into ArcGIS™ from version 9, and then the Python `win32com.client` module to access the R `StatConnector` (D)COM mechanism. The current release of ArcRstats uses the **sp**, **maptools**, and **rgdal** packages to interface spatial data, and **RODBC** to work with Access formated geodatabases, in addition to a wide range of analysis functions, including those from the **spatstat** package.

[25] `http://www.rpad.org/Rpad/`.
[26] `http://www.rpad.org/Rpad/InterruptionMap.Rpad`.
[27] `http://www.saga-gis.uni-goettingen.de/html/index.php`.
[28] `http://mapserver.gis.umn.edu/`.
[29] `http://www.env.duke.edu/geospatial/software/`.

The marine geospatial ecology tools project[30] follows up the work begun in ArcRstats, providing for execution in many environments, and using the Python route through the COM interface to ArcGIS™. It is not hard to write small Python scripts to interface R and ArcGIS™ through temporary files and the `system` function. This is illustrated by the **RPyGeo** package, which uses R to write Python scripts for the ArcGIS™ geoprocessor. The use of R Python interface is not as fruitful as it might be, because ArcGIS™ bundles its own usually rather dated version of Python, and ArcGIS™ itself only runs on Windows and is rather expensive, certainly compared to GRASS.

4.6 Installing rgdal

Because **rgdal** depends on external libraries, on GDAL and PROJ.4, and particular GDAL drivers may depend on further libraries, installation is not as easy as with self-contained R packages. Only the Windows binary package is self-contained, with a basic set of drivers available. For Linux/Unix and MacOSX, it is necessary to install **rgdal** from source, after first having installed the external dependencies. Users of open source GIS applications such as GRASS will already have GDAL and PROJ.4 installed anyway, because they are required for such applications.

In general, GDAL and PROJ.4 will install from source without difficulty, but care may be required to make sure that libraries needed for drivers are available and function correctly. If the programs `proj`, `gdalinfo`, and `ogrinfo` work correctly for data sources of interest after GDAL and PROJ.4 have been installed, then **rgdal** will also work correctly. Mac OSX users may find William Kyngesburye's frameworks[31] a useful place to start, if installation from source seems forbidding. More information is available on the 'maps' page at the Rgeo website,[32] and by searching the archives of the R-sig-geo mailing list.

Windows users needing other drivers, and for whom conversion using programs in the FWTools[33] binary for Windows is not useful, may choose to install **rgdal** from source, compiling the **rgdal** DLL with VC++ and linking against the FWTools DLLs – see the `inst/README.windows` file in the source package for details.

[30] `http://code.env.duke.edu/projects/mget.`
[31] `http://www.kyngchaos.com/software/unixport/frameworks.`
[32] `http://www.r-project.org/Rgeo.`
[33] `http://fwtools.maptools.org.`

5

Further Methods for Handling Spatial Data

This chapter is concerned with a more detailed explanation of some of the methods that are provided for working with the spatial classes described in Chap. 2. We first consider the question of the spatial support of observations, going on to look at overlay and sampling methods for a range of classes of spatial objects. Following this, we cover combining the data stored in the `data` slot of `Spatial*DataFrame` objects with additional data stored as vectors and data frames, as well as the combination of spatial objects. We also apply some of the functions that are available for handling and checking polygon topologies, including the dissolve operation.

5.1 Support

In data analysis in general, the relationship between the abstract constructs we try to measure and the operational procedures used to make the measurements is always important. Very often substantial metadata volumes are generated to document the performance of the instruments used to gather data. Naturally, the same applies to spatial data. Positional data need as much care in documenting their collection as other kinds of data. When approximations are used, they need to be recorded as such. Part of the issue is the correct recording of projection and datum, which are covered in Chap. 4. The time-stamping of observations is typically useful, for example when administrative boundaries change over time.

The recording of position for surveying, for example for a power company, involves the inventorying of cables, pylons, transformer substations, and other infrastructure. Much GIS software was developed to cater for such needs, inventory rather than analysis. For inventory, arbitrary decisions, such as placing the point coordinate locating a building by the right-hand doorpost facing the door from the outside, have no further consequences. When, however, spatial data are to be used for analysis and inference, the differences between arbitrary assumptions made during observation and other possible spatial representations of the phenomena of interest will feed through to the conclusions. The adopted representation is known as its support, and is discussed by Waller and Gotway (2004, pp. 38–39). The point support of a dwelling may be taken as the point location of its right-hand doorpost, a soil sample may have point support of a coordinate surveyed traditionally or by GPS. But the dwelling perhaps should have polygonal support, and in collecting soil samples, most often the point represents a central position in the circle or square used to gather a number of different samples, which are then bagged together for measurement.

An example of the effects of support is the impact of changes in voting district boundaries in election systems, which are not strictly proportional. The underlying voting behaviour is fixed, but different electoral results can be achieved by tallying results in different configurations or aggregations of the voters' dwellings.[1] When carried out to benefit particular candidates or parties, this is known as gerrymandering. The aggregations are arbitrary polygons, because they do not reflect a political entity as such. This is an example of change of support, moving from the position of the dwelling of the voter to some aggregation. Change of support is a significant issue in spatial data analysis, and is introduced in Schabenberger and Gotway (2005, pp. 284–285). A much more thorough treatment is given by Gotway and Young (2002), who show how statistical methods can be used to carry through error associated with change of support to further steps in analysis. In a very similar vein, it can be argued that researchers in particular subject domains should consider involving statisticians from the very beginning of their projects, to allow sources of potential uncertainty to be instrumented if possible. One would seek to control error propagation when trying to infer from the data collected later during the analysis and reporting phase (Guttorp, 2003; Wikle, 2003). An example might be marine biologists and oceanographers not collecting data at the same place and time, and hoping that data from different places and times could be readily combined without introducing systematic error.

One of the consequences of surveying as a profession being overtaken by computers, and of surveying data spreading out to non-surveyor users, is that understanding of the imprecision of positional data has been diluted. Some of the imprecision comes from measurement error, which surveyors know from their training and field experience. But a digital representation of a coordinate looks very crisp and precise, deceptively so. Surveying and cartographic representations are just a summary from available data. Where no data were collected, the actual values are guesswork and can go badly wrong, as users of maritime charts routinely find. Further, support is routinely changed for purposes of visualisation: contours or filled contours representing a grid make the

[1] The CRAN **BARD** package for automated redistricting and heuristic exploration of redistricter revealed preference is an example of the use of R for studying this problem.

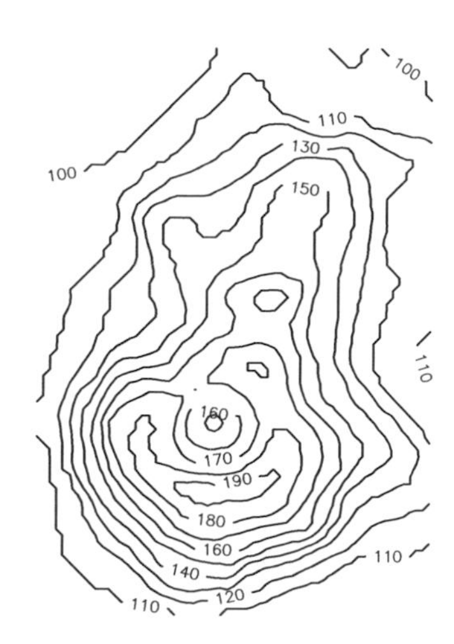

Fig. 5.1. Image plot and contour plot representations of Maunga Whau from the standard R `volcano` data set, for the same elevation class intervals (rotated to put north at the top)

data look much less 'chunky' than an image plot of the same data, as Fig. 5.1 shows. In fact, the data were digitised from a paper map by Ross Ihaka, as much other digital elevation data have been, and the paper map was itself a representation of available data, not an exact reproduction of the terrain. Even SRTM data can realistically be used only after cleaning; the 3 arcsec data used in Sect. 2.7 were re-sampled from noisier 1 arcsec data using a specific re-sampling and cleaning algorithm. A different algorithm would yield a slightly different digital elevation model.

While we perhaps expect researchers wanting to use R to analyse spatial data to be applied domain scientists, it is worth noting that geographical information science, the field of study attempting to provide GIS with more consistent foundations, is now actively incorporating error models into position measurement, and into spatial queries based on error in observed values. Say we are modelling crop yield based on soil type and other variables, and our spatial query at point i returns `"sand"`, when in fact the correct value at that location is `"clay"`, our conclusions will be affected. The general application of uncertainty to spatial data in a GIS context is reviewed by Worboys and Duckham (2004, pp. 328–358), and attribute error propagation is discussed by Heuvelink (1998). In an open computing environment like R, it is quite possible to think of 'uncertain' versions of the 'crisp' classes dealt with so far, in which, for example point position could be represented as a value drawn from a statistical model, allowing the impact of positional uncertainty on other methods of analysis to be assessed (see for example Leung et al., 2004).

5.2 Overlay

Accepting that moving from one spatial representation to another is a typical operation performed with spatial data, `overlay` methods are provided for a number of pairs of spatial data object types. Overlay methods involve combining congruent or non-congruent spatial data objects, and only some are provided directly, chiefly for non-congruent objects. Overlay operations are mentioned by Burrough and McDonnell (1998, pp. 52–53) and covered in much more detail by O'Sullivan and Unwin (2003, pp. 284–314) and Unwin (1996), who show how many of the apparently deterministic inferential problems in overlay are actually statistical in nature, as noted earlier. The basic approach is to query one spatial data object using a second spatial data object of the same or of a different class. The query result will typically be another spatial data object or an object pointing to one of the input objects. Overlaying a `SpatialPoints` object with a `SpatialPolygons` object returns a vector of numbers showing which `Polygons` object each coordinate in the `SpatialPoints` object falls within – this is an extension of the point-in-polygon problem to multiple points and polygons.

To continue the SRTM elevation data example, we can query `auck_el1`, which is a `SpatialGridDataFrame` object, using the transect shown in Fig. 2.7, stored as `SpatialPoints` object `transect_sp`. Using an overlay method, we obtain the elevation values retrieved from querying the grid cells as a `SpatialPointsDataFrame` object.

```
> summary(transect_sp)

Object of class SpatialPoints
Coordinates:
                 min       max
coords.x1 174.45800 175.29967
coords.x2 -37.03625 -37.03625
Is projected: FALSE
proj4string : [+proj=longlat +ellps=WGS84]
Number of points: 1011

> transect_el1 <- overlay(auck_el1, transect_sp)
> summary(transect_el1)

Object of class SpatialPointsDataFrame
Coordinates:
                 min       max
coords.x1 174.45800 175.29967
coords.x2 -37.03625 -37.03625
Is projected: FALSE
proj4string :
[+proj=longlat +ellps=WGS84 +datum=WGS84 +no_defs]
Number of points: 1011
```

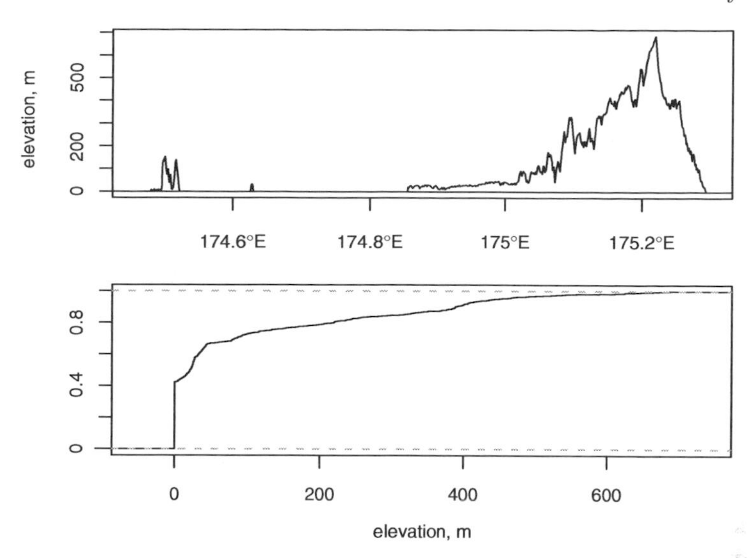

Fig. 5.2. Elevation values along a west–east transect, and a plot of the empirical cumulative distribution function values for elevation on the transect

```
Data attributes:
   Min. 1st Qu.  Median    Mean 3rd Qu.    Max.
      0       0      22     101     128     686
```

Figure 5.2 shows two views of the resulting data object, first a cross-section of elevation along the transect and below that a plot of the empirical cumulative distribution function for the transect data. This is effectively the same as the diagram termed the hypsometric curve by geomorphologists, the cumulative height frequency curve, with the axes swapped.

Spatial queries of this kind are very common, reading off raster data for sample points of known events or phenomena of interest. Following modelling, the same data are then used to predict the value of the phenomenon of interest for unobserved raster cells. The study reported by Wang and Unwin (1992) on landslide distribution on loess soils is an example, involving the spatial querying of slope and aspect raster layers calculated from a digital elevation model, and lithology classes. As Unwin (1996, pp. 132–136) points out, there are two key issues in the analysis. First, there is considerable error present in the input raster data, and that the only field data collected are for sites at which landslides are known to have taken place. This means that some landslides may not have been included. Second, no matching data are available for other locations at which no landslides have been observed. This context of no control being available on the phenomena of interest is quite common in applied environmental research.

5.3 Spatial Sampling

One way of trying to get control over data in a research setting like the one described might be to sample points from the total study area, to be able to examine whether the observed phenomena seem to be associated with particular ranges of values of the supposed environmental 'drivers'. Sample design is not typically paid much attention in applied spatial data analysis, very often for practical reasons, for example the need to handle incoming data as they flow in, rather than being able to choose which data to use. In the case of veterinary epidemiology, it is not easy to impose clear control because of the time pressure to offer policy advice. Schemes for spatial sampling have been given in the literature, for example by Ripley (1981, pp. 19–27), and they are available in **sp** using generic method `spsample`. Five sampling schemes are available: `"random"`, which places the points at random within the sampling area; `"regular"`, termed a *centric systematic sample* by Ripley and for which the grid offset can be set, and `"stratified"` and `"nonaligned"`, which are implemented as variations on the `"regular"` scheme – `"stratified"` samples one point at random in each cell, and `"nonaligned"` is a systematic masked scheme using combinations of random x and y to yield a single coordinate in each cell. The fifth scheme samples on a hexagonal lattice. The spatial data object passed to the `spsample` method can be simply a `Spatial` object, in which case sampling is carried out within its bounding box. It can be a line object, when samples are taken along the line or lines. More typically, it is a polygon object or a grid object, providing an observation window defining the study area or areas.

Above, we examined SRTM elevation values along a transect crossing the highest point in the region. We can get an impression of which parts of the distribution of elevation values differ from those of the region as a whole by sampling. In the first case, we sample within the GSHHS shoreline polygons shown in Fig. 2.7, using the `"random"` sampling scheme; this scheme drops points within lakes in polygons. The `spsample` methods may return `SpatialPoints` objects with a different number of points than requested. The second and third samples are taken from the `SpatialPixelsDataFrame` object omitting the `NA` values offshore. They differ in using `"random"` and `"regular"` sampling schemes, respectively. The selected points are shown in Fig. 5.3.

```
> set.seed(9876)
> polygon_random <- spsample(auck_gshhs, 1000, type = "random")
> polygon_random_el1 <- overlay(auck_el1, polygon_random)
> grid_random <- spsample(auck_el2, 1000, type = "random")
> grid_random_el1 <- overlay(auck_el1, grid_random)
> grid_regular <- spsample(auck_el2, 1000, type = "regular")
> grid_regular_el1 <- overlay(auck_el1, grid_regular)
```

```
            minimum lower-hinge median upper-hinge maximum    n
transect          0           0     22       128.0     686 1011
```

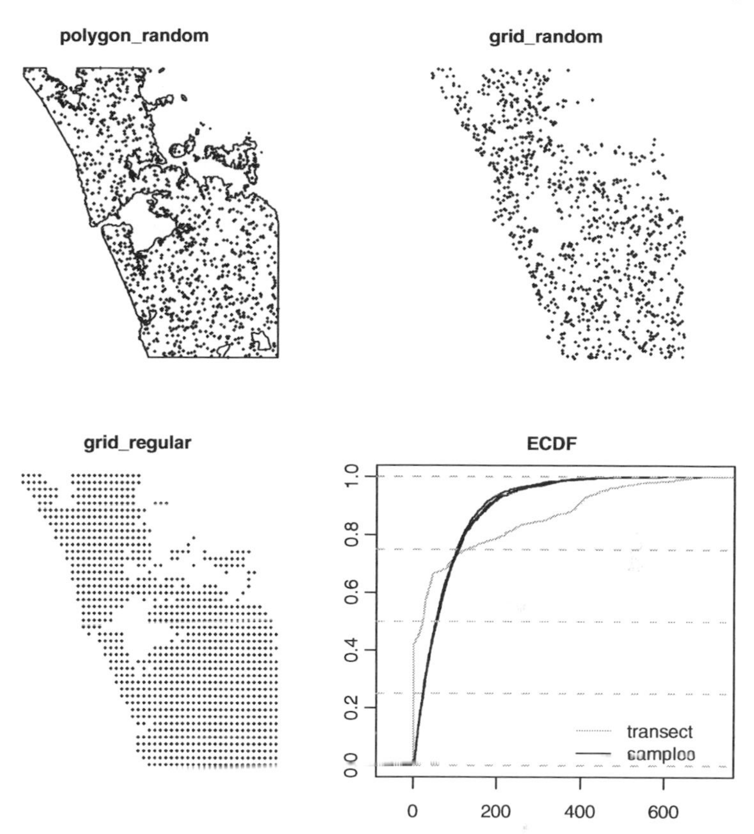

Fig. 5.3. Use of the `spsample` method: three sets of `SpatialPoints` objects and empirical cumulative distribution functions for elevation values for the sample points over-plotted on the transect values shown in Fig. 5.2

```
polygon_random    0    24    56    103.0    488 1000
grid_random       1    22    54    108.0    595  985
grid_regular      2    23    53    108.5    507 1020
```

Once again, we overlay our `SpatialPoints` objects created by sampling on the elevation `SpatialGridDataFrame` object. In this way we read off the elevation values for our sampled points, and can tabulate their five number summaries with that of the transect through the highest point. Both from the output and from the over-plotted empirical cumulative distribution function values shown in Fig. 5.3, we can see that the transect begins to diverge from the region as a whole above the 40th percentile. There is little difference between the three samples.

Alternative sampling schemes are contained in the **spatstat** package, with many point process generating functions for various window objects. The **spsurvey** package, using **sp** classes amongst other data representations, supplements these with some general procedures and the US Environmental Protection Agency Aquatic Resources Monitoring Generalized Random Tessellation Stratified procedure.[2]

5.4 Checking Topologies

In this section and the next, we look at a practical example involving the cleaning of spatial objects originally read into R from shapefiles published by the US Census. We then aggregate them up to metropolitan areas using a text table also from the US Census.

The data in this case are for polygons representing county boundaries in 1990 of North Carolina, South Carolina, and Virginia, as shown in Fig. 5.4. The attribute data for each polygon are the standard polygon identifiers, state and county identifiers, and county names. All the spatial objects have the same number of columns of attribute data of the same types and with the same names. The files are provided without coordinate reference systems as shapefiles; the metadata are used for choosing the `CRS` values.

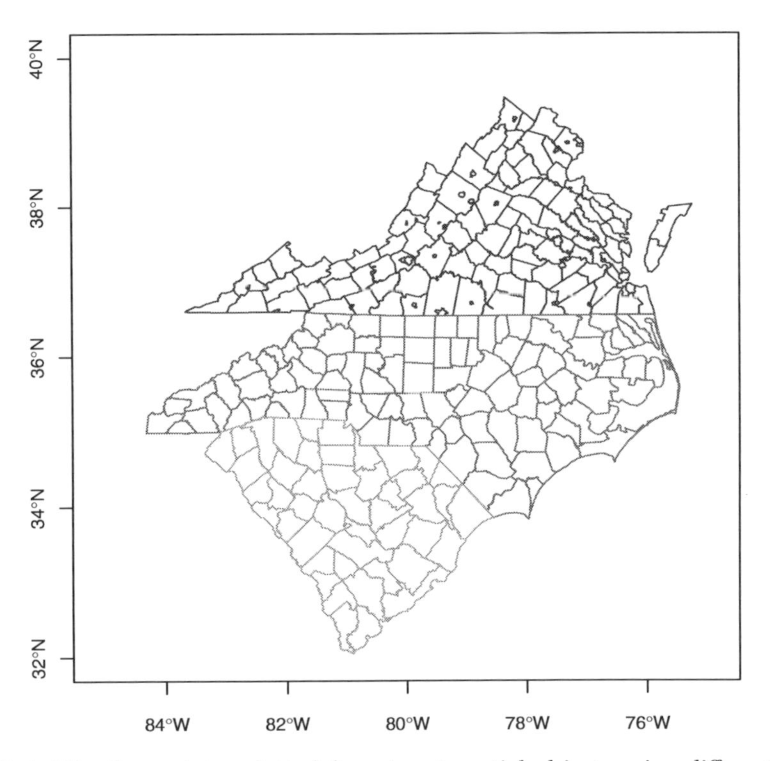

Fig. 5.4. The three states plotted from input spatial objects using different grey colours for county boundaries

[2] `http://www.epa.gov/nheerl/arm/index.htm`.

```
> library(rgdal)
> nc90 <- readOGR(".", "co37_d90")
> proj4string(nc90) <- CRS("+proj=longlat +datum=NAD27")
> sc90 <- readOGR(".", "co45_d90")
> proj4string(sc90) <- CRS("+proj=longlat +datum=NAD27")
> va90 <- readOGR(".", "co51_d90")
> proj4string(va90) <- CRS("+proj=longlat +datum=NAD27")
```

As read in, shapefiles usually have the polygon IDs set to the external file feature sequence number from zero to one less than the number of features. In our case, wanting to combine three states, we need to change the ID values so that they are unique across the study area. We can use the FIPS code (Federal Information Processing Standards Publication 6-4), which is simply the two-digit state FIPS code placed in front of the three-digit within-state FIPS county code, ending up with a five-digit string uniquely identifying each county. We can also drop the first four attribute data columns, two of which (area and perimeter) are misleading for objects in geographical coordinates, and the other two are internal ID values from the software used to generate the shapefiles, replicating the original feature IDs. We can start with the data set of South Carolina (sc90):

```
> library(maptools)

> names(sc90)

[1] "AREA"        "PERIMETER"  "CO45_D90_"  "CO45_D90_I" "ST"
[6] "CO"          "NAME"

> sc90a <- spChFIDs(sc90, paste(sc90$ST, sc90$CO, sep = ""))
> sc90a <- sc90a[, -(1:4)]
> names(sc90a)

[1] "ST"   "CO"   "NAME"
```

5.4.1 Dissolving Polygons

When we try the same sequence of commands for North Carolina, we run into difficulties:

```
> names(nc90)

[1] "AREA"        "PERIMETER"  "CO37_D90_"  "CO37_D90_I" "ST"
[6] "CO"          "NAME"

> nc90a <- spChFIDs(nc90, paste(nc90$ST, nc90$CO, sep = ""))

Error in spChFIDs(SP, x) : duplicate IDs
```

Tabulating the frequencies of polygons per unique county ID, we can see that 98 of North Carolina's counties are represented by single polygons, while one has two polygons, and one (on the coast) has four.

```
> table(table(paste(nc90$ST, nc90$CO, sep = "")))

 1  2  4
98  1  1
```

One reason for spatial data being structured in this way is that it is following the OpenGIS®[3] Simple Features Specification, which allows polygons to have one and only one external boundary ring, and an unlimited number of internal boundaries – holes. This means that multiple external boundaries – such as a county made up of several islands – are represented as multiple polygons. In the specification, they are linked to attribute data through a look-up table pointing to the appropriate attribute data row.

We need to restructure the SpatialPolygons object such that the Polygon objects belonging to each county belong to the same Polygons object. To do this, we use a function[4] in the **maptools** package also used for dissolving or merging polygons, but which can be used here to re-package the original features, so that each Polygons object corresponds to one and only one county:

```
> nc90a <- unionSpatialPolygons(nc90, IDs = paste(nc90$ST,
+     nc90$CO, sep = ""))
```

The function uses the IDs argument to set the ID slots of the output SpatialPolygons object. Having sorted out the polygons, we need to remove the duplicate rows from the data frame and put the pieces back together again:

```
> nc90_df <- as(nc90, "data.frame")[!duplicated(nc90$CO),
+     -(1:4)]
> row.names(nc90_df) <- paste(nc90_df$ST, nc90_df$CO, sep = "")
> nc90b <- SpatialPolygonsDataFrame(nc90a, nc90_df)
```

5.4.2 Checking Hole Status

Looking again at Fig. 5.4, we can see that while neither North Carolina nor South Carolina has included boroughs within counties, these are frequently found in Virginia. While data read from external sources are expected to be structured correctly, with the including polygon having an outer edge and an inner hole, into which the outer edge of the included borough fits, as described in Sect. 2.6.2, we can also check and correct the settings of the hole slot in Polygon objects. The checkPolygonsHoles function takes a Polygons object as its argument, and, if multiple Polygon objects belong to it, checks them for hole status using functions from the **gpclib** package:

```
> va90a <- spChFIDs(va90, paste(va90$ST, va90$CO, sep = ""))
> va90a <- va90a[, -(1:4)]
```

[3] See http://www.opengeospatial.org/.

[4] This function requires that the **gpclib** package is also installed.

```
> va90_pl <- slot(va90a, "polygons")
> va90_pla <- lapply(va90_pl, checkPolygonsHoles)
> p4sva <- CRS(proj4string(va90a))
> vaSP <- SpatialPolygons(va90_pla, proj4string = p4sva)
> va90b <- SpatialPolygonsDataFrame(vaSP, data = as(va90a,
+     "data.frame"))
```

Here we have changed the `Polygons` ID values as before, and then processed each `Polygons` object in turn for internal consistency, finally re-assembling the cleaned object. So we now have three spatial objects with mutually unique IDs, and with data slots containing data frames with the same numbers and kinds of columns with the same names.

5.5 Combining Spatial Data

It is quite often desirable to combine spatial data of the same kind, in addition to combining positional data of different kinds as discussed earlier in this chapter. There are functions `rbind` and `cbind` in R for combining objects by rows or columns, and `rbind` methods for `SpatialPixels` and `SpatialPixelsDataFrame` objects, as well as a `cbind` method for `SpatialGridDataFrame` objects are included in **sp**. In addition, methods with slightly different names to carry out similar operations are included in the **maptools** package.

5.5.1 Combining Positional Data

The `spRbind` method combines positional data, such as two `SpatialPoints` objects or two `SpatialPointsDataFrame` objects with matching column names and types in their data slots. The method is also implemented for `SpatialLines` and `SpatialPolygons` objects and their `*DataFrame` extensions. The methods do not check for duplication or overlapping of the spatial objects being combined, but do reject attempts to combine objects that would have resulted in non-unique IDs.

Because the methods only take two arguments, combining more than two involves repeating calls to the method:

```
> nc_sc_va90 <- spRbind(spRbind(nc90b, sc90a), va90b)
> FIPS <- sapply(slot(nc_sc_va90, "polygons"), function(x) slot(x,
+     "ID"))
> str(FIPS)

chr [1:282] "37001" "37003" ...

> length(slot(nc_sc_va90, "polygons"))

[1] 282
```

5.5.2 Combining Attribute Data

Here, as very often found in practice, we need to combine data for the same spatial objects from different sources, where one data source includes the geometries and an identifying index variable, and other data sources include the same index variable with additional variables. They often include more observations than our geometries, sometimes have no data for some of our geometries, and not are infrequently sorted in a different order. The data cleaning involved in getting ready for analysis is a little more tedious with spatial data, as we see, but does not differ in principle from steps taken with non-spatial data.

The text file provided by the US Census tabulating which counties belonged to each metropolitan area in 1990 has a header, which has already been omitted, a footer with formatting information, and many blank columns. We remove the footer and the blank columns first, and go on to remove rows with no data – the metropolitan areas are separated in the file by empty lines. The required rows and column numbers were found by inspecting the file before reading it into R:

```
> t1 <- read.fwf("90mfips.txt", skip = 21, widths = c(4,
+     4, 4, 4, 2, 6, 2, 3, 3, 1, 7, 5, 3, 51), colClasses = "character")

> t2 <- t1[1:2004, c(1, 7, 8, 14)]
> t3 <- t2[complete.cases(t2), ]
> cnty1 <- t3[t3$V7 != "  ", ]
> ma1 <- t3[t3$V7 == "  ", c(1, 4)]
> cnty2 <- cnty1[which(!is.na(match(cnty1$V7, c("37", "45",
+     "51")))), ]
> cnty2$FIPS <- paste(cnty2$V7, cnty2$V8, sep = "")
```

We next break out an object with metro IDs, state and county IDs, and county names (`cnty1`), and an object with metro IDs and metro names (`ma1`). From there, we subset the counties to the three states, and add the FIPS string for each county, to make it possible to combine the new data concerning metro area membership to our combined county map. We create an object (`MA_FIPS`) of county metro IDs by matching the `cnty2` FIPS IDs with those of the counties on the map, and then retrieving the metro area names from `ma1`. These two variables are then made into a data frame, the appropriate row names inserted and combined with the county map, with method `spCbind`. At last we are ready to dissolve the counties belonging to metro areas and to discard those not belonging to metro areas, using `unionSpatialPolygons`:

```
> MA_FIPS <- cnty2$V1[match(FIPS, cnty2$FIPS)]
> MA <- ma1$V14[match(MA_FIPS, ma1$V1)]
> MA_df <- data.frame(MA_FIPS = MA_FIPS, MA = MA, row.names = FIPS)
> nc_sc_va90a <- spCbind(nc_sc_va90, MA_df)
> ncscva_MA <- unionSpatialPolygons(nc_sc_va90a, nc_sc_va90a$MA_FIPS)
```

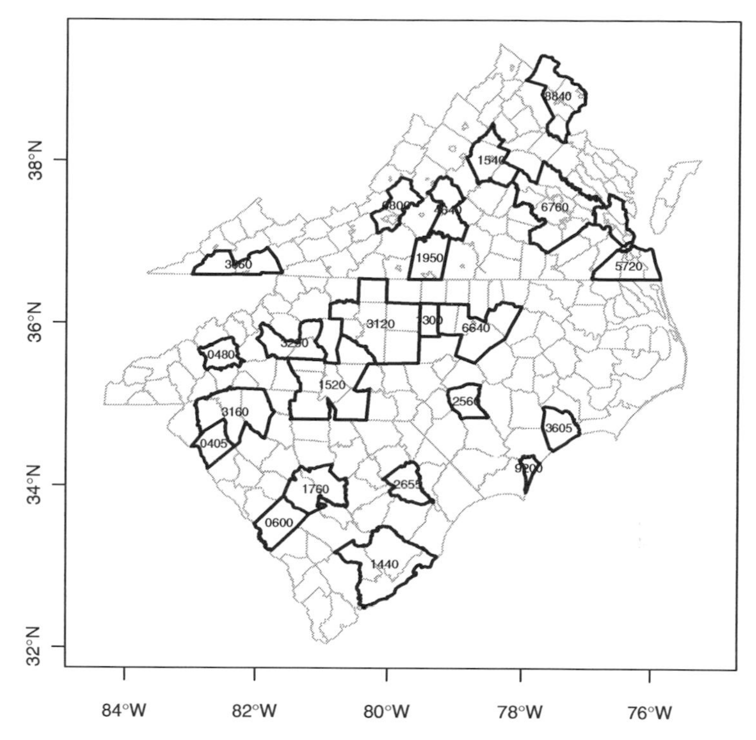

Fig. 5.5. The three states with county boundaries plotted in grey, and Metropolitan area boundaries plotted in black; Metro area standard IDs are shown

Figure 5.5 shows the output object plotted on top of the cleaned input county boundaries. There does appear to be a problem, however, because one of the output boundaries has no name – it is located between 6760 and 5720 in eastern Virginia. If we do some more matching, to extract the names of the metropolitan areas, we can display the name of the area with multiple polygons:

```
> np <- sapply(slot(ncscva_MA, "polygons"), function(x) length(slot(x,
+     "Polygons")))
> table(np)

np
 1  2
22  1

> MA_fips <- sapply(slot(ncscva_MA, "polygons"), function(x) slot(x,
+     "ID"))
> MA_name <- ma1$V14[match(MA_fips, ma1$V1)]
> data.frame(MA_fips, MA_name)[np > 1, ]

   MA_fips                                         MA_name
18    5720  Norfolk-Virginia Beach-Newport News, VA MSA
```

The Norfolk-Virginia Beach-Newport News, VA MSA is located on both sides of Hampton Roads, and the label has been positioned at the centre point of the largest member polygon.

5.6 Auxiliary Functions

New functions and methods are added to **maptools** quite frequently, often following suggestions and discussions on the R-sig-geo mailing list mentioned in Chap. 1. When positions are represented by geographical coordinates, it is often useful to be able to find the azimuth between them. The `gzAzimuth` function is a simple implementation of standard algorithms for this purpose, and gives azimuths calculated on the sphere between a matrix of points and a single point.[5] The `gcDestination` function returns the geographical coordinates of points at a given distance and bearing from given starting points.

A set of methods for matrices or `SpatialPoints` objects in geographical coordinates has been contributed to give timings for sunrise, sunset, and other solar measures for dates given as `POSIXct` objects:

```
> hels <- matrix(c(24.97, 60.17), nrow = 1)
> p4s <- CRS("+proj=longlat +datum=WGS84")
> Hels <- SpatialPoints(hels, proj4string = p4s)
> d041224 <- as.POSIXct("2004-12-24", tz = "EET")
> sunriset(Hels, d041224, direction = "sunrise", POSIXct.out = TRUE)

   day_frac                time
1 0.3924249 2004-12-24 09:25:05
```

Finally, `elide` methods have been provided for translating, rotating, and disguising coordinate positions in **sp** vector classes such as `SpatialPoints`. The geometries can be shifted in two dimensions, scaled such that the longest dimension is scaled $[0, 1]$, flipped, reflected, and rotated, if desired in relation to the bounding box of a different `Spatial` object. The methods can be used for standardising displays, for example in point pattern analysis, or for obfuscating position to meet in part privacy considerations. Since obscuring position was a reason for providing the methods, they have been given a suitably obscure name.

The methods discussed in this chapter are intended to provide ways for manipulating spatial objects to help in structuring analytical approaches to support problems amongst others. These are not the only ways to organise spatial data, do try to make it easier to concentrate on exploring and analysing the data, rather than dealing with the intricacies of particular representations peculiar to specific software or data providers.

[5] The function is based with permission on work by S. Abdali: The Correct Qibla, `http://patriot.net/users/abdali/ftp/qibla.pdf`.

6

Customising Spatial Data Classes and Methods

Although the classes defined in the **sp** package cover many needs, they do not go far beyond the most typical GIS data models. In applied research, it often happens that customised classes would suit the actual data coming from the instruments better. Since `S4` classes have mechanisms for inheritance, it may be attractive to build on the **sp** classes, so as to utilise their methods where appropriate. Here, we demonstrate a range of different settings in which **sp** classes can be extended. Naturally, this is only useful for researchers with specific and clear needs, so our goal is to show how (relatively) easy it may be to prototype classes extending **sp** classes for specific purposes.

6.1 Programming with Classes and Methods

This section explains the elementary basics of programming with classes and methods in R. The `S` language (implemented in R and `S-PLUS`™) contains two mechanisms for creating classes and methods: the traditional `S3` system and the more recent `S4` system (see Sect. 2.2, in which classes were described for the useR – here they are described for the developeR). This chapter is not a full introduction to R programming (see Braun and Murdoch (2007) for more details), but it will try to give some feel of how the `Spatial` classes in package **sp** can be extended to be used for wider classes of problems. For full details, the interested reader is referred to, for example, Venables and Ripley (2000) and Chambers (1998), the latter being a reference for new-style `S4` classes and methods. Example code is, for example, to be found in the source code for package **sp**, available from CRAN.

Suppose we define myfun as

```
> myfun <- function(x) {
+     x + 2
+ }
```

then, calling it with the numbers 1, 2, and 3 results in

```
> myfun(1:3)
```

```
[1] 3 4 5
```

or alternatively using a named argument:

```
> myfun(x = 1:3)
```

```
[1] 3 4 5
```

The return value of the function is the last expression evaluated. Often, we want to wrap existing functions, such as a plot function:

```
> plotXplus2Yminus3 <- function(x, y, ...) {
+     plot(x = x + 2, y = y - 3, ...)
+ }
```

In this case, the ... is used to pass information to the `plot` function without explicitly anticipating what it will be: named arguments `x` and `y` or the first two arguments if they are unnamed are processed, remaining arguments are passed on. The plot function is a generic method, with an instance that depends on the class of its first (S3) or first n arguments (S4). The available instances of `plot` are shown for S3-type methods by

```
> methods("plot")
```

```
 [1] plot.acf*            plot.data.frame*    plot.Date*
 [4] plot.decomposed.ts* plot.default        plot.dendrogram*
 [7] plot.density        plot.ecdf           plot.factor*
[10] plot.formula*       plot.hclust*        plot.histogram*
[13] plot.HoltWinters*   plot.isoreg*        plot.lm
[16] plot.medpolish*     plot.mlm            plot.POSIXct*
[19] plot.POSIXlt*       plot.ppr*           plot.prcomp*
[22] plot.princomp*      plot.profile.nls*   plot.spec
[25] plot.spec.coherency plot.spec.phase     plot.stepfun
[28] plot.stl*           plot.table*         plot.ts
[31] plot.tskernel*      plot.TukeyHSD

   Non-visible functions are asterisked
```

and for S4-type methods by

```
> library(sp)
> showMethods("plot")
```

```
Function: plot (package graphics)
x="ANY", y="ANY"
x="SpatialLines", y="missing"
x="Spatial", y="missing"
x="SpatialPoints", y="missing"
x="SpatialPolygons", y="missing"
```

where we first loaded **sp** to make sure there are some S4 plot methods to show.

6.1.1 S3-Style Classes and Methods

In Chap. 2, we presented R classes and methods from the perspective of a useR; here we shift perspective to that of a developeR. Building S3-style classes is simple. Suppose we want to build an object of class `foo`:

```
> x <- rnorm(10)
> class(x) <- "foo"
> x

 [1] -1.59553650 -1.17102368  0.80900393  0.63390826  0.01971040
 [6] -0.69330839 -1.56896726 -0.22350820  0.20268852  0.96951209
attr(,"class")
[1] "foo"
```

If we plot this object, for example by `plot(x)`, we get the same plot as when we would not have set the class to `foo`. If we know, however, that objects of class `foo` need to be plotted without symbols but with connected lines, we can write a plot method for this class:

```
> plot.foo <- function(x, y, ...) {
+     plot.default(x, type = "l", ...)
+ }
```

after which `plot(x)` will call this particular method, rather than a default plot method.

Class inheritance is obtained in S3 when an object is given multiple classes, as in

```
> class(x) <- c("foo", "bar")

> plot(x)
```

For this plot, first function `plot.foo` will be looked for, and if not found the second option `plot.bar` will be looked for. If none of them is found, the default `plot.default` will be used.

The S3 class mechanism is simple and powerful. Much of R works with it, including key functions such as `lm`.

```
> data(meuse)
> class(meuse)

[1] "data.frame"

> class(lm(log(zinc) ~ sqrt(dist), meuse))

[1] "lm"
```

There is, however, no checking that a class with a particular name does indeed contain the elements that a certain method for it expects. It also has design flaws, as method specification by dot separation is ambiguous in case of names such as `as.data.frame`, where one cannot tell whether it means that the method `as.data` acts on objects of class `frame`, or the method `as` acts on objects of class `data.frame`, or none of them (the answer is: none). For such reasons, S4-style classes and methods were designed.

6.1.2 S4-Style Classes and Methods

S4-style classes are formally defined using `setClass`. As an example, somewhat simplified versions of classes `CRS` and `Spatial` in **sp** are

```
> setClass("CRS", representation(projargs = "character"))
> setClass("Spatial", representation(bbox = "matrix",
+     proj4string = "CRS"), validity <- function(object) {
+     bb <- bbox(object)
+     if (!is.matrix(bb))
+         return("bbox should be a matrix")
+     n <- dimensions(object)
+     if (n < 2)
+         return("spatial.dimension should be 2 or more")
+     if (any(is.na(bb)))
+         return("bbox should never contain NA values")
+     if (any(!is.finite(bb)))
+         return("bbox should never contain infinite values")
+     if (any(bb[, "max"] < bb[, "min"]))
+         return("invalid bbox: max < min")
+     TRUE
+ })
```

The command `setClass` defines a class name as a formal class, gives the names of the class elements (called slots), and their type–type checking will happen upon construction of an instance of the class. Further checking, for example on valid dimensions and data ranges can be done in the `validity` function. Here, the validity function retrieves the bounding box using the generic `bbox` method. Generics, if not defined in the base R system, for example

```
> isGeneric("show")
[1] TRUE
```

can be defined with `setGeneric`. Defining a specific instance of a generic is done by `setMethod`:

```
> setGeneric("bbox", function(obj) standardGeneric("bbox"))
> setMethod("bbox", signature = "Spatial", function(obj) obj@bbox)
```

where the signature tells the class of the first (or first n) arguments. Here, the `@` operator is used to access the `bbox` slot in an S4 object, not to be confused with the `$` operator to access list elements.

We now illustrate this mechanism by providing a few examples of classes, building on those available in package **sp**.

6.2 Animal Track Data in Package Trip

CRAN Package **trip**, written by Michael Sumner (Kirkwood et al., 2006; Page et al., 2006), provides a class for animal tracking data. Animal tracking data consist of sets of (x, y, t) stamps, grouped by an identifier pointing to an individual animal, sensor, or perhaps isolated period of monitoring. A strategy for this (slightly simplified from that of **trip**) is to extend the

`SpatialPointsDataFrame` class by a length 2 character vector carrying the names of the time column and the trip identifier column in the `SpatialPoints DataFrame` attribute table.

Package **trip** does a lot of work to read and analyse tracking data from data formats typical for tracking data (Argos DAT), removing duplicate observations and validating the objects, for example checking that time stamps increase and movement speeds are realistic. We ignore this and stick to the bare bones.

We now define a class called `trip` that extends `SpatialPointsDataFrame`:

```
> library(sp)
> setClass("trip", representation("SpatialPointsDataFrame",
+     TOR.columns = "character"), validity <- function(object) {
+     if (length(object@TOR.columns) != 2)
+         stop("Time/id column names must be of length 2")
+     if (!all(object@TOR.columns %in% names(object@data)))
+         stop("Time/id columns must be present in attribute table")
+     TRUE
+ })

[1] "trip"

> showClass("trip")

Slots:

Name:  TOR.columns         data  coords.nrs       coords         bbox
Class:   character  data.frame     numeric       matrix       matrix

Name:  proj4string
Class:         CRS

Extends:
Class "SpatialPointsDataFrame", directly
Class "SpatialPoints", by class "SpatialPointsDataFrame", distance 2
Class "Spatial", by class "SpatialPointsDataFrame", distance 3
```

which checks, upon creation of objects, that indeed two variable names are passed and that these names refer to variables present in the attribute table.

6.2.1 Generic and Constructor Functions

It would be nice to have a constructor function, just like `data.frame` or `SpatialPoints`, and so we now create it and set it as the generic function to be called in case the first argument is of class `SpatialPointsDataFrame`.

```
> trip.default <- function(obj, TORnames) {
+     if (!is(obj, "SpatialPointsDataFrame"))
+         stop("trip only supports SpatialPointsDataFrame")
+     if (is.numeric(TORnames))
```

```
+         TORnames <- names(obj)[TORnames]
+     new("trip", obj, TOR.columns = TORnames)
+ }
> if (!isGeneric("trip")) setGeneric("trip", function(obj,
+     TORnames) standardGeneric("trip"))
```

```
[1] "trip"
```

```
> setMethod("trip", signature(obj = "SpatialPointsDataFrame",
+     TORnames = "ANY"), trip.default)
```

```
[1] "trip"
```

We can now try it out, with the turtle data of Chap. 2:

```
> turtle <- read.csv("seamap105_mod.csv")
```

```
> timestamp <- as.POSIXlt(strptime(as.character(turtle$obs_date),
+     "%m/%d/%Y %H:%M:%S"), "GMT")
> turtle <- data.frame(turtle, timestamp = timestamp)
> turtle$lon <- ifelse(turtle$lon < 0, turtle$lon + 360,
+     turtle$lon)
> turtle <- turtle[order(turtle$timestamp), ]
> coordinates(turtle) <- c("lon", "lat")
> proj4string(turtle) <- CRS("+proj=longlat +ellps=WGS84")
> turtle$id <- c(rep(1, 200), rep(2, nrow(coordinates(turtle)) -
+     200))
> turtle_trip <- trip(turtle, c("timestamp", "id"))
> summary(turtle_trip)
```

```
Object of class trip
Coordinates:
        min     max
lon 140.923 245.763
lat  21.574  39.843
Is projected: FALSE
proj4string : [+proj=longlat +ellps=WGS84]
Number of points: 394
Data attributes:
       id                          obs_date
 Min.   :1.000   01/02/1997 04:16:53:  1
 1st Qu.:1.000   01/02/1997 05:56:25:  1
 Median :1.000   01/04/1997 17:41:54:  1
 Mean   :1.492   01/05/1997 17:20:07:  1
 3rd Qu.:2.000   01/06/1997 04:31:13:  1
 Max.   :2.000   01/06/1997 06:12:56:  1
                 (Other)            :388
   timestamp
 Min.   :1996-08-11 01:15:00
 1st Qu.:1996-10-30 00:10:04
 Median :1997-01-24 23:31:01
 Mean   :1997-01-26 06:24:56
 3rd Qu.:1997-04-10 12:26:20
 Max.   :1997-08-13 20:19:46
```

6.2.2 Methods for Trip Objects

The summary method here is not defined for `trip`, but is the default summary inherited from class `Spatial`. As can be seen, nothing special about the trip features is mentioned, such as what the time points are and what the identifiers. We could alter this by writing a class-specific summary method

```
> summary.trip <- function(object, ...) {
+     cat("Object of class \"trip\"\nTime column: ")
+     print(object@TOR.columns[1])
+     cat("Identifier column: ")
+     print(object@TOR.columns[2])
+     print(summary(as(object, "Spatial")))
+     print(summary(object@data))
+ }
> setMethod("summary", "trip", summary.trip)

[1] "summary"

> summary(turtle_trip)

Object of class "trip"
Time column: [1] "timestamp"
Identifier column: [1] "id"
Object of class Spatial
Coordinates:
        min     max
lon 140.923 245.763
lat  21.574  39.843
Is projected: FALSE
proj4string : [+proj=longlat +ellps=WGS84]
       id                      obs_date
 Min.   :1.000   01/02/1997 04:16:53:  1
 1st Qu.:1.000   01/02/1997 05:56:25:  1
 Median :1.000   01/04/1997 17:41:54:  1
 Mean   :1.492   01/05/1997 17:20:07:  1
 3rd Qu.:2.000   01/06/1997 04:31:13:  1
 Max.   :2.000   01/06/1997 06:12:56:  1
                 (Other)            :388
   timestamp
 Min.   :1996-08-11 01:15:00
 1st Qu.:1996-10-30 00:10:04
 Median :1997-01-24 23:31:01
 Mean   :1997-01-26 06:24:56
 3rd Qu.:1997-04-10 12:26:20
 Max.   :1997-08-13 20:19:46
```

As `trip` extends `SpatialPointsDataFrame`, subsetting using `"["` and column selection or replacement using `"[["` or `"$"` all work, as these are inherited. Creating invalid trip objects can be prohibited by adding checks to the validity function in the class definition, for example will not work because the time and/or id column are not present any more.

A custom plot method for trip could be written, for example using colour to denote a change in identifier:

```
> setGeneric("lines", function(x, ...) standardGeneric("lines"))

[1] "lines"

> setMethod("lines", signature(x = "trip"), function(x,
+     ..., col = NULL) {
+     tor <- x@TOR.columns
+     if (is.null(col)) {
+         l <- length(unique(x[[tor[2]]]))
+         col <- hsv(seq(0, 0.5, length = l))
+     }
+     coords <- coordinates(x)
+     lx <- split(1:nrow(coords), x[[tor[2]]])
+     for (i in 1:length(lx)) lines(coords[lx[[i]], ],
+         col = col[i], ...)
+ })

[1] "lines"
```

Here, the `col` argument is added to the function header so that a reasonable default can be overridden, for example for black/white plotting.

6.3 Multi-Point Data: SpatialMultiPoints

One of the feature types of the OpenGeospatial Consortium (OGC) simple feature specification that has not been implemented in **sp** is the `MultiPoint` object. In a `MultiPoint` object, each feature refers to a *set of* points. The **sp** classes `SpatialPointsDataFrame` only provide reference to a single point. Instead of building a new class up from scratch, we try to re-use code and build a class `SpatialMultiPoint` from the `SpatialLines` class. After all, lines are just sets of ordered points.

In fact, the `SpatialLines` class implements the `MultiLineString` simple feature, where each feature can refer to multiple lines. A special case is formed if each feature only has a single line:

```
> setClass("SpatialMultiPoints", representation("SpatialLines"),
+     validity <- function(object) {
+         if (any(unlist(lapply(object@lines,
+             function(x) length(x@Lines))) !=
+             1))
+             stop("Only Lines objects with one Line element")
+         TRUE
+     })
```

```
[1] "SpatialMultiPoints"
```

```
> SpatialMultiPoints <- function(object) new("SpatialMultiPoints",
+     object)
```

As an example, we can create an instance of this class for two MultiPoint features each having three locations:

```
> n <- 5
> set.seed(1)
> x1 <- cbind(rnorm(n), rnorm(n, 0, 0.25))
> x2 <- cbind(rnorm(n), rnorm(n, 0, 0.25))
> x3 <- cbind(rnorm(n), rnorm(n, 0, 0.25))
> L1 <- Lines(list(Line(x1)), ID = "mp1")
> L2 <- Lines(list(Line(x2)), ID = "mp2")
> L3 <- Lines(list(Line(x3)), ID = "mp3")
> s <- SpatialLines(list(L1, L2, L3))
> smp <- SpatialMultiPoints(s)
```

If we now plot object `smp`, we get the same plot as when we plot `s`, showing the two lines. The `plot` method for a `SpatialLines` object is not suitable, so we write a new one:

```
> plot.SpatialMultiPoints <- function(x, ..., pch = 1:length(x@lines),
+     col = 1, cex = 1) {
+     n <- length(x@lines)
+     if (length(pch) < n)
+         pch <- rep(pch, length.out = n)
+     if (length(col) < n)
+         col <- rep(col, length.out = n)
+     if (length(cex) < n)
+         cex <- rep(cex, length.out = n)
+     plot(as(x, "Spatial"), ...)
+     for (i in 1:n) points(x@lines[[i]]@Lines[[1]]@coords,
+         pch = pch[i], col = col[i], cex = cex[i])
+ }
> setMethod("plot", signature(x = "SpatialMultiPoints",
+     y = "missing"), function(x, y, ...) plot.SpatialMultiPoints(x,
+     ...))
```

```
[1] "plot"
```

Here we chose to pass any named ... arguments to the plot method for a `Spatial` object. This function sets up the axes and controls the margins, aspect ratio, etc. All arguments that need to be passed to `points` (`pch` for symbol type, `cex` for symbol size, and `col` for symbol colour) need explicit naming and sensible defaults, as they are passed explicitly to the consecutive calls to `points`. According to the documentation of `points`, in addition to `pch`, `cex`, and `col`, the arguments `bg` and `lwd` (symbol fill colour and symbol line width) would need a similar treatment to make this plot method completely transparent with the base `plot` method – something an end user would hope for.

Having `pch`, `cex`, and `col` arrays, the length of the number of `MultiPoints` *sets* rather than the number of points to be plotted is useful for two reasons. First, the whole point of `MultiPoints` object is to distinguish *sets* of points. Second, when we extend this class to `SpatialMultiPointsDataFrame`, for example by

```
> cName <- "SpatialMultiPointsDataFrame"
> setClass(cName, representation("SpatialLinesDataFrame"),
+     validity <- function(object) {
+         lst <- lapply(object@lines, function(x) length(x@Lines))
+         if (any(unlist(lst) != 1))
+             stop("Only Lines objects with single Line")
+         TRUE
+     })
```

```
[1] "SpatialMultiPointsDataFrame"
```

```
> SpatialMultiPointsDataFrame <- function(object) {
+     new("SpatialMultiPointsDataFrame", object)
+ }
```

then we can pass symbol characteristics by (functions of) columns in the attribute table:

```
> df <- data.frame(x1 = 1:3, x2 = c(1, 4, 2), row.names = c("mp1",
+     "mp2", "mp3"))
> smp_df <- SpatialMultiPointsDataFrame(SpatialLinesDataFrame(smp,
+     df))
> setMethod("plot", signature(x = "SpatialMultiPointsDataFrame",
+     y = "missing"), function(x, y, ...) plot.SpatialMultiPoints(x,
+     ...))
```

```
[1] "plot"
```

```
> grys <- c("grey10", "grey40", "grey80")
```

```
> plot(smp_df, col = grys[smp_df[["x1"]]], pch = smp_df[["x2"]],
+     cex = 2, axes = TRUE)
```

for which the plot is shown in Fig. 6.1.

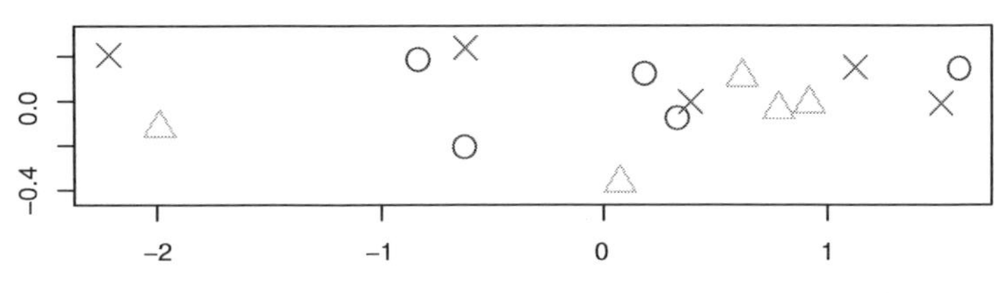

Fig. 6.1. Plot of the `SpatialMultiPointsDataFrame` object

6.4 Hexagonal Grids

Hexagonal grids are like square grids, where grid points are centres of matching hexagons, rather than squares. Package **sp** has no classes for hexagonal grids, but does have some useful functions for generating and plotting them. This could be used to build a class. Much of this code in **sp** is based on postings to the R-sig-geo mailing list by Tim Keitt, used with permission.

The spatial sampling method `spsample` has a method for sampling points on a hexagonal grid:

```
> data(meuse.grid)
> gridded(meuse.grid) = ~x + y
> xx <- spsample(meuse.grid, type = "hexagonal", cellsize = 200)
> class(xx)

[1] "SpatialPoints"
attr(,"package")
[1] "sp"
```

gives the points shown in the left side of Fig. 6.2. Note that an alternative hexagonal representation is obtained by rotating this grid 90°; we will not further consider that here.

```
> HexPts <- spsample(meuse.grid, type = "hexagonal", cellsize = 200)
> spplot(meuse.grid["dist"], sp.layout = list("sp.points",
+     HexPts, col = 1))
> HexPols <- HexPoints2SpatialPolygons(HexPts)
> df <- as.data.frame(meuse.grid)[overlay(meuse.grid, HexPts),
+     ]
> HexPolsDf <- SpatialPolygonsDataFrame(HexPols, df, match.ID = FALSE)
> spplot(HexPolsDf["dist"])
```

for which the plots are shown in Fig. 6.2.

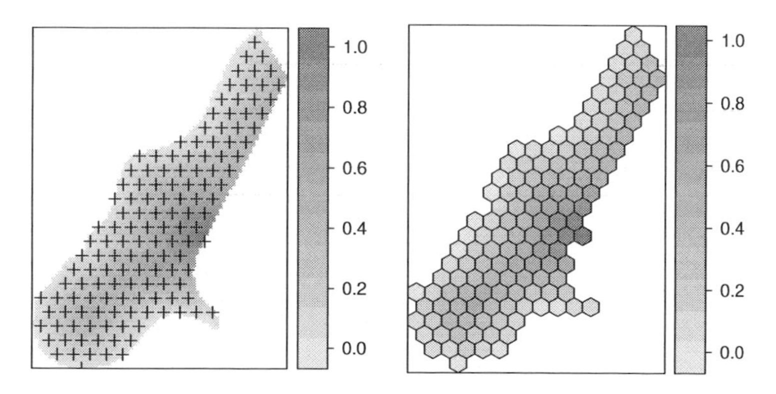

Fig. 6.2. Hexagonal points (*left*) and polygons (*right*)

We can now generate and plot hexagonal grids, but need to deal with two representations: as points and as polygons, and both representations do not tell by themselves that they represent a hexagonal grid.

For designing a hexagonal grid class we extend SpatialPoints, assuming that computation of the polygons can be done when needed without a prohibitive overhead.

```
> setClass("SpatialHexGrid", representation("SpatialPoints",
+     dx = "numeric"), validity <- function(object) {
+     if (object@dx <= 0)
+         stop("dx should be positive")
+     TRUE
+ })

[1] "SpatialHexGrid"

> setClass("SpatialHexGridDataFrame",
+     representation("SpatialPointsDataFrame",
+         dx = "numeric"), validity <- function(object) {
+         if (object@dx <= 0)
+             stop("dx should be positive")
+         TRUE
+     })

[1] "SpatialHexGridDataFrame"
```

Note that these class definitions do not check that instances actually do form valid hexagonal grids; a more robust implementation could provide a test that distances between points with equal y coordinate are separated by a multiple of dx, that the y-separations are correct and so on.

It might make sense to adapt the generic spsample method in package **sp** to return SpatialHexGrid objects; we can also add plot and spsample methods for them. Method overlay should work with a SpatialHexGrid as its first argument, by inheriting from SpatialPoints. Let us first see how to create the new classes. Without a constructor function we can use

```
> HexPts <- spsample(meuse.grid, type = "hexagonal", cellsize = 200)
> Hex <- new("SpatialHexGrid", HexPts, dx = 200)
> df <- as.data.frame(meuse.grid)[overlay(meuse.grid, Hex),
+     ]
> spdf <- SpatialPointsDataFrame(HexPts, df)
> HexDf <- new("SpatialHexGridDataFrame", spdf, dx = 200)
```

Because of the route taken to define both HexGrid classes, it is not obvious that the second extends the first. We can tell the S4 system this by setIs:

```
> is(HexDf, "SpatialHexGrid")

[1] FALSE

> setIs("SpatialHexGridDataFrame", "SpatialHexGrid")
> is(HexDf, "SpatialHexGrid")
```

```
[1] TRUE
```

to make sure that methods for `SpatialHexGrid` objects work as well for objects of class `SpatialHexGridDataFrame`.

When adding methods, several of them will need conversion to the polygon representation, so it makes sense to add the conversion function such that, for example `as(x, "SpatialPolygons")` will work:

```
> setAs("SpatialHexGrid", "SpatialPolygons",
+     function(from) HexPoints2SpatialPolygons(from,
+         from@dx))
> setAs("SpatialHexGridDataFrame", "SpatialPolygonsDataFrame",
+     function(from) SpatialPolygonsDataFrame(as(obj,
+         "SpatialPolygons"), obj@data,
+         match.ID = FALSE))
```

We can now add `plot`, `spplot`, `spsample`, and `overlay` methods for these classes:

```
> setMethod("plot", signature(x = "SpatialHexGrid", y = "missing"),
+     function(x, y, ...) plot(as(x, "SpatialPolygons"),\vspace*{-3pt}
+         ...))

[1] "plot"

> setMethod("spplot", signature(obj = "SpatialHexGridDataFrame"),
+     function(obj, ...) spplot(SpatialPolygonsDataFrame(as(obj,
+         "SpatialPolygons"), obj@data, match.ID = FALSE),
+         ...))

[1] "spplot"

> setMethod("spsample", "SpatialHexGrid", function(x, n,
+     type, ...) spsample(as(x, "SpatialPolygons"), n = n,
+     type = type, ...))

[1] "spsample"

> setMethod("overlay", c("SpatialHexGrid", "SpatialPoints"),
+     function(x, y, ...) overlay(as(x, "SpatialPolygons"),
+         y))

[1] "overlay"
```

After this, the following will work:

```
> spplot(meuse.grid["dist"], sp.layout = list("sp.points",
+     Hex, col = 1))
> spplot(HexDf["dist"])
```

Coercion to a data frame is done by

```
> as(HexDf, "data.frame")
```

Another detail not mentioned is that the bounding box of the hexgrid objects only match the grid centre points, not the hexgrid cells:

```
> bbox(Hex)

       min      max
x 178550.0 181450.0
y 329601.5 333585.3

> bbox(as(Hex, "SpatialPolygons"))

        min      max
r1 178450.0 181550.0
r2 329486.1 333700.7
```

One solution for this is to correct for this in a constructor function, and check for it in the validity test. Explicit coercion functions to the points representation would have to set the bounding box back to the points ranges. Another solution is to write a bbox method for the hexgrid classes, taking the risk that someone still looks at the incorrect bbox slot.

6.5 Spatio-Temporal Grids

Spatio-temporal data can be represented in different ways. One simple option is when observations (or model-results, or predictions) are given on a regular space–time grid.

Objects of class or extending `SpatialPoints`, `SpatialPixels`, and `SpatialGrid` do not have the constraint that they represent a two-dimensional space; they may have arbitrary dimension; an example for a three-dimensional grid is

```
> n <- 10
> x <- data.frame(expand.grid(x1 = 1:n, x2 = 1:n, x3 = 1:n),
+     z = rnorm(n^3))
> coordinates(x) <- ~x1 + x2 + x3
> gridded(x) <- TRUE
> fullgrid(x) <- TRUE
> summary(x)

Object of class SpatialGridDataFrame
Coordinates:
   min  max
x1 0.5 10.5
x2 0.5 10.5
x3 0.5 10.5
Is projected: NA
proj4string : [NA]
Number of points: 2
Grid attributes:
   cellcentre.offset cellsize cells.dim
x1                 1        1        10
x2                 1        1        10
```

```
x3                    1        1       10
Data attributes:
    Min.  1st Qu.    Median      Mean  3rd Qu.      Max.
-3.00800 -0.70630 -0.03970 -0.02012  0.68930  3.81000
```

We might assume here that the third dimension, x3, represents time. If we are happy with time somehow represented by a real number (in double precision), then we are done. A simple representation is that of decimal year, with, for example 1980.5 meaning the 183rd day of 1980, or, for example relative time in seconds after the start of some event.

When we want to use the POSIXct or POSIXlt representations, we need to do some more work to see the readable version. We now devise a simple three-dimensional space–time grid with the POSIXct representation.

```
> setClass("SpatialTimeGrid", "SpatialGrid",
+     validity <- function(object) {
+         stopifnot(dimensions(object) ==
+             3)
+         TRUE
+     })

[1] "SpatialTimeGrid"
```

Along the same line, we can extend the SpatialGridDataFrame for space-time:

```
> setClass("SpatialTimeGridDataFrame", "SpatialGridDataFrame",
+     validity <- function(object) {
+         stopifnot(dimensions(object) == 3)
+         TRUE
+     })

[1] "SpatialTimeGridDataFrame"

> setIs("SpatialTimeGridDataFrame", "SpatialTimeGrid")
> x <- new("SpatialTimeGridDataFrame", x)
```

A crude summary for this class could be written along these lines:

```
> summary.SpatialTimeGridDataFrame <- function(object,
+     ...) {
+     cat("Object of class SpatialTimeGridDataFrame\n")
+     x <- gridparameters(object)
+     t0 <- ISOdate(1970, 1, 1, 0, 0, 0)
+     t1 <- t0 + x[3, 1]
+     cat(paste("first time step:", t1, "\n"))
+     t2 <- t0 + x[3, 1] + (x[3, 3] - 1) * x[3, 2]
+     cat(paste("last time step: ", t2, "\n"))
+     cat(paste("time step:      ", x[3, 2], "\n"))
+     summary(as(object, "SpatialGridDataFrame"))
+ }
```

```
> setMethod("summary", "SpatialTimeGridDataFrame",
+     summary.SpatialTimeGridDataFrame)
```

```
[1] "summary"
```

```
> summary(x)
```

```
Object of class SpatialTimeGridDataFrame
first time step: 1970-01-01 00:00:01
last time step:  1970-01-01 00:00:10
time step:       1
Object of class SpatialGridDataFrame
Coordinates:
   min  max
x1 0.5 10.5
x2 0.5 10.5
x3 0.5 10.5
Is projected: NA
proj4string : [NA]
Number of points: 2
Grid attributes:
   cellcentre.offset cellsize cells.dim
x1                 1        1        10
x2                 1        1        10
x3                 1        1        10
Data attributes:
    Min.  1st Qu.   Median     Mean  3rd Qu.
-3.00800 -0.70630 -0.03970 -0.02012  0.68930
    Max.
 3.81000
```

Next, suppose we need a subsetting method that selects on the time. When the first subset argument is allowed to be a time range, this is done by

```
> subs.SpatialTimeGridDataFrame <- function(x, i, j, ...,
+     drop = FALSE) {
+     t <- coordinates(x)[, 3] + ISOdate(1970, 1, 1, 0,
+         0, 0)
+     if (missing(j))
+         j <- TRUE
+     sel <- t %in% i
+     if (!any(sel))
+         stop("selection results in empty set")
+     fullgrid(x) <- FALSE
+     if (length(i) > 1) {
+         x <- x[i = sel, j = j, ...]
+         fullgrid(x) <- TRUE
+         as(x, "SpatialTimeGridDataFrame")
+     }
+     else {
+         gridded(x) <- FALSE
```

```
+           x <- x[i = sel, j = j, ...]
+           cc <- coordinates(x)[, 1:2]
+           p4s <- CRS(proj4string(x))
+           SpatialPixelsDataFrame(cc, x@data, proj4string = p4s)
+       }
+ }
> setMethod("[", c("SpatialTimeGridDataFrame", "POSIXct",
+     "ANY"), subs.SpatialTimeGridDataFrame)
```

```
[1] "["
```

```
> t1 <- as.POSIXct("1970-01-01 0:00:03", tz = "GMT")
> t2 <- as.POSIXct("1970-01-01 0:00:05", tz = "GMT")
> summary(x[c(t1, t2)])
```

```
Object of class SpatialTimeGridDataFrame
first time step: 1970-01-01 00:00:03
last time step:  1970-01-01 00:00:05
time step:       2
Object of class SpatialGridDataFrame
Coordinates:
   min  max
x1 0.5 10.5
x2 0.5 10.5
x3 2.0  6.0
Is projected: NA
proj4string : [NA]
Number of points: 2
Grid attributes:
   cellcentre.offset cellsize cells.dim
x1                 1        1        10
x2                 1        1        10
x3                 3        2         2
Data attributes:
      Min.    1st Qu.      Median        Mean    3rd Qu.        Max.
-3.0080000 -0.6764000 -0.0002298 -0.0081510  0.6546000  3.8100000
```

```
> summary(x[t1])
```

```
Object of class SpatialPixelsDataFrame
Coordinates:
   min  max
x1 0.5 10.5
x2 0.5 10.5
Is projected: NA
proj4string : [NA]
Number of points: 100
Grid attributes:
   cellcentre.offset cellsize cells.dim
x1                 1        1        10
x2                 1        1        10
```

```
Data attributes:
   Min. 1st Qu.  Median    Mean 3rd Qu.    Max.
-2.8890 -0.4616  0.1353  0.0773  0.7779  2.6490
```

The reason to only convert back to `SpatialTimeGridDataFrame` when multiple time steps are present is that the time step ('cell size' in time direction) cannot be found when there is only a single step. In that case, the current selection method returns an object of class `SpatialPixelsDataFrame` for that time slice.

Plotting a set of slices could be done using levelplot or writing another `spplot` method:

```
> spplot.stgdf <- function(obj, zcol = 1, ..., format = NULL) {
+     if (length(zcol) != 1)
+         stop("can only plot a single attribute")
+     if (is.null(format))
+         format <- "%Y-%m-%d %H:%M:%S"
+     cc <- coordinates(obj)
+     df <- unstack(data.frame(obj[[zcol]], cc[, 3]))
+     ns <- as.character(coordinatevalues(getGridTopology(obj))[[3]] +
+         ISOdate(1970, 1, 1, 0, 0, 0), format = format)
+     cc2d <- cc[cc[, 3] == min(cc[, 3]), 1:2]
+     obj <- SpatialPixelsDataFrame(cc2d, df)
+     spplot(obj, names.attr = ns, ...)
+ }
> setMethod("spplot", "SpatialTimeGridDataFrame", spplot.stgdf)

[1] "spplot"
```

Now, the result of

```
> library(lattice)
> trellis.par.set(canonical.theme(color = FALSE))
> spplot(x, format = "%H:%M:%S", as.table = TRUE, cuts = 6,
+     col.regions = grey.colors(7, 0.55, 0.95, 2.2))
```

is shown in Fig. 6.3. The format argument passed controls the way time is printed; one can refer to the help of

```
> `?`(as.character.POSIXt)
```

for more details about the `format` argument.

6.6 Analysing Spatial Monte Carlo Simulations

Quite often, spatial statistical analysis results in a large number of spatial realisations of a random field, using some Monte Carlo simulation approach. Regardless whether individual values refer to points, lines, polygons, or grid cells, we would like to write some methods or functions that aggregate over these simulations, to get summary statistics such as the mean value, quantiles, or cumulative distributions values. Such aggregation can take place in two

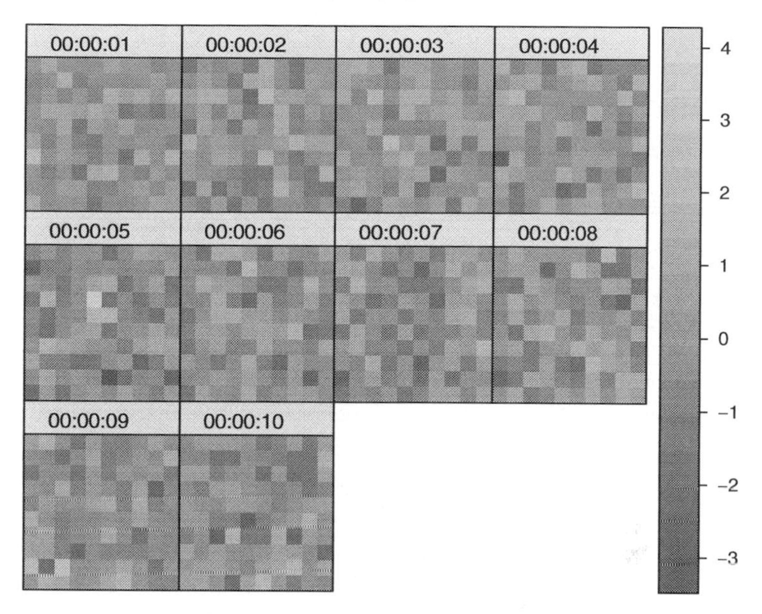

Fig. 6.3. `spplot` for an object of class `SpatialTimeGridDataFrame`, filled with random numbers

ways. Either we aggregate over the probability space and compute summary statistics for each geographical feature over the set of realisations (i.e. the rows of the attribute table), or for each realisation we aggregate over the complete geographical layer or a subset of it (i.e. aggregate over the columns of the attribute table).

Let us first generate, as an example, a set of 100 conditional Gaussian simulations for the zinc variable in the meuse data set:

```
> library(gstat)
> data(meuse)
> coordinates(meuse) <- ~x + y
> v <- vgm(0.5, "Sph", 800, 0.05)
> sim <- krige(log(zinc) ~ 1, meuse, meuse.grid, v, nsim = 100,
+     nmax = 30)
```

```
drawing 100 GLS realisations of beta...
[using conditional Gaussian simulation]
```

```
> sim@data <- exp(sim@data)
```

where the last statement back-transforms the simulations from the log scale to the observation scale. A quantile method for Spatial object attributes can be written as

```
> quantile.Spatial <- function(x, ..., byLayer = FALSE) {
+     stopifnot("data" %in% slotNames(x))
+     apply(x@data, ifelse(byLayer, 2, 1), quantile, ...)
+ }
```

after which we can find the sample below and above 95% confidence limits by

```
> sim$lower <- quantile.Spatial(sim[1:100], probs = 0.025)
> sim$upper <- quantile.Spatial(sim[1:100], probs = 0.975)
```

To get the sample distribution of the areal median, we can aggregate over layers:

```
> medians <- quantile.Spatial(sim[1:100], probs = 0.5,
+     byLayer = TRUE)

> hist(medians)
```

It should be noted that in these particular cases, the quantities computed by simulations could have been obtained faster and exactly by working analytically with ordinary (block) kriging and the normal distribution (Sect. 8.7.2).

A statistic that cannot be obtained analytically is the sample distribution of the area fraction that exceeds a threshold. Suppose that 500 is a crucial threshold, and we want to summarise the sampling distribution of the area fraction where 500 is exceeded:

```
> fractionBelow <- function(x, q, byLayer = FALSE) {
+     stopifnot(is(x, "Spatial") || !("data" %in%
+         slotNames(x)))
+     apply(x@data < q, ifelse(byLayer,
+         2, 1), function(r) sum(r)/length(r))
+ }

> over500 <- 1 - fractionBelow(sim[1:100], 200, byLayer = TRUE)
> summary(over500)

   Min. 1st Qu.  Median    Mean 3rd Qu.    Max. 
 0.6007  0.6768  0.6911  0.6897  0.7071  0.7448 

> quantile(over500, c(0.025, 0.975))

     2.5%     97.5% 
0.6460361 0.7336529 
```

For space–time data, we could write methods that aggregate over space, over time, or over space and time.

6.7 Processing Massive Grids

Up to now we have made the assumption that gridded data can be completely read and are kept by R in memory. In some cases, however, we need to process grids that exceed the memory capacity of the computers available. A method for analysing grids without fully loading them into memory then seems useful. Note that package **rgdal** allows for partial reading of grids, for example

```
> x <- readGDAL("70042108.tif", output.dim = c(120, 132))
> x$band1[x$band1 <= 0] <- NA
> spplot(x, col.regions = bpy.colors())
```

reads a downsized grid, where 1% of the grid cells remained. Another option is reading certain rectangular sections of a grid, starting at some offset.

Yet another approach is to use the low-level opening routines and then subset:

```
> library(rgdal)
> x <- GDAL.open("70042108.tif")
> class(x)

[1] "GDALReadOnlyDataset"
attr(,"package")
[1] "rgdal"

> x.subs <- x[1:100, 1:100, 1]
> class(x.subs)

[1] "SpatialGridDataFrame"
attr(,"package")
[1] "sp"

> gridparameters(x.subs)

  cellcentre.offset     cellsize cells.dim
x         174.20042 0.0008333333       100
y         -36.58292 0.0008333333       100
```

An object of class `GDALReadOnlyDataset` contains only a file handle. The subset method `"["` for it does not, as it quite often does, return an object of the same class but actually reads the data requested, with arguments interpreted as *rows*, *columns*, and raster *bands*, and returns a `SpatialGridDataFrame`. We now extend this approach to allow partial writing through `"["` as well. As the actual code is rather lengthy and involves a lot of administration, it will not all be shown and details can be found in the **rgdal** source code.

We define two classes,

```
> setClass("SpatialGDAL", representation("Spatial",
+     grid = "GridTopology", grod = "GDALReadOnlyDataset",
+     name = "character"))

[1] "SpatialGDAL"

> setClass("SpatialGDALWrite", "SpatialGDAL")

[1] "SpatialGDALWrite"
```

that derive from `Spatial`, contain a `GridTopology`, and a file handle in the `grod` slot. Next, we can define a function `open.SpatialGDAL` to open a raster file, returning a `SpatialGDAL` object and a function `copy.SpatialGDAL` that returns a writable copy of the opened raster. Note that some GDAL drivers allow only copying, some only writing, and some both.

```
> x <- open.SpatialGDAL("70042108.tif")
> nrows <- GDALinfo("70042108.tif")["rows"]
> ncols <- GDALinfo("70042108.tif")["columns"]
> xout <- copy.SpatialGDAL(x, "70042108out.tif")
> bls <- 20
> for (i in 1:(nrows/bls - 1)) {
+     r <- 1 + (i - 1) * bls
+     for (j in 1:(ncols/bls - 1)) {
+         c <- 1 + (j - 1) * bls
+         x.in <- x[r:(r + bls), c:(c + bls)]
+         xout[r:(r + bls), c:(c + bls)] <- x.in$band1 +
+             10
+     }
+     cat(paste("row-block", i, "\n"))
+ }
> close(x)
> close(xout)
```

This requires the functions `"["` and `"[<-"` to be present. They are set by

```
> setMethod("[", "SpatialGDAL", function(x, i, j, ...,
+     drop = FALSE) x@grod[i = i, j = j, ...])
> setReplaceMethod("[", "SpatialGDALWrite", function(x,
+     i, j, ..., value) {
+     ...
+ })
```

where, for the latter, the implementation details are omitted here. It should be noted that single rows or columns cannot be read this way, as they cannot be converted sensibly to a grid.

It should be noted that flat binary representations such as the Arc/Info Binary Grid allow much faster random access than ASCII representations or compressed formats such as jpeg varieties. Also, certain drivers in the GDAL library suggest an optimal block size for partial access (e.g. typically a single row), which is not used here[1].

This chapter has sketched developments beyond the base **sp** classes and methods used otherwise in this book. Although we think that the base classes cater for many standard kinds of spatial data analysis, it is clear that specific research problems will call for specific solutions, and that the R environment provides the high-level abstractions needed to help busy researchers get their work done.

[1] An attempt to use this block size is, at time of writing, found in the `blockApply` code, found in the THK CVS branch of the **rgdal** project on SourceForge.

Part II

Analysing Spatial Data

Analysing Spatial Data

The analysis of spatial data is usually undertaken to make inferences, that is to try to draw conclusions about a hypothesised data generating process or to use an estimated process to predict values at locations for which observations are unavailable. In some cases, the conclusions are sufficient in themselves, and in others, they are carried through to other hierarchical layers in the model under scrutiny. Haining (2003, pp. 184–185) and Bivand (2002, p. 409) suggest (following Tukey, 1977) that our understanding of the data may be partitioned into

$$\text{data} = \text{smooth} + \text{rough}.$$

If the data are spatial, we can see that there is room for another term, irrespective of whether we are more interested in the fit of the model itself or in calibrating the model in order to predict for new data:

$$\text{data} = \text{smooth} + \text{spatial smooth} + \text{rough}.$$

The added term constitutes the 'added value' of spatial data analysis, bringing better understanding or predictive accuracy at the price of using specialised methods for fitting the spatial smooth term. We will here be concerned with methods for finding out whether fitting this term is worth the effort, and, if so, how we might choose to go about doing so.

Before rushing off to apply such specialised methods, it is worth thinking through the research problem thoroughly. We have already mentioned the importance of the Distributed Statistical Computing conference in Vienna in 2003 for our work. At that meeting, Bill Venables presented a fascinating study of a real research problem in the management of tiger prawn fisheries. The variable of interest was the proportion by weight of two species of tiger prawn in the logbook on a given night at a given location. In a very careful treatment of the context available, the 'location' was not simply taken as a point in space with geographical coordinates:

'Rather than use latitude and longitude directly as predictors, we find it more effective to represent station locations using the following two predictors:

- The shortest distance from the station to the coast (variable R_{land}), and
- The distance from an origin in the west to the nearest point to the station along an arbitrary curve running nearly parallel to the coast (variable R_{dist}).

[...] Rather than use R_{dist} itself as a predictor, we use a natural spline basis that allows the fitted linear predictor to depend on the variable in a flexible curvilinear way.

[...] Similarly, we choose a natural spline term with four internal knots at the quantiles of the corresponding variable for the logbook data for the "distance from dry land" variable, R_{land}.

The major reason to use this system, which is adapted to the coastline, is that interactions between R_{land} and R_{dist} are more likely to be negligible than for latitude and longitude, thus simplifying the model. The fact that they do not form a true co-ordinate system equivalent to latitude and longitude is no real disadvantage for the models we propose.' Venables and Dichmont (2004, pp. 412–413)

The paper deserves to be required reading in its entirety for all spatial data analysts, not least because of its sustained focus on the research problem at hand. It also demonstrates that because applied spatial data analysis builds on and extends applied data analysis, specifically spatial methods should be used when the problem cannot be solved with general methods. Consequently, familiarity with the modelling chapters of textbooks using R for analysis will be of great help in distinguishing between situations calling for spatial solutions, and those that do not, even though the data are spatial. Readers will benefit from having one or more of Fox (2002), Dalgaard (2002), Faraway (2004, 2006), or Venables and Ripley (2002) to refer to in seeking guidance on making often difficult research decisions.

In introducing this part of the book – covering specialised spatial methods but touching in places on non-spatial methods – we use the classification of Cressie (1993) of spatial statistics into three areas, *spatial point patterns*, covered here in Chap. 7, *geostatistical data* in Chap. 8, and *lattice data*, here termed areal data, in Chaps. 9–11. In Chap. 1, we mentioned a number of central books on spatial statistics and spatial data analysis; Table II.1 shows very roughly which of our chapters contain material that illustrates some of the methods presented in more recent spatial statistics books, including treatments of all three areas of spatial statistics discussed earlier (see p. 13).

The coverage here is uneven, because only a limited number of the topics covered in these books could be accommodated; the specialised literature within the three areas will be referenced directly in the relevant chapters. On the other hand, the implementations discussed below may be extended to cover

Table II.1. Thematic cross-tabulation of chapters in this book with chapters and sections of chosen books on spatial statistics and spatial data analysis

Chapter	Cressie (1993)	Schabenberger and Gotway (2005)	Waller and Gotway (2004)	Fortin and Dale (2005)	O'Sullivan and Unwin (2003)
7	8	3	5	2.1–2.2	4–5
8	2–3	4–5	8	3.5	8–9
9–11	6–7	1, 6	6, 7, 9	3.1–3.4, 5	7

alternative methods; for example, the use of WinBUGS with R is introduced in Chap. 11 in general forms capable of extension. The choice of contributed packages is also uneven; we have used the packages that we maintain, but this does not constitute a recommendation of these rather than other approaches (see Fig. 1.1). Note that coloured versions of figures may be found on the book website together with complete code examples, data sets, and other support material.

7

Spatial Point Pattern Analysis

7.1 Introduction

The analysis of point patterns appears in many different areas of research. In ecology, for example, the interest may be focused on determining the spatial distribution (and its causes) of a tree species for which the locations have been obtained within a study area. Furthermore, if two or more species have been recorded, it may also be of interest to assess whether these species are equally distributed or competition exists between them. Other factors which force each species to spread in particular areas of the study region may be studied as well. In spatial epidemiology, a common problem is to determine whether the cases of a certain disease are clustered. This can be assessed by comparing the spatial distribution of the cases to the locations of a set of controls taken at random from the population.

In this chapter, we describe how the basic steps in the analysis of point patterns can be carried out using R. When introducing new ideas and concepts we have tried to follow Diggle (2003) as much as possible because this text offers a comprehensive description of point processes and applications in many fields of research. The examples included in this chapter have also been taken from that book and we have tried to reproduce some of the examples and figures included there.

In general, a point process is a stochastic process in which we observe the locations of some *events* of interest within a bounded region A. Diggle (2003) defines a point process as a 'stochastic mechanism which generates a countable set of events'. Diggle (2003) and Möller and Waagepetersen (2003) give proper definitions of different types of a point process and their main properties. The locations of the events generated by a point process in the area of study A will be called a *point pattern*. Sometimes, additional covariates may have been recorded and they will be attached to the locations of the observed events.

Other books covering this subject include Schabenberger and Gotway (2005, Chap. 3), Waller and Gotway (2004, Chaps. 5 and 6) and O'Sullivan and Unwin (2003, Chaps. 4 and 5).

7.2 Packages for the Analysis of Spatial Point Patterns

There are a number of packages for R which implement different functions for the analysis of spatial point patterns. The **spatial** package provides functions described in Venables and Ripley (2002, pp. 430–434), and **splancs** (Rowlingson and Diggle, 1993) and **spatstat** (Baddeley and Turner, 2005) provide other implementations and additional methods for the analysis of different types of point processes. The Spatial Task View contains a complete list of all the packages available in R for the analysis of point patterns. Other packages worth mentioning include **spatialkernel**, which implements different kernel functions and methods for the analysis of multivariate point processes. Given that most of the examples included in this chapter have been computed using **splancs** and **spatstat**, we focus particularly on these packages.

These packages use different data structures to store the information of a point pattern. Given that it would be tedious to rewrite all the code included in these packages to use **sp** classes, we need a simple mechanism to convert between formats. Package **maptools** offers some functions to convert between `ppp` objects representing two-dimensional point patterns (from **spatstat**, which uses old-style classes, see p. 24) and **sp** classes. Note that, in addition to the point coordinates, `ppp` objects include the boundary of the region where the point data have been observed, whilst **sp** classes do not, and it has to be stored separately. Data types used in **splancs** are based on a two-column matrix for the coordinates of the point pattern plus a similar matrix to store the boundary; the package was written before old-style classes were introduced. Function `as.points` is provided to convert to this type of structure. Hence, it is very simple to convert the coordinates from **sp** classes to use functions included in **splancs**.

Section 2.4 describes different types of **sp** classes to work with point data. They are `SpatialPoints`, for simple point data, and `SpatialPointsDataFrame`, when additional covariates are recorded. More information and examples can be found in the referred section. Hence, it should not be difficult to have the data available in the format required for the analysis whatever package is used.

To illustrate the use of some of the different techniques available for the analysis of point patterns, we have selected some examples from forest ecology, biology, and spatial epidemiology. The point patterns in Fig. 7.1 show the spatial distribution of cell centres (left), California redwood trees (right), and Japanese black pine (middle). All data sets have been re-scaled to fit into a one-by-one square. These data sets are described in Ripley (1977), Strauss (1975), Numata (1961) and all of them have been re-analysed in Diggle (2003).

These data sets are available in package **spatstat**. This package uses `ppp` objects to store point patterns, but package **maptools** provides some functions to convert between `ppp` objects and `SpatialPoints`, as shown in the following example. First we take the Japanese black pine saplings example, measured in a square sampling region in a natural forest, reading in the data provided with **spatstat**.

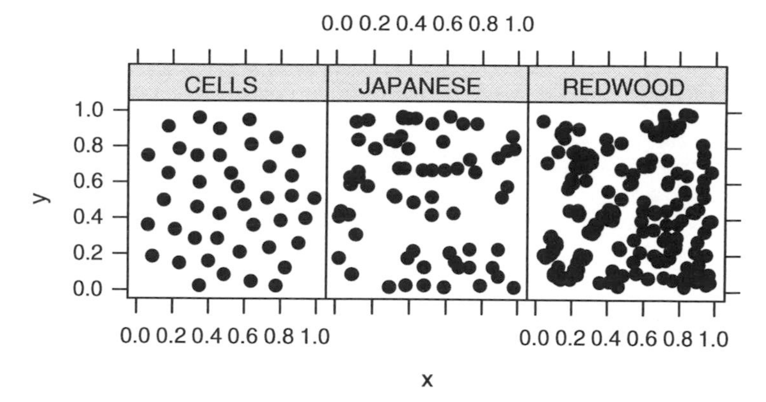

Fig. 7.1. Example of three point patterns re-scaled to fit in the unit square. On the left, spatial distribution of the location of cell centres (Ripley, 1977); in the middle, Japanese black pine saplings (Numata, 1961); and on the right, saplings of California redwood trees (Strauss, 1975)

```
> library(spatstat)
> data(japanesepines)

> summary(japanesepines)

Planar point pattern: 65 points
Average intensity 65 points per square unit (one unit = 5.7 metres)

Window: rectangle = [0, 1] x [0, 1] units
Window area =  1 square unit
Unit of length: 5.7 metres
```

The summary shows the average intensity in the region of interest; this region, known as a window, is also reported in the summary; windows are stored in objects of class `owin`. In this case, the points have been scaled to the unit square already, but the size of the sampling square can be used to retrieve the actual measurements. Note that **spatstat** windows may be of several forms, here the window is a rectangle. When we coerce a `ppp` object with a rectangular window to a `SpatialPoints` object, the point coordinates will by default be re-scaled to their original values.

```
> library(maptools)

> spjpines <- as(japanesepines, "SpatialPoints")
> summary(spjpines)

Object of class SpatialPoints
Coordinates:
     min max
[1,]   0 5.7
[2,]   0 5.7
```

```
Is projected: NA
proj4string : [NA]
Number of points: 65
```

We can get back to the unit square using the `elide` methods discussed in Chap. 5 as the summary of the output object shows.

```
> spjpines1 <- elide(spjpines, scale = TRUE, unitsq = TRUE)
> summary(spjpines1)
```

```
Object of class SpatialPoints
Coordinates:
     min max
[1,]   0   1
[2,]   0   1
Is projected: NA
proj4string : [NA]
Number of points: 65
```

Getting back to a `ppp` object is also done by coercing, but if we want to preserve the actual dimensions, we have to manipulate the `owin` object belonging to the `ppp` object directly. We return later to see how `SpatialPolygons` objects may be coerced into `owin` objects, and how **spatstat** `im` objects can interface with `SpatialGrid` objects.

```
> pppjap <- as(spjpines1, "ppp")
> summary(pppjap)
```

```
Planar point pattern: 65 points
Average intensity 65 points per square unit

Window: rectangle = [0, 1] x [0, 1] units
Window area =  1 square unit
```

These point patterns have been obtained by sampling in different regions, but it is not rare to find examples in which we have different types of events in the same region. In spatial epidemiology, for example, it is common to have two types of points: *cases* of a certain disease and *controls*, which usually reflect the spatial distribution of the population. In general, this kind of point pattern is called a *marked* point pattern because each point is assigned to a group and labelled accordingly.

The Asthma data set records the results of a case–control study carried out in 1992 on the incidence of asthma in children in North Derbyshire (United Kingdom). This data set has been studied by Diggle and Rowlingson (1994), Singleton et al. (1995), and Diggle (2003) to explore the relationship between asthma and the proximity to the main roads and three putative pollution sources (a coking works, chemical plant, and waste treatment centre). In the study, a number of relevant covariates were also collected by means of a questionnaire that was completed by the parents of the children attending 10 schools in the region. Children having suffered from asthma will act as cases whilst the remainder of the children included in the study will form the set of controls. Although this data set is introduced here, the spatial analysis of case–control data is described in the final part of this chapter.

The data set is available from Prof. Peter J. Diggle's website and comes in anonymised form. Barry Rowlingson provided some of the road lines. The original data were supplied by Dr. Joanna Briggs (University of Leeds, UK). To avoid computational problems in some of the methods described in this section, we have removed a very isolated point, which was one of the cases, and we have selected an appropriate boundary region.

The next example shows how to display the point pattern, including the boundary of the region (that we have created ourselves) and the location of the pollution sources using different **sp** layouts and function `spplot` (see Chap. 3 for more details). Given that the data set is a marked point pattern, we have converted it to a `SpatialPointsDataFrame` to preserve the type (case or control) of the events and all the other relevant information. In addition, we have created a `SpatialPolygons` object to store the boundary of the region and a `SpatialPointsDataFrame` object for the location of the three pollution sources. Given that the main roads are available, we have included them as well using a `SpatialLines` object. The final plot is shown in Fig. 7.2.

```
> library(rgdal)
> spasthma <- readOGR(".", "spasthma")
> spbdry <- readOGR(".", "spbdry")
> spsrc <- readOGR(".", "spsrc")
> sproads <- readOGR(".", "sproads")
```

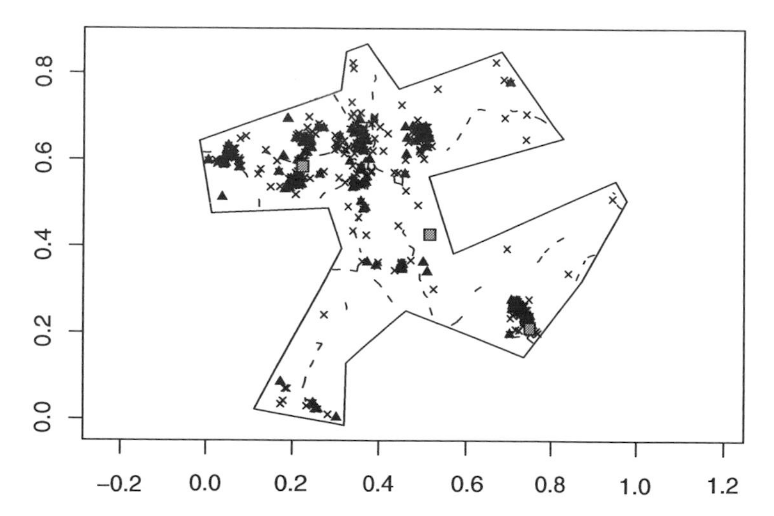

Fig. 7.2. Locations of the residence of asthmatic (cases, *filled triangle*) and non-asthmatic (controls, *cross*) in North Derbyshire, 1992 (Diggle and Rowlingson, 1994). The boundary has been taken to contain all points in the data set. The map shows the pollution sources (*grey filled square*) and the main roads (*dashed lines*)

7.3 Preliminary Analysis of a Point Pattern

The analysis of point patterns is focused on the spatial distribution of the observed events and making inference about the underlying process that generated them. In particular, there are two main issues of interest: the distribution of events in space and the existence of possible interactions between them. For a merely descriptive analysis, we would represent the locations of the point pattern in the study area. This will give us an idea of the distribution of the points, and it can lead to possible hypothesis about the spatial distribution of the events. Further statistical analyses can be done and they are described in this section.

7.3.1 Complete Spatial Randomness

When studying a point process, the most basic test that can be performed is that of *Complete Spatial Randomness* (CSR, henceforth). Intuitively, by CSR we mean that the events are distributed independently at random and uniformly over the study area. This implies that there are no regions where the events are more (or less) likely to occur and that the presence of a given event does not modify the probability of other events appearing nearby.

Informally, this can be tested by plotting the point pattern and observing whether the points tend to appear in clusters or, on the contrary, they follow a regular pattern. In any of these cases, the points are not distributed uniformly because they should be distributed filling all the space in the study area. Usually, clustered patterns occur when there is attraction (i.e. 'contagion') between points, whilst regular patterns occur when there is inhibition (i.e. 'competition') among points.

Figure 7.1 shows three examples of point patterns that have been generated by different biological mechanisms and seem to have different spatial distributions. In particular, plot 7.1 of the Japanese pine trees (middle) seems neither clustered nor regularly distributed, whilst the redwood seeds (right) show a clustered pattern and the cells (left) a regular one. Hence, only the spatial distribution of Japanese pine trees seems to be compatible with CSR.

To measure the degree of accomplishment of the CSR, several functions can be computed on the data. These are described in the following sections, together with methods to measure the uncertainty related to the observed pattern.

Testing for CSR is covered in Waller and Gotway (2004, pp. 118–126), O'Sullivan and Unwin (2003, pp. 88–92, including a discussion on pp. 108–112), and Schabenberger and Gotway (2005, pp. 86–99, including other methods not presented here).

7.3.2 G Function: Distance to the Nearest Event

The G function measures the distribution of the distances from an arbitrary event to its nearest event. If these distances are defined as $d_i = \min_j\{d_{ij}, \forall j \neq i\}$, $i = 1, \ldots, n$, then the G function can be estimated as

$$\hat{G}(r) = \frac{\#\{d_i : d_i \leq r, \forall i\}}{n},$$

where the numerator is the number of elements in the set of distances that are lower than or equal to d and n is the total number of points. Under CSR, the value of the G function is

$$G(r) = 1 - \exp\{-\lambda \pi r^2\},$$

where λ represents the mean number of events per unit area (or *intensity*).

The compatibility with CSR of the point pattern can be assessed by plotting the empirical function $\hat{G}(d)$ against the theoretical expectation. In addition, point-wise envelopes under CSR can be computed by repeatedly simulating a CSR point process with the same estimated intensity $\hat{\lambda}$ in the study region (Diggle, 2003, p. 13) and check whether the empirical function is contained inside. The next chunk of code shows how to compute this by using **spatstat** functions `Gest` and `envelope`. The results have been merged in a data frame in order to use conditional Lattice graphics.

```
> r <- seq(0, sqrt(2)/6, by = 0.005)
> envjap <- envelope(as(spjpines1, "ppp"), fun = Gest,
+      r = r, nrank = 2, nsim = 99)
> envred <- envelope(as(spred, "ppp"), fun = Gest, r = r,
+      nrank = 2, nsim = 99)
> envcells <- envelope(as(spcells, "ppp"), fun = Gest,
+      r = r, nrank = 2, nsim = 99)
> Gresults <- rbind(envjap, envred, envcells)
> Gresults <- cbind(Gresults, DATASET = rep(c("JAPANESE",
+      "REDWOOD", "CELLS"), each = length(r)))
```

Figure 7.3 shows the empirical function $\hat{G}(r)$ against $G(r)$ together with the 96% pointwise envelopes (because `nrank=2`) of the same point pattern examined using the G function. The plot is produced by taking the pairs $(G(r), \hat{G}(r))$ for a set of reasonable values of the distance r, so that in the x-axis we have the values of the theoretical value of $G(r)$ under CSR and in the y-axis the empirical function $\hat{G}(r)$. The results show that only the Japanese trees seem to be homogeneously distributed, whilst the redwood seeds show a clustered pattern (values of $\hat{G}(r)$ above the envelopes) and the location of the cells shows a more regular pattern (values of $\hat{G}(r)$ below the envelopes).

`envelope` is a very flexible function that can be used to compute Monte Carlo envelopes of a certain type of functions. Basically, it works by randomly simulating a number of point patterns so that the summary function is computed for all of them. The resulting values are then used to compute point-wise

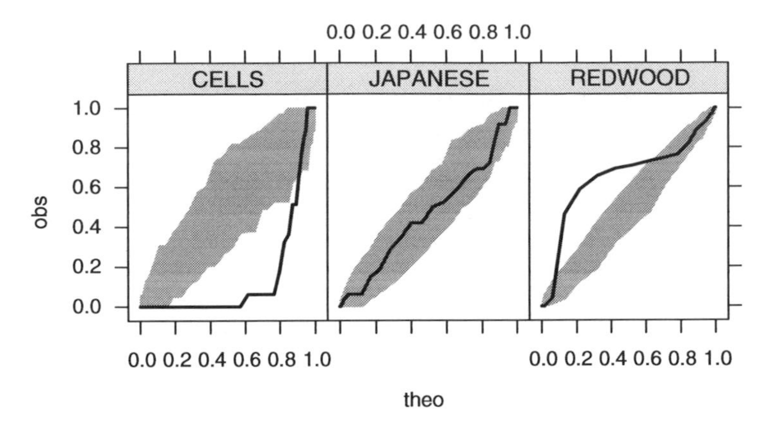

Fig. 7.3. Envelopes and observed values of the G function for three point patterns

(i.e. at different distances) or global Monte Carlo envelopes. `envelope` can be passed the way the point patterns are generated (by default, CSR). The reader is referred to the manual page for more information.

7.3.3 *F* Function: Distance from a Point to the Nearest Event

The F function measures the distribution of all distances from an arbitrary point of the plane to its nearest event. This function is often called the *empty space* function because it is a measure of the average space left between events. Under CSR, the expected value of the F function is

$$F(r) - 1 - \exp\{-\lambda\pi r^2\}.$$

Hence, we can compare the estimated value of the F function to its theoretical value and compute simulation envelopes as before.

```
> Fenvjap <- envelope(as(spjpines1, "ppp"), fun = Fest,
+     r = r, nrank = 2, nsim = 99)
> Fenvred <- envelope(as(spred, "ppp"), fun = Fest, r = r,
+     nrank = 2, nsim = 99)
> Fenvcells <- envelope(as(spcells, "ppp"), fun = Fest,
+     r = r, nrank = 2, nsim = 99)
> Fresults <- rbind(Fenvjap, Fenvred, Fenvcells)
> Fresults <- cbind(Fresults, DATASET = rep(c("JAPANESE",
+     "REDWOOD", "CELLS"), each = length(r)))
```

Figure 7.4 shows the empirical F functions and their associated 96% envelopes (because `nrank=2`) for the three data sets presented before. The Japanese data are compatible with the CSR hypothesis, whereas the cells point pattern shows a regular pattern ($\hat{F}(r)$ is above the envelopes) and the redwood points seem to be clustered, given the low values of $\hat{F}(r)$.

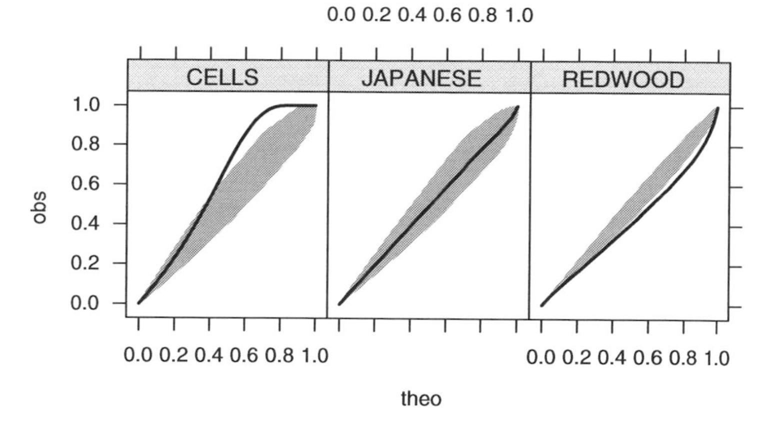

Fig. 7.4. Envelopes and observed values of the F function for three point patterns

7.4 Statistical Analysis of Spatial Point Processes

A first description of the point pattern can be done by estimating the spatial statistical density from the observed data. The spatial density has the same properties as a univariate density, but its domain is the study area where the point process takes place.

As an alternative function to measure the spatial distribution of the events, we can work with the *intensity* $\lambda(x)$ of the point process, which is proportional to its spatial density. The constant of proportionality is the expected number of events of the point process in the area A. That is, for two point processes with the same spatial density but different intensities, the number of events observed will be higher for the process with the highest intensity.

The intensity and spatial density are part of the *first-order* properties because they measure the distribution of events in the study region. Note that neither the intensity nor the spatial density give any information on the interaction between two arbitrary points. This is measured by *second-order properties*, which reflect any tendency of the events to appear clustered, independently, or regularly spaced.

First- and second-order properties are properly defined in, for example, Diggle (2003, p. 43) and Möller and Waagepetersen (2003, Chap. 4). We focus on the estimation of the intensity and the assessment of clustering, as explained in the following sections. Waller and Gotway (2004, pp. 130–146) and Schabenberger and Gotway (2005, 90–103, 110–112) discuss the estimation of the intensity of a point pattern and the assessment of clustering as well.

The separation between first- and second-order properties can be difficult to disentangle without further assumptions. For example, do groups of events appear at a specific location because the intensity is higher there or because events are clustered? In general, it is assumed that interaction between points occurs at small scale, while large-scale variation is reflected on the intensity

(Diggle, 2003, p. 143). Waller and Gotway (2004, 146–147) also discuss the roles of first and second-order properties.

In the remainder of this chapter, we focus on Poisson processes because they offer a simple approach to a wide range of problems. Loosely, we can distinguish between homogeneous and inhomogeneous Poisson point processes (HPP and IPP, respectively). Both HPP and IPP assume that the events occur independently and are distributed according to a given intensity. The main difference between the two point processes is that the HPP assumes that the intensity function is constant, while the intensity of an IPP varies spatially. In a sense, the IPP is a generalisation of the HPP or, inversely, the HPP can be regarded as an IPP with constant intensity. Poisson processes are also described in Schabenberger and Gotway (2005, pp. 81–86, 107–110) and Waller and Gotway (2004, pp. 126–130).

Note that other spatial processes may be required when more complex data sets are to be analysed. For example, when events are clustered, points do not occur independently of each other and a clustered process would be more appropriate. See Diggle (2003, Chap. 5) and Möller and Waagepetersen (2003) for a wider description of other spatial point processes. `spatstat` provides a number of functions to fit some of the models described therein.

7.4.1 Homogeneous Poisson Processes

A homogeneous Poisson process is characterised as representing the kind of point process in which all events are independently and uniformly distributed in the region A where the point process occurs. This means that the location of one point does not affect the probabilities of other points appearing nearby and that there are no regions where events are more likely to appear.

More formally, Diggle (2003) describes an HPP in a region A as fulfilling:

1. The number of events in A, with area $|A|$, is Poisson distributed with mean $\lambda|A|$, where λ is the constant intensity of the point process.
2. Given n observed events in region A, they are uniformly distributed in A.

The HPP is also *stationary* and *isotropic*. It is stationary because the intensity is constant and the second-order intensity depends only on the relative positions of two points (i.e. direction and distance). In addition, it is isotropic because the second-order intensity is invariant to rotation. Hence, the point process has constant intensity and its second-order intensity depends only on the distance between the two points, regardless of the relative positions of the points.

These constraints reflect that the intensity of the point process is constant, that is $\lambda(x) = \lambda > 0, \forall x \in A$, and that events appear independently of each other. Hence, the HPP is the formal definition of a point process which is CSR.

7.4.2 Inhomogeneous Poisson Processes

In most cases assuming that a point process under study is homogeneous is not realistic. Clear examples are the distribution of the population in a city or the location of trees in a forest. In both cases, different factors affect the spatial distribution. In the case of the population, it can be the type of housing, neighbourhood, etc., whilst in the case of the trees, it can be the environmental factors such as humidity, quality of the soil, slope and others.

The IPP is a generalisation of the HPP, which allows for a non-constant intensity. The same principle of independence between events holds, but now the spatial variation can be more diverse, with events appearing more likely in some areas than others. As a result, the intensity will be a generic function $\lambda(x)$ that varies spatially.

7.4.3 Estimation of the Intensity

As stated previously, the intensity of an HPP point process is constant. Hence, the problem of estimating the intensity is the problem of estimating a constant function λ such as the expected number of events in region A ($\int_A \lambda \, \mathrm{d}x$) is equal to the observed number of cases. This is the volume under the surface defined by the intensity in region A. Once we have observed the (homogeneous) point process, we have the locations of a set of n points. So, an unbiased estimator of the intensity is $n/|A|$, where $|A|$ is the area of region A. This ensures that the expected number of points is, in fact, the observed number of points.

For IPP, the estimation of the intensity can be done in different ways. It can be done non-parametrically by means of kernel smoothing or parametrically by proposing a specific function for the intensity whose parameters are estimated by maximising the likelihood of the point process. If we have observed n points $\{x_i\}_{i=1}^n$, the form of a kernel smoothing estimator is the following (Diggle, 1985; Berman and Diggle, 1989):

$$\hat{\lambda}(x) = \frac{1}{h^2} \sum_{i=1}^{n} \kappa\Big(\frac{||x - x_i||}{h}\Big) / q(||x||), \tag{7.1}$$

where $\kappa(u)$ is a bivariate and symmetrical kernel function. $q(||x||)$ is a border correction to compensate for the missing observations that occur when x is close to the border of the region A. Bandwidth h measures the level of smoothing. Small values will produce very peaky estimates, whilst large values will produce very smooth functions.

Silverman (1986) gives a detailed description of different kernel functions and their properties. In the examples included in this chapter, we have used the *quartic* kernel (also known as *biweight*), whose expression in two dimensions is

$$\kappa(u) = \begin{cases} \frac{3}{\pi}(1 - \|u\|^2)^2 & \text{if } u \in (-1, 1) \\ 0 & \text{Otherwise} \end{cases},$$

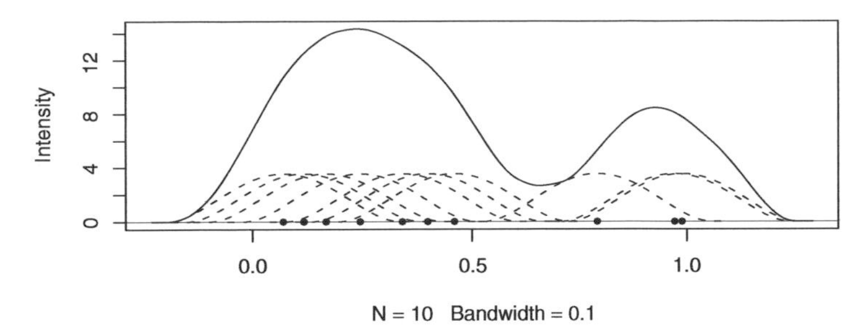

Fig. 7.5. Example of the contribution of the different points to the estimate of the intensity. Dashed lines represent the kernel around each observation, whilst the solid line is the estimate of the intensity

where $\|u\|^2$ denotes the squared norm of point $u = (u_1, u_2)$ equal to $u_1^2 + u_2^2$. Figure 7.5 shows an example of estimation of the intensity by kernel smoothing in a one-dimensional setting, but the same ideas are used in a spatial framework.

Methods for the selection of the bandwidth of kernel smoothers in a general setting are given by Silverman (1986). In the context of spatial analysis, a few proposals have been made so far, but it is not clear how to choose an optimal value for the bandwidth in the general case. It seems reasonable to use several values depending on the process under consideration, and choose a value that seems plausible.

Diggle (1985) and Berman and Diggle (1989) propose a criterion based on minimising the Mean Square Error (MSE) of the kernel smoothing estimator when the underlying point process in a stationary Cox process (see, e.g. p. 68 of Diggle (2003) for details). However, it can still be used as a general exploratory method and a guidance in order to choose the bandwidth. Kelsall and Diggle (1995a,b, 1998) propose and compare different methods for the selection of the bandwidth when a case–control point pattern is used. Clark and Lawson (2004) have compared these and other methods for disease mapping, including some methods for the automatic selection of the bandwidth.

We have applied the approach proposed by Berman and Diggle (1989), which is implemented in function `mse2d` to the redwood data set.

```
> library(splancs)
> mserw <- mse2d(as.points(coordinates(spred)), as.points(list(x = c(0,
+     1, 1, 0), y = c(0, 0, 1, 1))), 100, 0.15)
> bw <- mserw$h[which.min(mserw$mse)]
```

Figure 7.6 shows different values of the bandwidth and their associated values of the MSE. The value that minimises it is 0.039, but it should be noted that the curve is very flat around that point, which means that many other values of the bandwidth are plausible. This is a common problem in the analysis of real data sets.

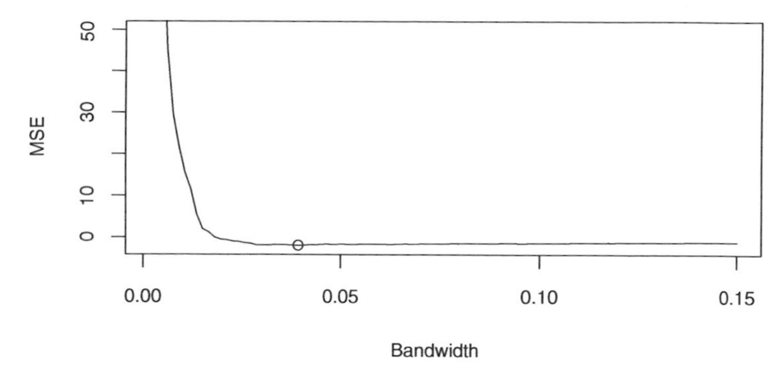

Fig. 7.6. Values of the mean square error for several values of the bandwidth using the redwood data set. The value that minimises it is 0.039 but many other values seem plausible, given the flatness of the curve

Fig. 7.7. Different estimates of the intensity of the redwood data set using a quartic kernel and different values of the bandwidth

It must be noted that when estimating the intensity by kernel smoothing, the key choice is not that of the kernel function but the bandwidth. Different kernels will produce very similar estimates for equivalent bandwidths, but the same kernel with different bandwidths will produce dramatically different results. An example of this fact is shown in Fig. 7.7, where four different bandwidths have been used to estimate the intensity of the redwood data.

```
> poly <- as.points(list(x = c(0, 0, 1, 1), y = c(0, 1,
+     1, 0)))
> sG <- Sobj_SpatialGrid(spred, maxDim = 100)$SG
> grd <- slot(sG, "grid")
> summary(grd)
> k0 <- spkernel2d(spred, poly, h0 = bw, grd)
> k1 <- spkernel2d(spred, poly, h0 = 0.05, grd)
> k2 <- spkernel2d(spred, poly, h0 = 0.1, grd)
```

```
> k3 <- spkernel2d(spred, poly, h0 = 0.15, grd)
> df <- data.frame(k0 = k0, k1 = k1, k2 = k2, k3 = k3)
> kernels <- SpatialGridDataFrame(grd, data = df)
> summary(kernels)
```

Package **spatstat** provides similar functions to estimate the intensity by kernel smoothing using an isotropic Gaussian kernel. We have empirically adjusted the value of the bandwidth to make the kernel estimates comparable. See Härdle et al. (2004, Sect. 3.4.2) for a full discussion. When calling `density` on a `ppp` object (which in fact calls `density.ppp`), we have used the additional arguments `dimxy` and `xy` to make sure that the grid used to compute the estimates is compatible with that stored in `kernels`. Finally, the kernel estimate is returned in an `im` class that is converted into a `SpatialGridDataFrame` and the values incorporated into `kernels`.

```
> xy <- list(x = coordinates(kernels)[, 1], y = coordinates(kernels)[,
+     2])
> k4 <- density(as(spred, "ppp"), 0.5 * bw, dimyx = c(100,
+     100), xy = xy)
> kernels$k4 <- as(k4, "SpatialGridDataFrame")$v
> k5 <- density(as(spred, "ppp"), 0.5 * 0.05, dimyx = c(100,
+     100), xy = xy)
> kernels$k5 <- as(k5, "SpatialGridDataFrame")$v
> k6 <- density(as(spred, "ppp"), 0.5 * 0.1, dimyx = c(100,
+     100), xy = xy)
> kernels$k6 <- as(k6, "SpatialGridDataFrame")$v
> k7 <- density(as(spred, "ppp"), 0.5 * 0.15, dimyx = c(100,
+     100), xy = xy)
> kernels$k7 <- as(k7, "SpatialGridDataFrame")$v
> summary(kernels)
```

7.4.4 Likelihood of an Inhomogeneous Poisson Process

The previous procedure to estimate the intensity is essentially non-parametric. Alternatively, a specific parametric or semi-parametric form for the intensity may be of interest (e.g. to include available covariates). Standard statistical techniques, such as the maximisation of the likelihood, can be used to estimate the parameters that appear in the expression of the intensity.

The expression of the likelihood can be difficult to work out for many point processes. However, in the case of the IPP (and, hence, the HPP) it has a very simple expression. The log-likelihood of a realisation of n independent events of an IPP with intensity $\lambda(x)$ is (Diggle, 2003, p. 104)

$$L(\lambda) = \sum_{i=1}^{n} \log \lambda(x_i) - \int_A \lambda(x)\,\mathrm{d}x,$$

where $\int_A \lambda(x)\,\mathrm{d}x$ is the expected number of cases of the IPP with intensity $\lambda(x)$ in region A.

When the intensity of the point process is estimated parametrically, the likelihood can be maximised to obtain the estimates of the parameters of the model. Diggle (2003, p. 104) suggests a log-linear model

$$\log \lambda(x) = \sum_{j=1}^{p} \beta_j z_j(x)$$

using covariates $z_j(x)$, $j = 1, \ldots, p$ measured at a location x. These models can be fit using standard numerical integration techniques.

The following example defines the log-intensity (`loglambda`) at a given point $x = (x_1, x_2)$ using the parametric specification given by

$$\log \lambda(x) = \alpha + \beta_1 x_1 + \beta_2 x_2 + \beta_3 x_1^2 + \beta_4 x_2^2 + \beta_5 x_1 * x_2. \tag{7.2}$$

This expression is in turn used to construct the likelihood of an IPP (`L`). Function `adapt` (from the package with the same name) is used to compute numerically the integral that appears in the expression of the likelihood.

```
> loglambda <- function(x, alpha, beta) {
+     l <- alpha + sum(beta * c(x, x * x, prod(x)))
+     return(l)
+ }
> L <- function(alphabeta, x) {
+     l <- apply(x, 1, loglambda, alpha = alphabeta[1],
+         beta = alphabeta[-1])
+     l <- sum(l)
+     intL <- adapt(2, c(0, 0), c(1, 1), functn = function(x,
+         alpha, beta) {
+         exp(loglambda(x, alpha, beta))
+     }, alpha = alphabeta[1], beta = alphabeta[-1])
+     l <- l - intL$value
+     return(l)
+ }
```

The following example uses the locations of maple trees from the Lansing Woods data set (Gerard, 1969) in order to show how to fit a parametric intensity using (7.2). The parameters are estimated by maximising the likelihood using function `optim`.

```
> library(adapt)
> data(lansing)
> x <- as.points(lansing[lansing$marks == "maple", ])

> optbeta <- optim(par = c(log(514), 0, 0, 0, 0, 0), fn = L,
+     control = list(maxit = 1000, fnscale = -1), x = x)
```

The values of the coefficients $\alpha, \beta_1, \ldots, \beta_5$ are 5.53, 5.64, –0.774, –5.01, –1.2, 0.645, for a value of the (maximised) likelihood of 2778.4. Figure 7.8 shows the location of the maple trees and the estimated intensity according

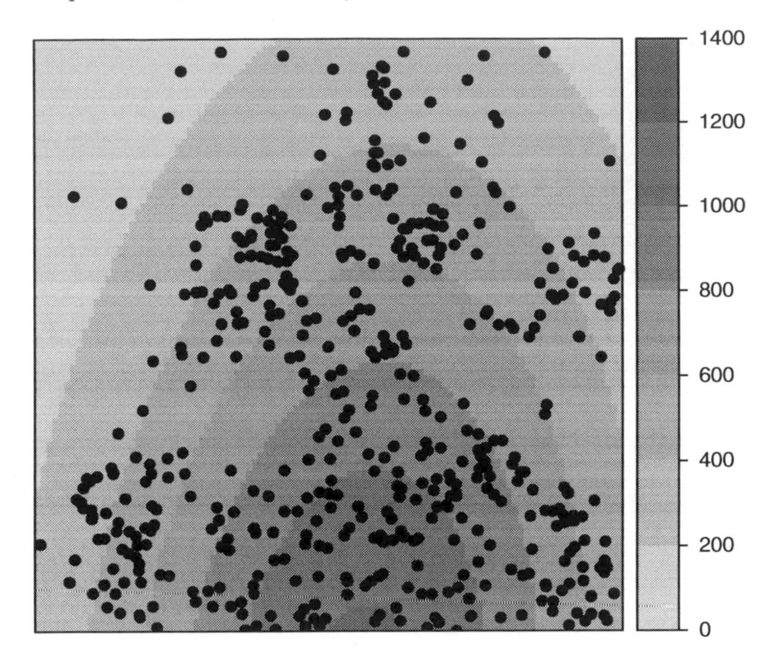

Fig. 7.8. Location of maple trees from the Lansing data set and their estimated parametric intensity using model (7.2)

to parametric model in (7.2). See Diggle (2003, Chap. 7) for a similar analysis using all the tree species in the Lansing Woods data set.

The same example can be run using function `ppm` from **spatstat** as follows (`x` and `y` representing the coordinates of the point pattern):

```
> lmaple <- lansing[lansing$marks == "maple", ]
> ppm(Q = lmaple, trend = ~x + y + I(x^2) + I(y^2) + I(x *
+     y))

Nonstationary multitype Poisson process
Possible marks:
blackoak hickory maple misc redoak whiteoak

Trend formula: ~x + y + I(x^2) + I(y^2) + I(x * y)

Fitted coefficients for trend formula:
(Intercept)           x           y      I(x^2)      I(y^2)
  3.7310742   5.6400643  -0.7663636  -5.0115142  -1.1983209
   I(x * y)
  0.6375824
```

As the authors mention in the manual page, `ppm` can be compared to `glm` because it can be used to fit a specific type of point process model to a particular point pattern. In this case, the `family` argument used in `glm` to define the model is substituted by `interaction`, which defines the point process to be fit. By default, a Poisson point process is used, but many other point processes can be fitted (see manual page for details).

7.4.5 Second-Order Properties

Second-order properties measure the strength and type of the interactions between events of the point process. Hence, they are particularly interesting if we are keen on studying clustering or competition between events.

Informally, the *second-order intensity* of two points x and y reflects the probability of any pair of events occurring in the vicinities of x and y, respectively. Diggle (2003, p. 43) and Möller and Waagepetersen (2003, Chap. 4) give a more formal description of the second-order intensity. Schabenberger and Gotway (2005, pp. 99–103) and Waller and Gotway (2004, pp. 137–147) also discuss second-order properties and the role of the K-function.

An alternative way of measuring second-order properties when the spatial process is HPP is by means of the K-function (Ripley, 1976, 1977). The K-function measures the number of events found up to a given distance of any particular event and it is defined as

$$K(s) = \lambda^{-1} E[N_0(s)],$$

where $E[.]$ denotes the expectation and $N_0(s)$ represents the number of further events up to a distance s around an arbitrary event. To compute this function, Ripley (1976) also proposed an unbiased estimate equal to

$$\hat{K}(s) = (n(n-1))^{-1}|A| \sum_{i=1}^{n} \sum_{j \neq i} w_{ij}^{-1} |\{x_j : d(x_i, x_j) \leq s\}|, \qquad (7.3)$$

where w_{ij} are weights equal to the proportion of the area inside the region A of the circle centred at x_i and radius $d(x_i, x_j)$, the distance between x_i and x_j.

The value of the K-function for an HPP is $K(s) = \pi s^2$. By comparing the estimated value $\hat{K}(s)$ to the theoretical value we can assess what kind of interaction exists. Usually, we assume that these interactions occur at small scales, and so will be interested in relatively small values of s. Values of $\hat{K}(s)$ higher than πs^2 are characteristic of clustered processes, whilst values smaller than that are found when there exists competition between events (regular pattern).

```
> Kenvjap <- envelope(as(spjpines1, "ppp"), fun = Kest,
+     r = r, nrank = 2, nsim = 99)
> Kenvred <- envelope(as(spred, "ppp"), fun = Kest, r = r,
+     nrank = 2, nsim = 99)
> Kenvcells <- envelope(as(spcells, "ppp"), fun = Kest,
+     r = r, nrank = 2, nsim = 99)
> Kresults <- rbind(Kenvjap, Kenvred, Kenvcells)
```

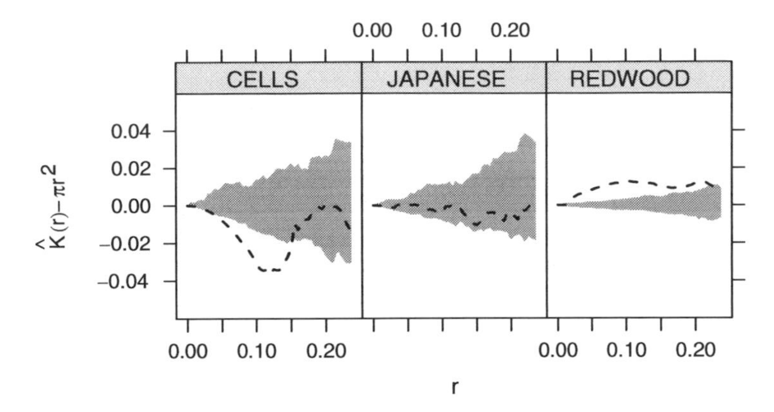

Fig. 7.9. Envelopes and observed values of Ripley's K-function for three point patterns

```
> Kresults <- cbind(Kresults, DATASET = rep(c("JAPANESE",
+     "REDWOOD", "CELLS"), each = length(r)))
```

Figure 7.9 shows the estimated K-function minus the theoretical value under CSR of the three point patterns that we have considered before. Note that the biological interpretations must be made cautiously because the underlying mechanisms are quite different and the scale of the interactions (if any) will probably be different for each point pattern. This is reflected in two ways: the width of the envelopes, which reflects the variability of the process under the null hypothesis of CSR, and the scale of the interaction. This seems to exist only for the cells, which follow a regular pattern, and the redwood seeds, which seem to be clustered. The Japanese trees point pattern is compatible with CSR because the estimated K-function is contained within the envelopes.

Inhomogeneous *K*-Function

Baddeley et al. (2000) propose a version of the K-function for non-homogeneous point processes, in particular, for the class of point processes which are second-order reweighted-stationary, which includes IPPs. This means that the second-order intensity of two points, divided by their respective intensities, is stationary. The inhomogeneous K-function is used in Sect. 7.5.5 in the analysis of case–control point patterns.

7.5 Some Applications in Spatial Epidemiology

In this section we focus on different applications of the analysis of point patterns in Spatial Epidemiology. Gatrell et al. (1996) and Diggle (2003) describe most of the methods contained here, but a comprehensive description of spatial methods for the analysis of epidemiological data can be found in

Elliott et al. (2000) and Waller and Gotway (2004). Furthermore, Chap. 11 describes the analysis of epidemiological data when they are aggregated.

The distribution of the cases of a certain disease can be regarded as the realisation of a point process, which reflects the underlying distribution of the population (which usually is not homogeneous) plus any other risk factors related to the disease and that are likely to depend on the subjects. Hence, we need to have accurate records of the locations of the disease cases, which can also include additional information on the individuals such as age, gender, and others.

In a spatial setting, the primary interest is on the spatial distribution of the cases, but any underlying risk factor that affects this spatial distribution should be taken into account. It is clear that looking solely at the spatial distribution of the cases in order to detect areas of high incidence is useless because the distribution of the cases will reflect that of the population. To overcome this problem, it would be necessary to have an estimate of the spatial distribution of the population so that it can be compared to that of the cases. For this reason, a set of controls can be randomly selected from the population at risk so that its spatial variation can be estimated (see, e.g. Prince et al., 2001).

Different authors have approached this problem in different ways. Diggle and Chetwynd (1991), for example compute the difference of the homogeneous K-function of cases and controls. Kelsall and Diggle (1995a,b) use non-parametric estimates of the distribution of the ratio between the intensities of cases and controls (i.e. the *relative* risk). Kelsall and Diggle (1998) propose a similar model and the use of binary regression and additive models to account for covariates and a smoothing term to model the residual spatial variation. More recently, Diggle et al. (2007) use the inhomogeneous K-function to compare the spatial distribution of cases and controls after accounting for the effect of relevant covariates.

Many of these methods are also covered, including new examples, and discussed in Schabenberger and Gotway (2005, pp. 103–122), Waller and Gotway (2004, Chap. 6) and O'Sullivan and Unwin (2003, see the discussion in Chap. 5).

7.5.1 Case–Control Studies

As we need to estimate the spatial distribution of the population, a number of individuals can be taken at random to make a set of controls. Controls are often selected using the population register or, if it is not available, the events of another non-related disease (Diggle, 1990). Furthermore, some strategies, such as stratification and matching (Jarner et al., 2002), can be done in order to account for other sources of confounding, such as age and sex. As discussed by Diggle (2000) when matching is used in the selection of the controls, the hypothesis of random selection from the population is violated and specific methods to handle this are required (Diggle et al., 2000; Jarner et al., 2002).

In general, we have a set of n_1 cases and n_0 controls. Conditioning on the number of cases and controls, we can assume that they are realisations of two IPP with intensities $\lambda_1(x)$ and $\lambda_0(x)$, respectively. In this setting, assuming that the distribution of cases and controls is the same means that the intensities $\lambda_1(x)$ and $\lambda_0(x)$ are equal up to a proportionality constant, which is equal to the ratio between n_1 and n_0: $\lambda_1(x) = \frac{n_1}{n_0}\lambda_0(x)$. Note that the ratio between cases and controls is determined only by the study design.

Spatial Variation of the Relative Risk

Kelsall and Diggle (1995a,b) consider the estimator of the disease risk given by the ratio between the intensity of the cases and controls $\rho(x) = \lambda_1(x)/\lambda_0(x)$ in order to assess the variation of the risk. Under the null hypothesis of equal spatial distribution, the ratio is a constant $\rho_0 = n_1/n_0$.

Alternatively, a risk estimate $r(x)$ can be estimated by working with the logarithm of the ratio of the densities of cases and controls:

$$r(x) = \log(f(x)/g(x)), \tag{7.4}$$

$f(x) = \lambda_1(x)/\int_A \lambda_1(x)\,\mathrm{d}x$ and $g(x) = \lambda_0(x)/\int_A \lambda_0(x)\,\mathrm{d}x$, respectively. In this case, the null hypothesis of equal spatial distributions becomes $r(x) = 0$. The advantage of this approach is that 0 is the reference value for equal spatial distribution without regarding the number of cases and controls. Unfortunately, this presents several computational problems because the intensity of the controls may be zero at some points, as addressed by, for example, Waller and Gotway (2004, pp. 165–166).

Kelsall and Diggle (1995a) propose the use of a kernel smoothing to estimate each intensity and evaluate different alternatives to estimate the optimum bandwidth for each kernel smoothing. They conclude that the best option is to select the bandwidth by cross-validation and use the same bandwidth in both cases.

They choose the bandwidth that minimises the following criterion:

$$CV(h) = -\int_A \hat{r}_h(x)^2\,\mathrm{d}x - 2n_1^{-1}\sum_{i=1}^{n_1}\hat{r}_h^{-i}(x_i)/\hat{f}_h^{-i}(x_i) + 2n_0^{-1}\sum_{i=n_1+1}^{n_1+n_0}\hat{r}_h^{-i}(x_i)/\hat{g}_h^{-i}(x_i),$$

where the superscript $-i$ means that the function is computed by removing the ith point.

This criterion is not currently implemented, but it can be done easily by using function `lambdahat` in package **spatialkernel**, which allows for the computing of the intensity at a set of point using a different set of points. Our implementation of the method (which does not use border correction to avoid computational instability) gave an optimal bandwidth of 0.275. However, as discussed in Diggle et al. (2007), these automatic methods should be used as a guidance when estimating the bandwidth. In this case, the value of 0.275 seems a bit high given the scale of the data and we have set it to 0.125.

```
> bwasthma <- 0.125
```

To avoid computational problems, we use the risk estimator $\rho(x)$. First of all, we need to create a grid over the study region where the risk ratio will be estimated, using the helper function `Sobj_SpatialGrid`.

```
> library(maptools)
> sG <- Sobj_SpatialGrid(spbdry, maxDim = 50)$SG
> gt <- slot(sG, "grid")
```

The risk ratio can be computed easily by estimating the intensity of cases and controls first, and then taking the ratio (as shown below) after using the `spkernel2d` function from **splancs**.

```
> pbdry <- slot(slot(slot(spbdry, "polygons")[[1]], "Polygons")[[1]],
+     "coords")
```

After unpacking the boundary coordinates of the study area, the point locations are divided between cases and controls and the intensities of each subset calculated for grid cells lying within the study area, using the chosen bandwidth. The **splancs** package uses a simple form of single polygon boundary, while **spatstat** can use multiple separate polygons (`SpatialPolygons` objects can be coerced to suitable `owin` objects).

```
> library(splancs)
> cases <- spasthma[spasthma$Asthma == "case", ]
> ncases <- nrow(cases)
> controls <- spasthma[spasthma$Asthma == "control", ]
> ncontrols <- nrow(controls)
> kcases <- spkernel2d(cases, pbdry, h0 = bwasthma, gt)
> kcontrols <- spkernel2d(controls, pbdry, h0 = bwasthma,
+     gt)
```

The results contain missing values for grid cells outside the study area, and so we first construct a `SpatialGridDataFrame` object to hold them and coerce to a `SpatialPixelsDataFrame` to drop the missing cells. The ratio is calculated, setting non-finite values from division by zero to missing.

```
> df0 <- data.frame(kcases = kcases, kcontrols = kcontrols)
> spkratio0 <- SpatialGridDataFrame(gt, data = df0)
> spkratio <- as(spkratio0, "SpatialPixelsDataFrame")
> spkratio$kratio <- spkratio$kcases/spkratio$kcontrols
> is.na(spkratio$kratio) <- !is.finite(spkratio$kratio)
> spkratio$logratio <- log(spkratio$kratio) - log(ncases/ncontrols)
```

To assess departure from the null hypothesis, they propose the following test statistic:

$$T = \int_A (\rho(x) - \rho_0)^2 \, \mathrm{d}x.$$

This integral can be estimated up to a proportionality constant by computing $\rho(x)$ on a regular grid of points $\{s_i, i = 1, \ldots, p\}$ and computing the sum of the values $\{(\rho(s_i) - \rho_0)^2, i = 1, \ldots, p\}$. Hence, an estimate of T is given by

$$\hat{T} = |c| \sum_{i=1}^{p} (\hat{\rho}(s_i) - \hat{\rho}_0)^2,$$

where $|c|$ is the area of the cells of the grid, $\hat{\rho}_0$ is n_1/n_0, and $\hat{\rho}(x)$ the estimate of the risk ratio.

Note that the former test is to assess whether there is constant risk all over the study region. However, risk is likely to vary spatially and another appropriate test can be done by substituting ρ_0 for $\hat{\rho}(x)$ (Kelsall and Diggle, 1995a). Now we are testing for significance of risk given that we assume that its variation is not homogeneous (i.e. equal to $\hat{\rho}(x)$) and the test statistic is

$$T = \int_A (\rho(x) - \hat{\rho}(x))^2 \, \mathrm{d}x.$$

Significance of the observed value of the test statistic can be computed by means of a Monte Carlo test (Kelsall and Diggle, 1995b). In this test, we compute k values of the test statistic T by re-labelling cases and controls (keeping n_1 and n_0 fixed) and calculating a new risk ratio $\hat{\rho}_i(x)$ $i = 1, \ldots, n$ for each new set of cases and controls. This will provide a series of values $T^1, \ldots, T^k$ under the null hypothesis. If we call T^0 the value of T for the observed data set, the significance (p-value) can be computed by taking $(t+1)/(k+1)$, where t is the number of values of $T^i, i = 1, \ldots, n$ greater than T^0.

The Monte Carlo test is based on the fact that cases and controls are equally distributed under the null hypothesis. In that case, if we change the label of a case to be a control (or viceversa), the new set of cases (or controls) still have the same spatial distribution and will have the same risk function $\rho(x)$. If that is not the case, then the re-labelling of cases and controls will produce different risk functions.

```
> idxinbdry <- overlay(sG, spbdry)
> idxna <- !is.na(idxinbdry)
```

We use the `overlay` method to find grid cells within the study area boundary, and use the number of included grid cells to set up objects to hold the results for the re-labelled cases and controls:

```
> niter <- 99
> ratio <- rep(NA, niter)
> pvaluemap <- rep(0, sum(idxna))
> rlabelratio <- matrix(NA, nrow = niter, ncol = sum(idxna))
```

The probability map is calculated by repeating the re-labelling process `niter` times, and tallying the number of times that the observed kernel density ratio is less than the re-labelled ratios. In the loop, the first commands carry out the re-labelling from the full set of points, and the remainder calculate the ratio and store the results:

```
> for (i in 1:niter) {
+     idxrel <- sample(spasthma$Asthma) == "case"
+     casesrel <- spasthma[idxrel, ]
+     controlsrel <- spasthma[!idxrel, ]
+     kcasesrel <- spkernel2d(casesrel, pbdry, h0 = bwasthma,
+         gt)
+     kcontrolsrel <- spkernel2d(controlsrel, pbdry, h0 = bwasthma,
+         gt)
+     kratiorel <- kcasesrel[idxna]/kcontrolsrel[idxna]
+     is.na(kratiorel) <- !is.finite(kratiorel)
+     rlabelratio[i, ] <- kratiorel
+     pvaluemap <- pvaluemap + (spkratio$kratio < kratiorel)
+ }
```

Figure 7.10 shows the kernel ratio of cases and controls, using a bandwidth of 0.125, as discussed before. We may have computational problems when estimating the intensity at points very close to the boundary of the study area and obtain `NA` instead of the value of the intensity. To avoid problems with this, we have filtered out these points using a new index called `idxna2`.

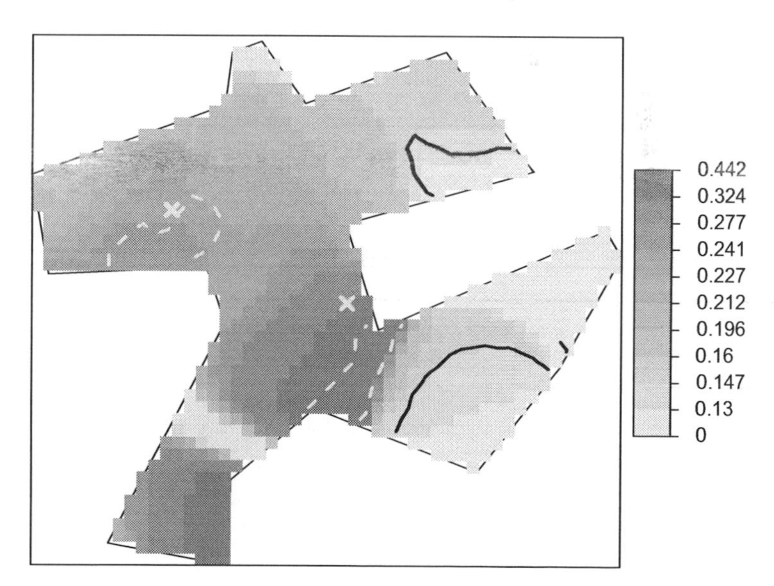

Fig. 7.10. Kernel ratio of the intensity of cases and controls. The continuous and dashed lines show the surfaces associated with 0.95 and 0.05 p-values, respectively, grey crosses mark the pollution sources. The value of $\hat{\rho}_0$ which marks a flat constant risk is 0.2

This will ensure that we use the same number of points when we estimate the value of the test statistic for the observed data and the permuted re-labelled sets of cases and controls.

```
> idxna2 <- apply(rlabelratio, 2, function(x) all(is.finite(x)))
> rhomean <- apply(rlabelratio[, idxna2], 2, mean)
> c <- prod(slot(gt, "cellsize"))
> ratiorho <- c * sum((spkratio$kratio[idxna2] - ncases/ncontrols)^2)
> ratio <- c * apply(rlabelratio[, idxna2], 1, function(X,
+     rho0) {
+     sum((X - rho0)^2)
+ }, rho0 = ncases/ncontrols)
> pvaluerho <- (sum(ratio > ratiorho) + 1)/(niter + 1)
```

The results for the test with null hypothesis $\rho = \hat{\rho}_0$ turned out to be non-significant (p-value of 0.69), which means that the observed risk ratio is consistent with a constant risk ratio. In principle, this agrees with the fact that Diggle and Rowlingson (1994) did not find a significant association with distance from main roads or two of the pollution sources and only a possible association with the remaining site, which should be further investigated. However, they found some relationship with other risk factors, but these were not of a spatial nature and, hence, this particular test is unable to detect it.

Had the p-value of the test been significant, 90% point confidence surfaces could be computed in a similar way to the envelopes shown before, but considering the different values of the estimates of $\rho(x)$ under random labelling and computing the p-value at each point. The procedure computes, for each point x_j in the grid, the proportion of values $\hat{\rho}_i(x_j)$ that are lower than $\hat{\rho}(x_j)$, where the $\hat{\rho}_i(x_j), i = 1, \ldots, R$ are the estimated ratios obtained by re-labelling cases and controls. Finally, the 0.05 and 0.95 contours of the p-value surface can be displayed on the plot of $\hat{\rho}(x)$ to highlight areas of significant low and high risk, respectively. This is shown in Fig. 7.10.

The contour lines at a given value can be obtained using function `contourLines`, which takes an `image` object. This will generate contour lines that can be converted to `SpatialLinesDataFrame` objects so that they can be added to a plot as a layout.

```
> spkratio$pvaluemap <- (pvaluemap + 1)/(niter + 1)
> imgpvalue <- as.image.SpatialGridDataFrame(spkratio["pvaluemap"])
> clpvalue <- contourLines(imgpvalue, levels = c(0, 0.05,
+     0.95, 1))
> cl <- ContourLines2SLDF(clpvalue)
```

7.5.2 Binary Regression Estimator

Kelsall and Diggle (1998) propose a binary regression estimator to estimate the probability of being a case at a given location, which can be easily extended to allow for the incorporation of covariates. In principle, the probabilities can be estimated by assuming that we have a variable Y_i, which labels cases ($y_i = 1$) and controls ($y_i = 0$) in a set of $n = n_1 + n_2$ events. Conditioning on the point locations, Y_i is a realisation of a Bernoulli variable Y_i with probability

$$P(Y_i = 1 | X_i = x_i) = p(x_i) = \frac{\lambda_1(x_i)}{\lambda_0(x_i) + \lambda_1(x_i)}.$$

In practise, the following Nadaraya–Watson kernel estimator can be used:

$$\hat{p}_h(x) = \frac{\sum_{i=1}^n h^{-2}\kappa_h((x - x_i)/h) y_i}{\sum_{i=1}^n h^{-2}\kappa_h((x - x_i)/h)}, \tag{7.5}$$

where $\kappa_h(u)$ is a kernel function. Note that $p(x)$ is related to the log-ratio relative risk $r(x)$ as follows:

$$\text{logit}(p(x)) = \log\left(\frac{p(x)}{1 - p(x)}\right) = \log\left(\frac{\lambda_1(x)}{\lambda_0(x)}\right) = r(x) + \log(n_1/n_0).$$

$\hat{p}_h(x)$ can be estimated as

$$\hat{p}_h(x) = \frac{\hat{\lambda}_1(x)}{\hat{\lambda}_1(x) + \hat{\lambda}_0(x)}.$$

To estimate the bandwidth that appears in this new estimator, Kelsall and Diggle (1998) suggest another cross-validation criterion based on the value of h that minimises

$$CV(h) = \left[\prod_{i=1}^n \hat{p}_h^{-i}(x_i)^{y_i}(1 - \hat{p}_h^{-i}(x_i))^{1-y_i}\right]^{-1/n}.$$

Using the new criterion we obtained a bandwidth of 0.225, which is very similar to the one obtained with the previous cross-validation criterion. However, we believe that this value would over-smooth the data and we have set it to 0.125. The estimator for $p(x)$ can be computed easily, as is shown below. Figure 7.11 shows the resulting estimate.

```
> bwasthmap <- 0.125

> lambda1 <- spkernel2d(cases, pbdry, h0 = bwasthmap, gt)
> lambda0 <- spkernel2d(controls, pbdry, h0 = bwasthmap,
+     gt)
> lambda1 <- lambda1[idxna]
> lambda0 <- lambda0[idxna]
> spkratio$prob <- lambda1/(lambda1 + lambda0)
> is.na(spkratio$prob) <- !is.finite(spkratio$prob)
```

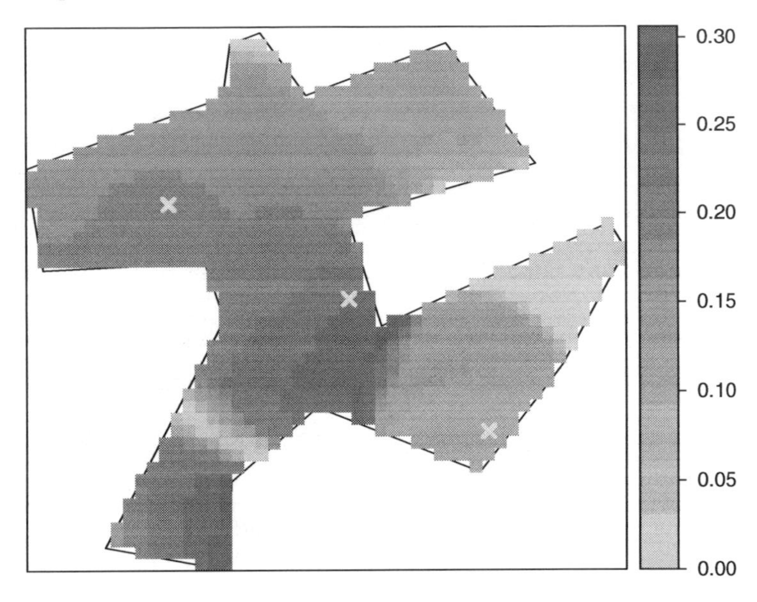

Fig. 7.11. Binary regression estimator using the probability of being a case at every grid cell in the study region

7.5.3 Binary Regression Using Generalised Additive Models

This formulation allows the inclusion of covariates in the model by means of standard logistic regression. In addition, the residual spatial variation can be modelled by including a smooth spatial function. In other words, if u is a vector of covariates observed at location x and $g(x)$ is a smooth function not dependent on the covariates, the formulation is

$$\text{logit}(p(x)) = u'\beta + g(x).$$

If the covariates are missing, the former expression is just another way of estimating the probability surface. Kelsall and Diggle (1998) estimate $g(x)$ using a kernel weighted regression. We have used package **mgcv** (Wood, 2006) to fit the Generalised Additive Model (GAM) models but, given that this package lacks the same non-parametric estimator used in Kelsall and Diggle (1998), we have preferred the use of a penalised spline instead.

The following example shows how to fit a GAM using the distance of the events to the pollution sources and main roads, and controlling for known and possible risk factors such as gender, age, previous events of hay fever, and having at least one smoker in the house. Rows have been filtered so that only children with a valid value of Gender (1 or 2) are used. We have included the distance as a proxy of the actual exposure to any risk factor caused by the pollution sources or the roads. Other models that consider a special modelling for the distance are considered later.

```
> spasthma$y <- as.integer(!as.integer(spasthma$Asthma) -
+     1)
> ccasthma <- coordinates(spasthma)
> spasthma$x1 <- ccasthma[, 1]
> spasthma$x2 <- ccasthma[, 2]
> spasthma$dist1 <- sqrt(spasthma$d2source1)
> spasthma$dist2 <- sqrt(spasthma$d2source2)
> spasthma$dist3 <- sqrt(spasthma$d2source3)
> spasthma$droads <- sqrt(spasthma$roaddist2)
> spasthma$smoking <- as.factor(as.numeric(spasthma$Nsmokers >
+     0))
> spasthma$Genderf <- as.factor(spasthma$Gender)
> spasthma$HayFeverf <- as.factor(spasthma$HayFever)
```

```
> library(mgcv)
> gasthma <- gam(y ~ 1 + dist1 + dist2 + dist3 + droads +
+     Genderf + Age + HayFeverf + smoking + s(x1, x2),
+     data = spasthma[spasthma$Gender == 1 | spasthma$Gender ==
+         2, ], family = binomial)
```

```
> summary(gasthma)
```

```
Family: binomial
Link function: logit

Formula:
y ~ 1 + dist1 + dist2 + dist3 + droads + Genderf + Age + HayFeverf +
    smoking + s(x1, x2)

Parametric coefficients:
              Estimate Std. Error z value Pr(>|z|)
(Intercept) -2.0326784  0.9195177  -2.211   0.0271 *
dist1        0.9822575  6.0714999   0.162   0.8715
dist2       -9.5790621  5.7708614  -1.660   0.0969 .
dist3       11.2247321  7.8724979   1.426   0.1539
droads       0.0001479  0.0001717   0.861   0.3890
Genderf2    -0.3476861  0.1562020  -2.226   0.0260 *
Age         -0.0679031  0.0382349  -1.776   0.0757 .
HayFeverf1   1.1881331  0.1875414   6.335 2.37e-10 ***
smoking1     0.1651210  0.1610362   1.025   0.3052
---
Signif. codes:  0 '***' 0.001 '**' 0.01 '*' 0.05 '.' 0.1 ' ' 1

Approximate significance of smooth terms:
            edf Est.rank Chi.sq p-value
s(x1,x2) 2.001        2  7.004  0.0301 *
---
Signif. codes:  0 '***' 0.001 '**' 0.01 '*' 0.05 '.' 0.1 ' ' 1

R-sq.(adj) =  0.0403   Deviance explained = 4.94%
UBRE score = -0.12348  Scale est. = 1         n = 1283
```

The results show that the significant variables are the presence of reported hay fever (p-value 2.4e-10) and gender (p-value 0.026). The coefficient of the second pollution source is marginally significant (p-value 0.097). The smoothed residual term using splines is significant (p-value 0.0301), which suggests that there may have been some residual spatial variation unexplained in the generalised linear model.

7.5.4 Point Source Pollution

In the previous model, we have shown how to consider the exposure to a number of pollution sources by including the distance as a covariate in the model. However, this approach does not allow for a more flexible parametric modelling of the exposure according to the distance to a pollution source. Diggle (1990) proposed the use of an IPP for the cases in which their intensity accounts for the distance to the pollution sources. In particular, the intensity is as follows:

$$\lambda_1(x) = \rho\lambda_0(x) f(x - x_0; \theta),$$

ρ measures the overall number of events per unit area, $\lambda_0(x)$ is the spatial variation of the underlying population (independent of the effect of the source), and $f(x - x_0; \theta)$ is a function of the distance from point x to the location of the source x_0 and has parameters θ. Diggle (1990) uses a decaying function with distance

$$f(x - x_0; \alpha, \beta) = 1 + \alpha \exp(-\beta||x - x_0||^2).$$

Parameters ρ, α, and β of $\lambda_1(x)$ can be estimated by maximising the likelihood of the IPP, assuming that $\lambda_0(x)$ is estimated by kernel smoothing taking a certain value h_0 of the bandwidth. That is, the value of h_0 is not obtained by the maximisation procedure, but choosing a reasonable value for h_0 can be difficult and it can have an important impact on the results.

A slightly different approach that does not require the choice of a bandwidth is considered in Diggle and Rowlingson (1994). It is based on the previous scenario, but conditioning on the location of cases and controls to model the probability of being a case at location x:

$$p(x) = \frac{\lambda_1(x)}{\lambda_1(x) + \lambda_0(x)} = \frac{\rho f(x - x_0; \alpha, \beta)}{1 + \rho f(x - x_0; \alpha, \beta)}.$$

As in the previous scenario, the remaining parameters of the model can be estimated by maximising the log-likelihood:

$$L(\rho, \theta) = \sum_{i=1}^{n_1} \log(p(x_i)) + \sum_{j=1}^{n_0} \log(1 - p(x_j)).$$

This model can be fitted using function `tribble` from package `splancs`. Given that $\lambda_0(x)$ vanishes we only need to pass the distances to the source and the labels of cases and controls.

To compare models that may include different sets of pollution sources or covariates, Diggle and Rowlingson (1994) compare the difference of the log-likelihoods by means of a chi-square test. The following example shows the results for the exposure model with distance to source two and another model with only the covariate hay fever.

```
> D2_mat <- as.matrix(spasthma$dist2)
> RHO <- ncases/ncontrols
> expsource2 <- tribble(ccflag = spasthma$y, vars = D2_mat,
+     rho = RHO, alphas = 1, betas = 1)

> print(expsource2)

Call:
tribble(ccflag = spasthma$y, vars = D2_mat, alphas = 1, betas = 1,
    rho = RHO)
Kcode = 2

Distance decay parameters:
        Alpha     Beta
[1,] 1.305824 25.14672

rho parameter : 0.163395847637003

     log-likelihood : -580.495955916672
null log-likelihood : -581.406203518987

        D = 2(L-Lo) : 1.82049520462942

> Hay_mat <- as.matrix(spasthma$HayFever)
> exphay <- tribble(ccflag = spasthma$y, rho = RHO, covars = Hay_mat,
+     thetas = 1)

> print(exphay)

Call:
tribble(ccflag = spasthma$y, rho = RHO, covars = Hay_mat, thetas = 1)
Kcode = 2

Covariate parameters:
[1] 1.103344

rho parameter : 0.163182953009353

     log-likelihood : -564.368250327802
null log-likelihood : -581.406203518987

        D = 2(L-Lo) : 34.0759063823702
```

As the output shows, the log-likelihood for the model with exposure to source 2 is −580.5, whilst for the model with the effect of hay fever is only −564.4. This means that there is a significant difference between the two models and that the model that accounts for the effect of hay fever is preferable. Even though the second source has a significant impact on the increase of the cases of asthma, its effect is not as important as the effect of having suffered from hay fever. However, another model could be proposed to account for both effects at the same time.

```
> expsource2hay <- tribble(ccflag = spasthma$y, vars = D2_mat,
+     rho = RHO, alphas = 1, betas = 1, covars = Hay_mat,
+     thetas = 1)
```

This new model (output not shown) has a log-likelihood of −563, with two more parameters than the model with hay fever. Hence, the presence of the second source has a small impact on the increase of cases of asthma after adjusting for the effect of hay fever, which can be regarded as the main factor related to asthma, and the model with hay fever only should be preferred. The reader is referred to Diggle and Rowlingson (1994) and Diggle (2003, p. 137) for more details on how the models can be compared and results for other models.

These types of models are extended by Diggle et al. (1997), who consider further options for the choice of the function $f(x - x_0, \alpha, \beta)$ to accommodate different spatial variants of the risk around the source.

In our experience, these models can be very sensitive to the initial values for certain data sets, especially if they are sparse. Hence, it is advised to fit the model using different values for the initial values to ensure that the algorithm is not trapped in a local maximum of the likelihood.

Assessment of General Spatial Clustering

As discussed by Diggle (2000), it is important to distinguish between spatial variation of the risk and clustering. Spatial variation occurs when the risk is not homogeneous in the study region (i.e. all individuals do not have the same risk) but cases appear independently of each other according to this risk surface, whilst clustering occurs when the occurrence of cases is not at random and the presence of a case increases the probability of other cases appearing nearby.

The former methods allow us to inspect a raised incidence in the number of cases around certain pre-specified sources. However, no such source is identified a priori, and a different type of test is required to assess clustering in the cases.

Diggle and Chetwynd (1991) propose a test based on the homogeneous K-function to assess clustering of the cases as compared to the controls. The null hypothesis is as before, that is cases and controls are two IPP that have the same intensities up to a proportionality constant. Hence, they will produce the same K-functions. Note that the inverse is not always true, that is two point processes with the same homogeneous K-function can be completely different (Baddeley and Silverman, 1984). Diggle and Chetwynd (1991) take the difference of the two K-functions to evaluate whether the cases tend to cluster after considering the inhomogeneous distribution of the population: $D(s) = K_1(s) - K_0(s)$, where $K_1(s)$ and $K_0(s)$ are the homogeneous K-functions of cases and controls, respectively.

The test statistic is

$$D = \int_A \frac{D(s)}{\mathrm{var}[D(s)]^{1/2}} \, \mathrm{d}s,$$

where $\mathrm{var}[D(s)]$ is the variance of $D(s)$ under the null hypothesis. Diggle and Chetwynd (1991) compute the value of this variance under random labelling of cases and controls so that the significance of the test statistic can be assessed. Note that under the null hypothesis the expected value of the test statistic D is zero. Finally, the integral is approximated in practice by a discrete sum at a set of finite distances, as the T statistic was computed before.

Significant departure from 0 means that there is a difference in the distribution of cases and controls, with clustering occurring at the range of those distances for which $D(s) > 0$. Furthermore, pointwise envelopes can be provided for the test statistic by the same Monte Carlo test so that the degree of clustering can be assessed. Function `Kenv.label` also provides envelopes for the difference of the K-functions but it does not carry out any test of significance.

```
> s <- seq(0, 0.15, by = 0.01)

> khcases <- khat(coordinates(cases), pbdry, s)
> khcontrols <- khat(coordinates(controls), pbdry, s)
> khcov <- khvmat(coordinates(cases), coordinates(controls),
+     pbdry, s)
> T0 <- sum(((khcases - khcontrols))/sqrt(diag(khcov)))

> niter <- 99
> T <- rep(NA, niter)

> khcasesrel <- matrix(NA, nrow = length(s), ncol = niter)
> khcontrolsrel <- matrix(NA, nrow = length(s), ncol = niter)
> for (i in 1:niter) {
+     idxrel <- sample(spasthma$Asthma) == "case"
+     casesrel <- coordinates(spasthma[idxrel, ])
+     controlsrel <- coordinates(spasthma[!idxrel, ])
+     khcasesrel[, i] <- khat(casesrel, pbdry, s)
+     khcontrolsrel[, i] <- khat(controlsrel, pbdry, s)
+     khdiff <- khcasesrel[, i] - khcontrolsrel[, i]
+     T[i] <- sum(khdiff/sqrt(diag(khcov)))
+ }

> pvalue <- (sum(T > T0) + 1)/(niter + 1)
```

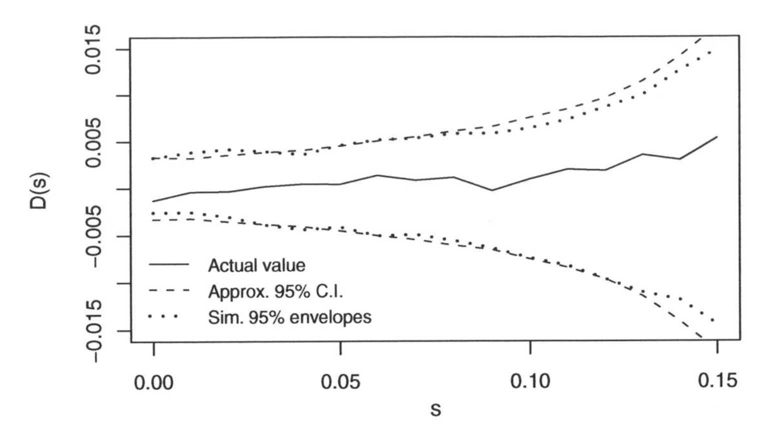

Fig. 7.12. Actual value of $D(s)$ with approximate 95% confidence intervals and 95% envelopes

The p-value for this data set is 0.36, meaning that there is no significant difference between the distribution of cases and controls.The outcome is consistent with the fact that the observed K-function is contained by the simulation envelopes and approximated 95% confidence intervals, as shown in Fig. 7.12.

7.5.5 Accounting for Confounding and Covariates

Diggle et al. (2007) propose a similar way of assessing clustering by means of the inhomogeneous K-function $K_{I,\lambda}(s)$ (Baddeley et al., 2000). For an IPP with intensity $\lambda(x)$, it can be estimated as

$$\hat{K}_{I,\lambda}(s) = |A|^{-1} \sum_{i=1}^{n} \sum_{j \neq i} w_{ij}^{-1} \frac{|\{x_j : d(x_i, x_j) \leq s\}|}{\lambda(x_i)\lambda(x_j)}.$$

Note that this estimator is a generalisation of the estimator of the homogeneous K-function from expression (7.3) and that in fact reduces to it when instead of an IPP we have an HPP (the intensity becomes $\lambda(x) = \lambda$). Similarly, the value of $K_{I,\lambda}(s)$ for an IPP with intensity $\lambda(s)$ is πs^2.

In practise the intensity $\lambda(x)$ needs to be estimated either parametrically or non-parametrically, so that the estimator that we use is

$$\hat{K}_{I,\hat{\lambda}}(s) = |A|^{-1} \sum_{i=1}^{n} \sum_{j \neq i} w_{ij}^{-1} \frac{|\{x_j : d(x_i, x_j) \leq s\}|}{\hat{\lambda}(x_i)\hat{\lambda}(x_j)}.$$

Values of $\hat{K}_{I,\hat{\lambda}}(s)$ higher than πs^2 will mean that the point pattern shows more aggregation than that shown by $\lambda(x)$ and values lower than πs^2 reflect more relative homogeneity.

To be able to account for confounding and risk factors, Diggle et al. (2007) propose the use of a semi-parametric estimator of the intensity in a case–control setting. The basic assumption is that controls are drawn from an IPP with spatially varying intensity $\lambda_0(x)$. The cases are assumed to appear as a result of the inhomogeneous distribution of the population, measured by $\lambda_0(x)$, plus other risk factors, measured by a set of spatially referenced covariates $z(x)$. Hence, the intensity of the cases is modelled as

$$\lambda_1(x) = \exp\{\alpha + \beta z(x)\}\lambda_0(x),$$

where α and β are the intercept and covariate coefficients of the model, respectively. When there are no covariates, the intensity of the cases reduces to

$$\lambda_1(x) = \frac{n_1}{n_0}\lambda_0(x).$$

Note that it is possible to use any generic non-negative function $f(z(x);\theta)$ to account for other types of effects

$$\lambda_1(x) = \lambda_0(x)f(z(x);\theta).$$

This way it is possible to model non-linear and additive effects.

To estimate the parameters that appear in the intensity of the cases, we can use the same working variables Y_i that we have used before (see the binary regression estimator in Sect. 7.5.2), with values 1 for cases and 0 for controls. Conditioning on the locations of cases and controls, Y_i is a realisation of a Bernoulli process with probability

$$P(Y_i = 1|x_i, z(x)) = p(x_i) = \frac{\lambda_1(x)}{\lambda_0(x) + \lambda_1(x)} = \frac{\exp\{\alpha + \beta z(x)\}}{1 + \exp\{\alpha + \beta z(x)\}}. \quad (7.6)$$

Hence, conditioning on the locations of cases and controls, the problem is reformulated as a logistic regression and α and β can be estimated using function `glm`.

Baddeley et al. (2000) estimate the intensity non-parametrically and use the same data to estimate both the intensity and the inhomogeneous K-function, but Diggle et al. (2007) show that this can give poor performance in detecting clustering. This problem arises from the difficulty of disentangling inhomogeneous spatial variation of process from clustering of the events (Cox, 1955). Another problem that appears in practise is that the intensities involved must be bounded away from zero. If kernel smoothing is used, a good alternative to the quartic kernel is a Gaussian bivariate kernel.

The following piece of code shows how to estimate the inhomogeneous K-function both without covariates and accounting for hay fever.

```
> glmasthma <- glm(y ~ HayFeverf, data = spasthma, family = "binomial")
> prob <- fitted(glmasthma)
> weights <- exp(glmasthma$linear.predictors)
> library(spatialkernel)
> setkernel("gaussian")
> lambda0 <- lambdahat(coordinates(controls), bwasthma,
+     coordinates(cases), pbdry, FALSE)$lambda
> lambda1 <- weights[spasthma$Asthma == "case"] * lambda0
> ratiocc <- ncases/ncontrols
> kihnocov <- kinhat(coordinates(cases), ratiocc * lambda0,
+     pbdry, s)$k
> kih <- kinhat(coordinates(cases), lambda1, pbdry, s)$k
```

To assess for any residual clustering left after adjusting for covariates, Diggle et al. (2007) suggest the following test statistic:

$$D = \int_0^{s_0} \frac{\hat{K}_{I,\hat{\lambda}_1}(s) - E[s]}{\mathrm{var}(K_{I,\lambda}(s))^{1/2}} \, \mathrm{d}s,$$

$E[s]$ is the expectation of $\hat{K}_{I,\hat{\lambda}_1}(s)$ under the null hypothesis. In principle, it should be πs^2, but when kernel estimators are used in the computation of the intensity, the estimate of $K_{I,\lambda}(s)$ may be biased. $E[s]$ can be computed as the average of all the estimates $\hat{K}_{I,\hat{\lambda}_1}(s)$, which have been obtained during the Monte Carlo simulations (as explained below). $\mathrm{var}(K_{I,\lambda}(s))$ can be computed in a similar way.

The Monte Carlo test proposed by Diggle et al. (2007) is similar to the one that we used in the homogeneous case (see Sect. 7.5.4), with the difference that the re-labelling must be done taking into account the effects of the covariates. That is, when we relabel cases and controls, the probability of being a case will not be the same for all points but it will depend on the values of $z(x)$. In particular, these probabilities are given by (7.6). The values of the covariates are fixed to the values obtained by fitting the model with the observed data set (i.e. they are not re-estimated when the points are re-labelled) because we are only interested in testing for the spatial variation and not that related to the estimation of the coefficients of the covariates.

```
> niter <- 99
> kinhomrelnocov <- matrix(NA, nrow = length(s), ncol = niter)
> kinhomrel <- matrix(NA, nrow = length(s), ncol = niter)
> for (i in 1:niter) {
+     idxrel <- sample(spasthma$Asthma, prob = prob) ==
+         "case"
+     casesrel <- coordinates(spasthma[idxrel, ])
+     controlsrel <- coordinates(spasthma[!idxrel, ])
+     lambda0rel <- lambdahat(controlsrel, bwasthma, casesrel,
+         pbdry, FALSE)$lambda
+     lambda1rel <- weights[idxrel] * lambda0rel
+     kinhomrelnocov[, i] <- kinhat(casesrel, ratiocc *
```

```
+         lambda0rel, pbdry, s)$k
+     kinhomrel[, i] <- kinhat(casesrel, lambda1rel, pbdry,
+         s)$k
+ }

> kinhsdnocov <- apply(kinhomrelnocov, 1, sd)
> kihmeannocov <- apply(kinhomrelnocov, 1, mean)
> D0nocov <- sum((kihnocov - kihmeannocov)/kinhsdnocov)
> Dnocov <- apply(kinhomrelnocov, 2, function(X) {
+     sum((X - kihmeannocov)/kinhsdnocov)
+ })
> pvaluenocov <- (sum(Dnocov > D0nocov) + 1)/(niter + 1)

> kinhsd <- apply(kinhomrel, 1, sd)
> kihmean <- apply(kinhomrel, 1, mean)
> D0 <- sum((kih - kihmean)/kinhsd)
> D <- apply(kinhomrel, 2, function(X) {
+     sum((X - kihmean)/kinhsd)
+ })
> pvalue <- (sum(D > D0) + 1)/(niter + 1)
```

Figure 7.13 shows the estimated values of the inhomogeneous K-function plus 95% envelopes under the null hypothesis. In both cases there are no signs of spatial clustering. The p-values are 0.14 (no covariates) and 0.18 (with hay fever). The increase in the p-value when hay fever is used to modulate the intensity shows how it accounts for some spatial clustering. This is consistent with the plots in Fig. 7.13.

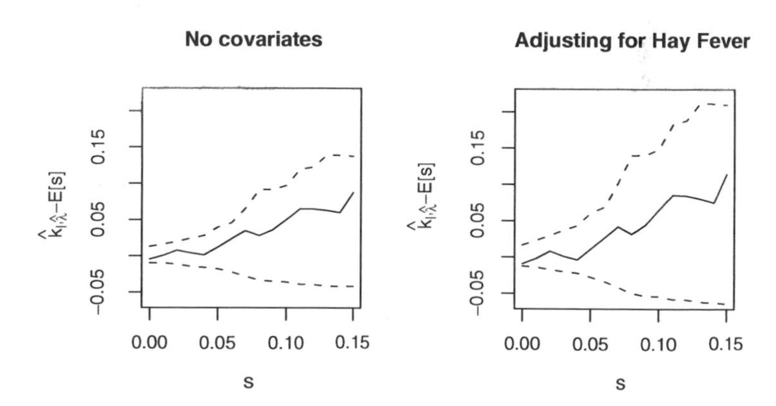

Fig. 7.13. Results of the test based on the inhomogeneous K-function for the asthma data set. The intensity has been modulated to account for the effect of suffering from hay fever

7.6 Further Methods for the Analysis of Point Patterns

In this chapter we have just covered some key examples but the analysis of point patterns with R goes beyond this. Other important problems that we have not discussed here are the analysis of marked point processes (Schabenberger and Gotway 2005, pp. 118–122; Diggle 2003, pp. 82–85), spatio-temporal analysis (see Schabenberger and Gotway 2005, pp. 442–445; Diggle 2006), and complex model fitting and simulation from different point processes (as extensively discussed in Möller and Waagepetersen, 2003). Baddeley et al. (2005) provide a recent compendium of theoretical problems and applications of the analysis of point patterns, including a description of package **spatstat**. Some of the examples described therein should be reproducible using the contents of this chapter.

The Spatial Task View contains a list of other packages for the analysis and visualisation of point patterns. The reader is referred there for updated information.

8

Interpolation and Geostatistics

8.1 Introduction

Geostatistical data are data that could in principle be measured anywhere, but that typically come as measurements at a limited number of observation locations: think of gold grades in an ore body or particulate matter in air samples. The pattern of observation *locations* is usually not of primary interest, as it often results from considerations ranging from economical and physical constraints to being 'representative' or random sampling varieties. The interest is usually in inference of aspects of the variable that have not been measured such as maps of the estimated values, exceedence probabilities or estimates of aggregates over given regions, or inference of the process that generated the data. Other problems include monitoring network optimisation: where should new observations be located or which observation locations should be removed such that the operational value of the monitoring network is maximised.

Typical spatial problems where geostatistics are used are the following:

- The estimation of ore grades over mineable units, based on drill hole data
- Interpolation of environmental variables from sample or monitoring network data (e.g. air quality, soil pollution, ground water head, hydraulic conductivity)
- Interpolation of physical or chemical variables from sample data
- Estimation of spatial averages from continuous, spatially correlated data

In this chapter we use the Meuse data set used by Burrough and McDonnell (1998). The notation we use follows mostly that of Christensen (1991), as this text most closely links geostatistics to linear model theory. Good texts on geostatistics are Chilès and Delfiner (1999), Christensen (1991), Cressie (1993), and Journel and Huijbregts (1978). More applied texts are, for example Isaaks and Strivastava (1989), Goovaerts (1997), and Deutsch and Journel (1992).

Geostatistics deals with the analysis of random fields $Z(s)$, with Z random and s the non-random spatial index. Typically, at a limited number of sometimes arbitrarily chosen sample locations, measurements on Z are available, and prediction (interpolation) of Z is required at non-observed locations s_0, or the mean of Z is required over a specific region B_0. Geostatistical analysis involves estimation and modelling of spatial correlation (covariance or semivariance), and evaluating whether simplifying assumptions such as stationarity can be justified or need refinement. More advanced topics include the conditional simulation of $Z(s)$, for example over locations on a grid, and model-based inference, which propagates uncertainty of correlation parameters through spatial predictions or simulations.

Much of this chapter will deal with package **gstat**, because it offers the widest functionality in the geostatistics curriculum for R: it covers variogram cloud diagnostics, variogram modelling, everything from global simple kriging to local universal cokriging, multivariate geostatistics, block kriging, indicator and Gaussian conditional simulation, and many combinations. Other R packages that provide additional geostatistical functionality are mentioned where relevant, and discussed at the end of this chapter.

8.2 Exploratory Data Analysis

Spatial exploratory data analysis starts with the plotting of maps with a measured variable. To express the observed value, we can use colour or symbol size,

```
> library(lattice)
> library(sp)
> data(meuse)
> coordinates(meuse) <- c("x", "y")
> spplot(meuse, "zinc", do.log = T)
> bubble(meuse, "zinc", do.log = T, key.space = "bottom")
```

to produce plots with information similar to that of Fig. 3.8.

The evident structure here is that zinc concentration is larger close to the river Meuse banks. In case of an evident spatial trend, such as the relation between top soil zinc concentration and distance to the river here, we can also plot maps with fitted values and with residuals (Cleveland, 1993), as shown in Fig. 8.1, obtained by

```
> xyplot(log(zinc) ~ sqrt(dist), as.data.frame(meuse))
> zn.lm <- lm(log(zinc) ~ sqrt(dist), meuse)
> meuse$fitted.s <- predict(zn.lm, meuse) - mean(predict(zn.lm,
+     meuse))
> meuse$residuals <- residuals(zn.lm)
> spplot(meuse, c("fitted.s", "residuals"))
```

where the formula `y ~ x` indicates dependency of `y` on `x`. This figure reveals that although the trend removes a large part of the variability, the residuals do not appear to behave as spatially unstructured or white noise: residuals with

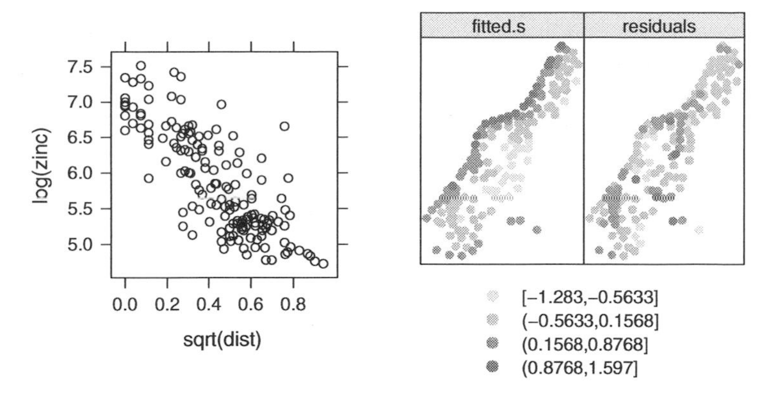

Fig. 8.1. Zinc as a function of distance to river (*left*), and fitted-residual maps (fitted.s: mean subtracted) for the linear regression model of log zinc and square-root transformed distance to the river

a similar value occur regularly close to another. More exploratory analysis will take place when we further analyse these data in the context of geostatistical models; first we deal with simple, non-geostatistical interpolation approaches.

8.3 Non-Geostatistical Interpolation Methods

Usually, interpolation is done on a regular grid. For the Meuse data set, coordinates of points on a regular grid are already defined in the `meuse.grid` data.frame, and are converted into a `SpatialPixelsDataFrame` by

```
> data(meuse.grid)
> coordinates(meuse.grid) <- c("x", "y")
> meuse.grid <- as(meuse.grid, "SpatialPixelsDataFrame")
```

Alternatively, we could interpolate to individual points, sets of irregularly distributed points, or to averages over square or irregular areas (Sect. 8.5.6).

8.3.1 Inverse Distance Weighted Interpolation

Inverse distance-based weighted interpolation (IDW) computes a weighted average,

$$\hat{Z}(s_0) = \frac{\sum_{i=1}^{n} w(s_i) Z(s_i)}{\sum_{i=1}^{n} w(s_i)},$$

where weights for observations are computed according to their distance to the interpolation location,

$$w(s_i) = ||s_i - s_0||^{-p},$$

with $||\cdot||$ indicating Euclidean distance and p an inverse distance weighting power, defaulting to 2. If s_0 coincides with an observation location, the observed value is returned to avoid infinite weights.

The inverse distance power determines the degree to which the nearer point(s) are preferred over more distant points; for large values IDW converges to the one-nearest-neighbour interpolation. It can be tuned, for example using cross validation (Sect. 8.6.1). IDW can also be used within local search neighbourhoods (Sect. 8.5.5).

```
> library(gstat)
> idw.out <- idw(zinc ~ 1, meuse, meuse.grid, idp = 2.5)
[inverse distance weighted interpolation]
> as.data.frame(idw.out)[1:5, ]
  var1.pred var1.var      x      y
1  701.9621       NA 181180 333740
2  799.9616       NA 181140 333700
3  723.5780       NA 181180 333700
4  655.3131       NA 181220 333700
5  942.0218       NA 181100 333660
```

The output variable is called `var1.pred`, and the `var1.var` values are `NA` because inverse distance does not provide prediction error variances.

Inverse distance interpolation results usually in maps that are very similar to kriged maps when a variogram with no or a small nugget is used. In contrast to kriging, by only considering distances to the prediction location it ignores the spatial configuration of observations; this may lead to undesired effects if the observation locations are strongly clustered. Another difference is that weights are guaranteed to be between 0 and 1, resulting in interpolated values never outside the range of observed values.

8.3.2 Linear Regression

For spatial prediction using simple linear models, we can use the R function `lm`:

```
> zn.lm <- lm(log(zinc) ~ sqrt(dist), meuse)
> meuse.grid$pred <- predict(zn.lm, meuse.grid)
> meuse.grid$se.fit <- predict(zn.lm, meuse.grid, se.fit = TRUE)$se.fit
```

Alternatively, the `predict` method used here can provide the prediction or confidence intervals for a given confidence level. Alternatively, we can use the function `krige` in **gstat** for this,

```
> meuse.lm <- krige(log(zinc) ~ sqrt(dist), meuse, meuse.grid)
[ordinary or weighted least squares prediction]
```

that in this case does not *krige* as no variogram is specified, but uses linear regression.

Used in this form, the result is identical to that of `lm`. However, it can also be used to predict with regression models that are refitted within local neighbourhoods around a prediction location (Sect. 8.5.5) or provide mean predicted values for spatial areas (Sect. 8.5.6). The variance it returns is the prediction error variance when predicting for points or the estimation error variance when used for blocks.

A special form of linear regression is obtained when polynomials of spatial coordinates are used for predictors, for example for a second-order polynomial

```
> meuse.tr2 <- krige(log(zinc) ~ 1, meuse, meuse.grid,
+     degree = 2)

[ordinary or weighted least squares prediction]
```

This form is called trend surface analysis.

It is possible to use `lm` for trend surface analysis, for example for the second-order trend with a formula using `I` to treat powers and products 'as is':

```
> lm(log(zinc) ~ I(x^2) + I(y^2) + I(x * y) + x + y, meuse)
```

or the short form

```
> lm(log(zinc) ~ poly(x, y, 2), meuse)
```

It should be noted that for `lm`, the first form does not standardise coordinates, which often yields huge numbers when powered, and that the second form does standardise coordinates in such a way that it cannot be used in a subsequent `predict` call with different coordinate ranges. Also note that trend surface fitting is highly sensitive to outlying observations. Another place to look for trend surface analysis is function `surf.ls` in package **spatial**.

8.4 Estimating Spatial Correlation: The Variogram

In geostatistics the spatial correlation is modelled by the variogram instead of a correlogram or covariogram, largely for historical reasons. Here, the word variogram will be used synonymously with semivariogram. The variogram plots semivariance as a function of distance.

In standard statistical problems, correlation can be estimated from a scatterplot, when several data pairs $\{x, y\}$ are available. The spatial correlation between two observations of a variable $z(s)$ at locations s_1 and s_2 cannot be estimated, as only a single pair is available. To estimate spatial correlation from observational data, we therefore need to make stationarity assumptions before we can make any progress. One commonly used form of stationarity is *intrinsic stationarity*, which assumes that the process that generated the samples is a random function $Z(s)$ composed of a mean and residual

$$Z(s) = m + e(s), \tag{8.1}$$

with a constant mean

$$\mathrm{E}(Z(s)) = m \tag{8.2}$$

and a variogram defined as

$$\gamma(h) = \frac{1}{2}\mathrm{E}(Z(s) - Z(s+h))^2. \tag{8.3}$$

Under this assumption, we basically state that the variance of Z is constant, and that spatial correlation of Z does not depend on location s, but only on separation distance h. Then, we can form *multiple* pairs $\{z(s_i), z(s_j)\}$ that have (nearly) identical separation vectors $h = s_i - s_j$ and estimate correlation from them. If we further assume *isotropy*, which is direction independence of semivariance, we can replace the vector h with its length, $||h||$.

Under this assumption, the variogram can be estimated from N_h sample data pairs $z(s_i)$, $z(s_i + h)$ for a number of distances (or distance intervals) $\tilde{h}_j$ by

$$\hat{\gamma}(\tilde{h}_j) = \frac{1}{2N_h}\sum_{i=1}^{N_h}(Z(s_i) - Z(s_i+h))^2, \quad \forall h \in \tilde{h}_j \tag{8.4}$$

and this estimate is called the *sample* variogram.

A wider class of models is obtained when the mean varies spatially, and can, for example be modelled as a linear function of known predictors $X_j(s)$, as in

$$Z(s) = \sum_{j=0}^{p} X_j(s)\beta_j + e(s) = X\beta + e(s), \tag{8.5}$$

with $X_j(s)$ the known spatial regressors and β_j unknown regression coefficients, usually containing an intercept for which $X_0(s) \equiv 1$. The $X_j(s)$ form the columns of the $n \times (p+1)$ design matrix X, β is the column vector with $p+1$ unknown coefficients.

For varying mean models, stationarity properties refer to the residual $e(s)$, and the sample variogram needs to be computed from estimated residuals.

8.4.1 Exploratory Variogram Analysis

A simple way to acknowledge that spatial correlation is present or not is to make scatter plots of pairs $Z(s_i)$ and $Z(s_j)$, grouped according to their separation distance $h_{ij} = ||s_i - s_j||$. This is done for the `meuse` data set in Fig. 8.2, by

```
> hscat(log(zinc) ~ 1, meuse, (0:9) * 100)
```

where the strip texts indicate the distance classes, and sample correlations are shown in each panel.

A second way to explore spatial correlation is by plotting the variogram and the variogram cloud. The variogram cloud is obtained by plotting all possible squared differences of observation pairs $(Z(s_i) - Z(s_j))^2$ against their separation distance h_{ij}. One such variogram cloud, obtained by

```
> library(gstat)
> variogram(log(zinc) ~ 1, meuse, cloud = TRUE)
```

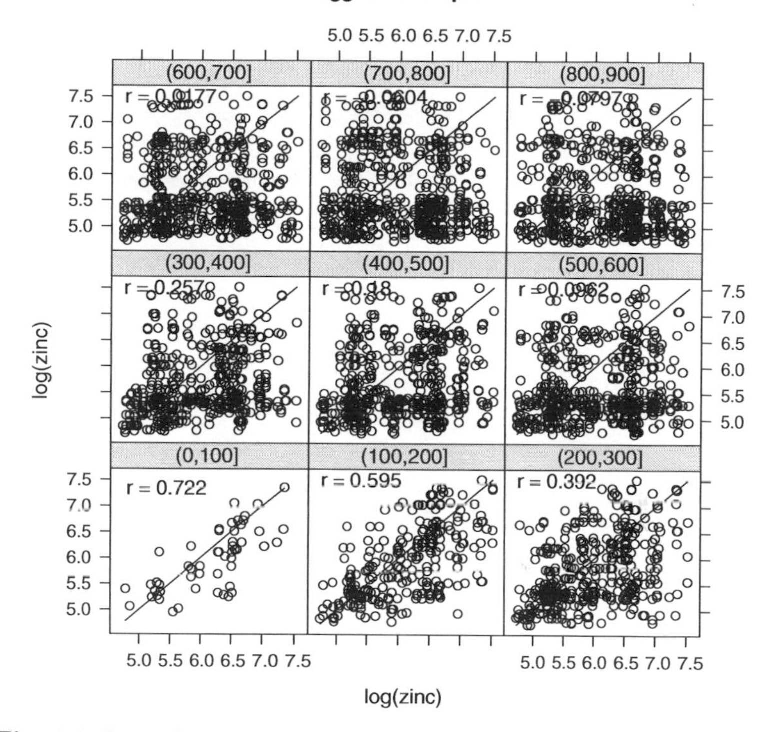

Fig. 8.2. Lagged scatter plot for the log-zinc data in the `meuse` data set

is plotted in Fig. 8.3 (top). The plot shows a lot of scatter, as could be expected: when $Z(s)$ follows a Gaussian distribution, $(Z(s_i) - Z(s_j))^2$ follows a $\chi^2(1)$ distribution. It does show, however, some increase of maximum values for distances increasing up to 1,000 m.

Essentially, the sample variogram plot of (8.4) obtained by

```
> plot(variogram(log(zinc) ~ 1, meuse))
```

is nothing but a plot of averages of semivariogram cloud values over distance intervals h; it is shown in Fig. 8.3, bottom. It smooths the variation in the variogram cloud and provides an estimate of semivariance (8.3), although Stein (1999) discourages this approach.

The `~ 1` defines a single constant predictor, leading to a spatially constant mean coefficient, in accordance with (8.2); see p. 26 for a presentation of formula objects.

As any point in the variogram cloud refers to a pair of points in the data set, the variogram cloud can point us to areas with unusual high or low variability. To do that, we need to select a subset of variogram cloud points. In

```
> sel <- plot(variogram(zinc ~ 1, meuse, cloud = TRUE),
+     digitize = TRUE)
> plot(sel, meuse)
```

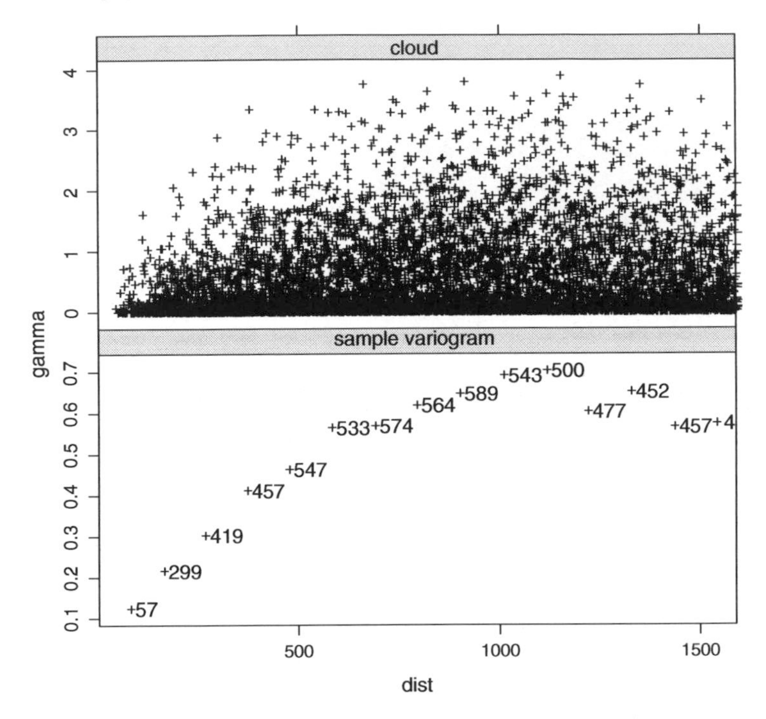

Fig. 8.3. Variogram cloud (*top*) and sample variogram (*bottom*) for log-zinc data; numbers next to symbols refer to the value N_h in (8.4)

the user is asked to digitise an area in the variogram cloud plot after the first command. The second command plots the selected point pairs on the observations map. Figure 8.4 shows the output plots of such a session. The point pairs with largest semivariance at short distances were selected, because they indicate the areas with the strongest gradients. The map shows that these areas are not spread randomly: they connect the maximum values closest to the Meuse river with values usually more inland. This can be an indication of non-stationarity or of anisotropy. Log-transformation or detrending may remove this effect, as we see later.

In case of outlying observations, extreme variogram cloud values are easily identified to find the outliers. These may need removal, or else robust measures for the sample variogram can be computed by passing the logical argument `cressie=TRUE` to the `variogram` function call (Cressie, 1993).

A sample variogram $\hat{\gamma}(h)$ always contains a signal that results from the true variogram $\gamma(h)$ and a sampling error, due to the fact that N_h and s are not infinite. To verify whether an increase in semivariance with distance *could* possibly be attributed to chance, we can compute variograms from the same data, after randomly re-assigning measurements to spatial locations. If the sample variogram falls within the (say 95%) range of these random variograms, complete spatial randomness of the underlying process *may be* a

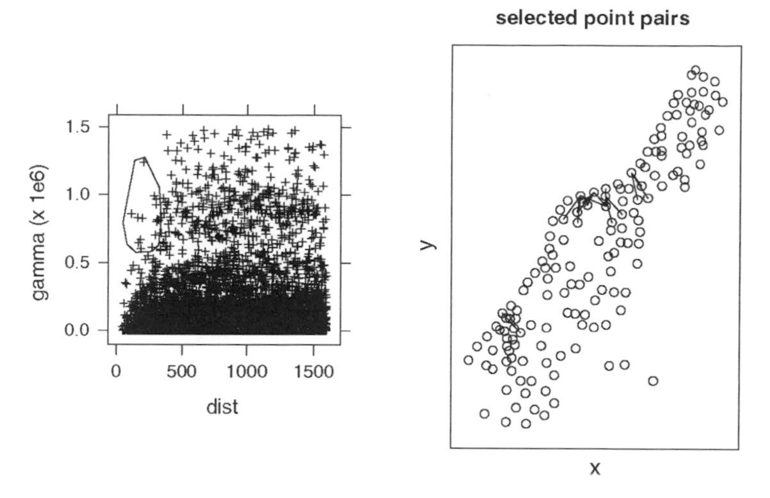

Fig. 8.4. Interactively selected point pairs on the variogram cloud (*left*) and map of selected point pairs (*right*)

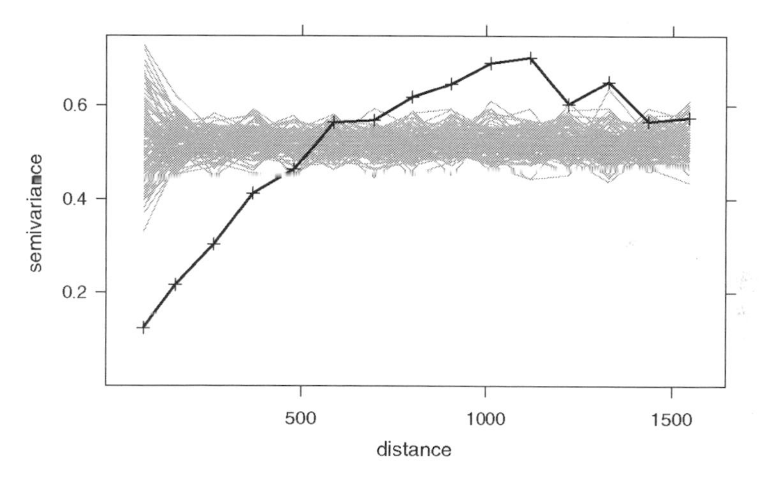

Fig. 8.5. Sample variogram (*bold*) compared to 100 variograms for randomly re-allocated data (*grey lines*)

plausible hypothesis. Figure 8.5 shows an example of such a plot for the log zinc data; here the hypothesis of absence of spatial correlation seems unlikely. In general, however, concluding or even assuming that an underlying process is *completely* spatially uncorrelated is quite unrealistic for real, natural processes. A common case is that the spatial correlation is difficult to infer from sample data, because of their distribution, sample size, or spatial configuration. In certain cases spatial correlation is nearly absent.

8.4.2 Cutoff, Lag Width, Direction Dependence

Although the command

```
> plot(variogram(log(zinc) ~ 1, meuse))
```

simply computes and plots the sample variogram, it does make a number of decisions by default. It decides that direction is ignored: point pairs are merged on the basis of distance, not direction. An alternative is, for example to look in four different angles, as in

```
> plot(variogram(log(zinc) ~ 1, meuse, alpha = c(0, 45,
+     90, 135))),
```

see Fig. 8.7. Directions are now divided over their principal direction, e.g., any point pair between 22.5° and 67.5° is used for the 45° panel. You might want to split into a finer direction subdivision, for example passing `alpha = seq(0,170,10)`, but then the noise component of resulting sample variograms will increase, as the number of point pairs for each separate estimate decreases.

A similar issue is the cutoff distance, which is the maximum distance up to which point pairs are considered and the width of distance interval over which point pairs are averaged in bins.

The default value **gstat** uses for the cutoff value is one third of the largest diagonal of the bounding box (or cube) of the data. Just as for time series data autocorrelations are never computed for lags farther than half the series length, there is little point in computing semivariances for long distances other than mere curiosity: wild oscillations usually show up that reveal little about the process under study. Good reasons to decrease the cutoff may be when a local prediction method is foreseen, and only semivariances up to a rather small distance are required. In this case, the modelling effort, and hence the computing of sample variograms should be limited to this distance (e.g. twice the radius of the planned search neighbourhood).

For the interval width, **gstat** uses a default of the cutoff value divided by 15. Usually, these default values will result in some initial overview of the spatial correlation. Choosing a smaller interval width will result in more detail, as more estimates of $\gamma(h)$ appear, but also in estimates with more noise, as N_h inevitably decreases. It should be noted that apparent local fluctuations of consecutive $\hat{\gamma}(h)$ values may still be attributed to sampling error. The errors $\hat{\gamma}(h_i) - \gamma(h_i)$ and $\hat{\gamma}(h_j) - \gamma(h_j)$ will be correlated, because $\gamma(\hat{h}_i)$ and $\gamma(\hat{h}_j)$ usually share a large number of common points used to form pairs.

The default cutoff and interval width values may not be appropriate at all, and can be overridden, for example by

```
> plot(variogram(log(zinc) ~ 1, meuse, cutoff = 1000, width =
50))
```

The distance vector does not have to be cut in regular intervals; one can specify each interval by

```
> variogram(log(zinc) ~ 1, meuse, boundaries = c(0, 50,
+     100, seq(250, 1500, 250)))
```

which is especially useful for data sets that have much information on short distance variability: it allows one to zoom in on the short distance variogram without revealing irrelevant details for the longer distances.

8.4.3 Variogram Modelling

The variogram is often used for spatial prediction (interpolation) or simulation of the observed process based on point observations. To ensure that predictions are associated with non-negative prediction variances, the matrix with semivariance values between all observation points and any possible prediction point needs to be non-negative definite. For this, simply plugging in sample variogram values from (8.4) is not sufficient. One common way is to infer a parametric variogram *model* from the data. A non-parametric way, using smoothing and cutting off negative frequencies in the spectral domain, is given in Yao and Journel (1998); it will not be discussed here.

The traditional way of finding a suitable variogram model is to fit a parametric model to the sample variogram (8.4). An overview of the basic variogram models available in **gstat** is obtained by

```
> show.vgms()
> show.vgms(model = "Mat", kappa.range = c(0.1, 0.2, 0.5,
+     1, 2, 5, 10), max = 10)
```

where the second command gives an overview of various models in the Matérn class.

In **gstat**, valid variogram models are constructed by using one or combinations of two or more basic variogram models. Variogram models are derived from `data.frame` objects, and are built as follows:

```
> vgm(1, "Sph", 300)

  model psill range
1   Sph     1   300

> vgm(1, "Sph", 300, 0.5)

  model psill range
1   Nug   0.5     0
2   Sph   1.0   300

> v1 <- vgm(1, "Sph", 300, 0.5)
> v2 <- vgm(0.8, "Sph", 800, add.to = v1)
> v2

  model psill range
1   Nug   0.5     0
2   Sph   1.0   300
3   Sph   0.8   800
```

```
> vgm(0.5, "Nug", 0)

  model psill range
1   Nug   0.5     0
```

and so on. Each component (row) has a model type ('Nug', 'Sph', ...), followed by a partial sill (the vertical extent of the model component) and a range parameter (the horizontal extent). Nugget variance can be defined in two ways, because it is almost always present. It reflects usually measurement error and/or micro-variability. Note that **gstat** uses range *parameters*, for example for the exponential model with partial sill c and range parameter a

$$\gamma(h) = c(1 - \mathrm{e}^{-h/a}).$$

This implies that for this particular model the *practical range*, the value at which this model reaches 95% of its asymptotic value, is $3a$; for the Gaussian model the practical range is $\sqrt{3}a$. A list of model types is obtained by

```
> vgm()

   short                      long
1    Nug              Nug (nugget)
2    Exp         Exp (exponential)
3    Sph           Sph (spherical)
4    Gau            Gau (gaussian)
5    Exc Exclass (Exponential class)
6    Mat              Mat (Matern)
7    Cir            Cir (circular)
8    Lin              Lin (linear)
9    Bes              Bes (bessel)
10   Pen       Pen (pentaspherical)
11   Per            Per (periodic)
12   Hol                Hol (hole)
13   Log         Log (logarithmic)
14   Pow               Pow (power)
15   Spl              Spl (spline)
16   Err    Err (Measurement error)
17   Int          Int (Intercept)
```

Not all of these models are equally useful, in practice. Most practical studies have so far used exponential, spherical, Gaussian, Matérn, or power models, with or without a nugget, or a combination of those.

For weighted least squares fitting a variogram model to the sample variogram (Cressie, 1985), we need to take several steps:

1. Choose a suitable model (such as exponential, ...), with or without nugget
2. Choose suitable initial values for partial sill(s), range(s), and possibly nugget
3. Fit this model, using one of the fitting criteria.

For the variogram obtained by

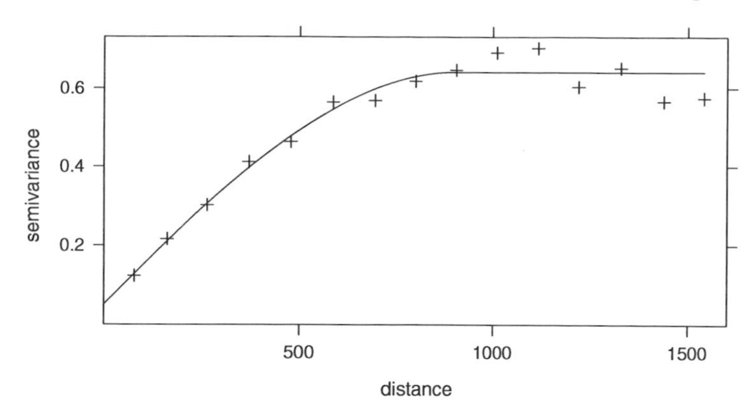

Fig. 8.6. Sample variogram (*plus*) and fitted model (*dashed line*)

```
> v <- variogram(log(zinc) ~ 1, meuse)
> plot(v)
```

and shown in Fig. 8.6, the spherical model looks like a reasonable choice. Initial values for the variogram fit are needed for `fit.variogram`, because for the spherical model (and many other models) fitting the range parameter involves non-linear regression. The following fit works:

```
> fit.variogram(v, vgm(1, "Sph", 800, 1))

  model      psill    range
1   Nug 0.05065923   0.0000
2   Sph 0.59060463 896.9976
```

but if we choose initial values too far off from reasonable values, as in

```
> fit.variogram(v, vgm(1, "Sph", 10, 1))

  model psill range
1   Nug     1     0
2   Sph     1    10
```

the fit will not succeed. To stop execution in an automated fitting script, a construct like

```
> v.fit <- fit.variogram(v, vgm(1, "Sph", 10, 1))
> if (attr(v.fit, "singular")) stop("singular fit")
```

will halt the procedure on this fitting problem.

The fitting method uses non-linear regression to fit the coefficients. For this, a weighted sum of square errors $\sum_{j=1}^{p} w_j(\gamma(h) - \hat{\gamma}(h))^2$, with $\gamma(h)$ the value according to the parametric model, is minimised. The optimisation routine alternates the following two steps until convergence: (i) a direct fit over the partial sills, and (ii) non-linear optimising of the range parameter(s) given the last fit of partial sills. The minimised criterion is available as

```
> attr(v.fit, "SSErr")

[1] 9.011194e-06
```

Table 8.1. Values for argument `fit.method` in function `fit.variogram`

`fit.method`	Weight
1	N_j
2	$N_j/\{\gamma(h_j)\}^2$
6	1
7	N_j/h_j^2

Different options for the weights w_j are given in Table 8.1, the default value chosen by **gstat** is 7. Two things should be noted: (i) for option 2, the weights change after each iteration, which may confuse the optimisation routine, and (ii) for the linear variogram with no nugget, option 7 is equivalent to option 2. Option 7 is default as it seems to work in many cases; it will, however, give rise to spurious fits when a sample semivariance estimate for distance (very close to) zero gives rise to an almost infinite weight. This may happen when duplicate observations are available.

An alternative approach to fitting variograms is by visual fitting, the so-called eyeball fit. Package **geoR** provides a graphical user interface for interactively adjusting the parameters:

```
> library(geoR)
> v.eye <- eyefit(variog(as.geodata(meuse["zinc"]), max.dist = 1500))
> ve.fit <- as.vgm.variomodel(v.eye[[1]])
```

The last function converts the model saved in `v.eye` to a form readable by **gstat**.

Typically, visual fitting will minimise $|\gamma(h) - \hat{\gamma}(h)|$ with emphasis on short distance/small $\gamma(h)$ values, as opposed to a weighted squared difference, used by most numerical fitting. An argument to prefer visual fitting over numerical fitting may be that the person who fits has knowledge that goes beyond the information in the data. This may for instance be related to information about the nugget effect, which may be hard to infer from data when sample locations are regularly spread. Information may be borrowed from other studies or derived from measurement error characteristics for a specific device. In that case, one could, however, also consider partial fitting, by keeping, for example the nugget to a fixed value.

Partial fitting of variogram coefficients can be done with **gstat**. Suppose we know for some reason that the partial sill for the nugget model (i.e. the nugget variance) *is* 0.06, and we want to fit the remaining parameters, then this is done by

```
> fit.variogram(v, vgm(1, "Sph", 800, 0.06), fit.sills = c(FALSE,
+     TRUE))

  model     psill    range
1   Nug 0.0600000   0.0000
2   Sph 0.5845836 923.0066
```

Alternatively, the range parameter(s) can be fixed using argument `fit.ranges`.

Maximum likelihood fitting of variogram models does not need the sample variogram as intermediate form, as it fits a model directly to a quadratic form of the data, that is the variogram cloud. REML (restricted maximum likelihood) fitting of only partial sills, not of ranges, can be done using **gstat** function `fit.variogram.reml`:

```
> fit.variogram.reml(log(zinc) ~ 1, meuse, model = vgm(0.6,
+     "Sph", 800, 0.06))

  model      psill range
1   Nug 0.02011090     0
2   Sph 0.57116195   800
```

Full maximum likelihood or restricted maximum likelihood fitting of variogram models, including the range parameters, can be done using function `likfit` in package **geoR**, or with function `fitvario` in package **RandomFields**. Maximum likelihood fitting is optimal under the assumption of a Gaussian random field, and can be very time consuming for larger data sets.

8.4.4 Anisotropy

Anisotropy may be modelled by defining a range *ellipse* instead of a circular or spherical range. In the following example

```
> v.dir <- variogram(log(zinc) ~ 1, meuse, alpha = (0:3) *
+     45)
> v.anis <- vgm(0.6, "Sph", 1600, 0.05, anis = c(45, 0.3))
> plot(v.dir, v.anis)
```

the result of which is shown in Fig. 8.7, for four main directions. The fitted model has a range in the principal direction (45°, NE) of 1,600, and of $0.3 \times 1{,}600 = 480$ in the minor direction (135°).

When more measurement information is available, one may consider plotting a *variogram map*, as in

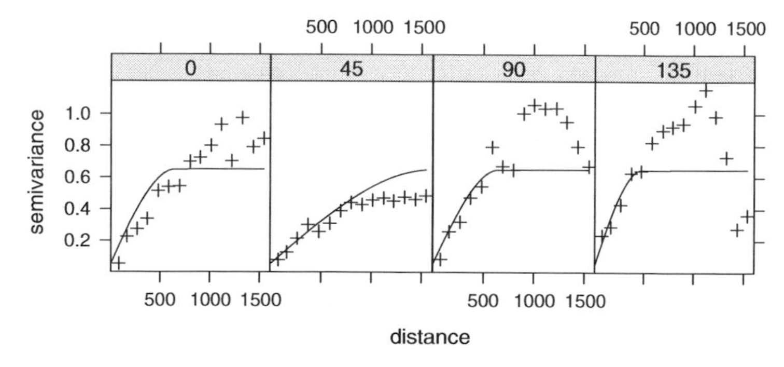

Fig. 8.7. Directional sample variogram (*plus*) and fitted model (*dashed line*) for four directions (0 is North, 90 is East)

```
> plot(variogram(log(zinc) ~ 1, meuse, map = TRUE, cutoff = 1000,
+     width = 100))
```

which bins h vectors in square grid cells over x and y, meaning that distance and direction are shown in much more detail. Help is available for the plotting function `plot.variogramMap`.

Package **gstat** does not provide automatic fitting of anisotropy parameters. Function `likfit` in **geoR** does, by using (restricted) maximum likelihood.

8.4.5 Multivariable Variogram Modelling

We use the term multivariable geostatistics here for the case where multiple dependent spatial variables are analysed jointly. The case where the trend of a single dependent variable contains more than a constant only is not called multivariable in this sense, and will be treated in Sect. 8.5.

The main tool for estimating semivariances between different variables is the cross variogram, defined for collocated[1] data as

$$\gamma_{ij}(h) = \mathrm{E}[(Z_i(s) - Z_i(s+h))(Z_j(s) - Z_j(s+h))]$$

and for non-collocated data as

$$\gamma_{ij}(h) = \mathrm{E}[(Z_i(s) - m_i)(Z_j(s) - m_j)],$$

with m_i and m_j the means of the respective variables. Sample cross variograms are the obvious sums over the available pairs or cross pairs, in the line of (8.4).

As multivariable analysis may involve numerous variables, we need to start organising the available information. For that reason, we collect all the observation data specifications in a `gstat` object, created by the function `gstat`. This function does nothing else than ordering (and actually, copying) information needed later in a single object. Consider the following definitions of four heavy metals:

```
> g <- gstat(NULL, "logCd", log(cadmium) ~ 1, meuse)
> g <- gstat(g, "logCu", log(copper) ~ 1, meuse)
> g <- gstat(g, "logPb", log(lead) ~ 1, meuse)
> g <- gstat(g, "logZn", log(zinc) ~ 1, meuse)
> g

data:
logCd : formula = log(cadmium)`~`1 ; data dim = 155 x 14
logCu : formula = log(copper)`~`1 ; data dim = 155 x 14
logPb : formula = log(lead)`~`1 ; data dim = 155 x 14
logZn : formula = log(zinc)`~`1 ; data dim = 155 x 14

> vm <- variogram(g)
> vm.fit <- fit.lmc(vm, g, vgm(1, "Sph", 800, 1))

> plot(vm, vm.fit)
```

[1] Each observation location has all variables measured.

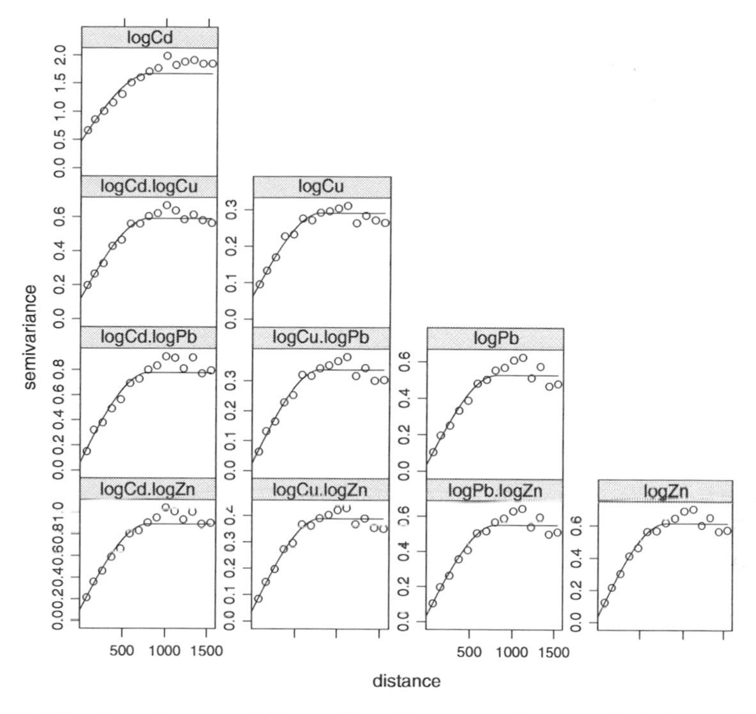

Fig. 8.8. Direct variograms (*diagonal*) and cross variograms (*off-diagonal*) along with fitted linear model of coregionalization (*dashed line*)

the plot of which is shown in Fig. 8.8. By default, `variogram` when passing a `gstat` object computes all direct and cross variograms, but this can be turned off. The function `fit.lmc` fits a linear model of co-regionalization, which is a particular model that needs to have identical model components (here nugget, and spherical with range 800), and needs to have positive definite partial sill matrices, to ensure non-negative prediction variances when used for spatial prediction (cokriging).

As the variograms in Fig. 8.8 indicate, the variables have a strong cross correlation. Because these variables are collocated, we could compute direct correlations:

```
> cor(as.data.frame(meuse)[c("cadmium", "copper", "lead",
+     "zinc")])
```

```
          cadmium    copper      lead      zinc
cadmium 1.0000000 0.9254499 0.7989466 0.9162139
copper  0.9254499 1.0000000 0.8183069 0.9082695
lead    0.7989466 0.8183069 1.0000000 0.9546913
zinc    0.9162139 0.9082695 0.9546913 1.0000000
```

which confirm this, but ignore spatial components. For non-collocated data, the direct correlations may be hard to compute.

The `fit.lmc` function fits positive definite coefficient matrices by first fitting models individually (while fixing the ranges) and then replacing non-positive definite coefficient matrices by their nearest positive definite approximation, taking out components that have a negative eigenvalue. When eigenvalues of exactly zero occur, a small value may have to be added to the direct variogram sill parameters; use the `correct.diagonal` argument for this.

Variables do not need to have a constant mean but can have a trend function specified, as explained in Sect. 8.4.6.

8.4.6 Residual Variogram Modelling

Residual variograms are calculated by default when a more complex model for the trend is used, for example as in

```
> variogram(log(zinc) ~ sqrt(dist), meuse)
```

where the trend is simple linear (Fig. 8.1), for example reworking (8.5) to

$$\log(Z(s)) = \beta_0 + \sqrt{D(s)}\beta_1 + e(s),$$

with $D(s)$ the distance to the river. For defining trends, the full range of R formulas can be used: the right-hand side may contain factors, in which case trends are calculated with respect to the factor level means, and may contain interactions of all sorts; see p. 26 for explanation on S formula syntax.

By default, the residuals **gstat** uses are ordinary least squares residuals (i.e. regular regression residuals), meaning that for the sake of estimating the trend, observations are considered independent. To honour a dependence structure present, generalised least squares residuals can be calculated instead. For this, a variogram model to define the covariance structure is needed. In the following example

```
> f <- log(zinc) ~ sqrt(dist)
> vt <- variogram(f, meuse)
> vt.fit <- fit.variogram(vt, vgm(1, "Exp", 300, 1))
> vt.fit

  model      psill    range
1   Nug 0.05712231   0.0000
2   Exp 0.17641559 340.3201

> g.wls <- gstat(NULL, "log-zinc", f, meuse, model = vt.fit,
+     set = list(gls = 1))
> (variogram(g.wls)$gamma - vt$gamma)/mean(vt$gamma)

 [1]  1.133887e-05 -6.800894e-05 -1.588582e-04 -2.520913e-04
 [5] -5.461007e-05 -1.257573e-04  2.560629e-04  1.509185e-04
 [9]  4.812184e-07 -5.292472e-05 -2.998868e-04  2.169712e-04
[13] -1.771773e-04  1.872195e-04  3.095021e-05
```

it is clear that the difference between the two approaches is marginal, but this does not need to be the case in other examples.

For multivariable analysis, **gstat** objects can be formed where the trend structure can be specified uniquely for each variable. If multivariable residuals are calculated using weighted least squares, this is done on a per-variable basis, ignoring cross correlations for trend estimation.

8.5 Spatial Prediction

Spatial prediction refers to the prediction of unknown quantities $Z(s_0)$, based on sample data $Z(s_i)$ and assumptions regarding the form of the trend of Z and its variance and spatial correlation.

Suppose we can write the trend as a linear regression function, as in (8.5). If the predictor values for s_0 are available in the $1 \times p$ row-vector $x(s_0)$, V is the covariance matrix of $Z(s)$ and v the covariance vector of $Z(s)$ and $Z(s_0)$, then the best linear unbiased predictor of $Z(s_0)$ is

$$\hat{Z}(s_0) = x(s_0)\hat{\beta} + v'V^{-1}(Z(s) - X\hat{\beta}), \qquad (8.6)$$

with $\hat{\beta} = (X'V^{-1}X)^{-1}X'V^{-1}Z(s)$ the generalized least squares estimate of the trend coefficients and where X' is the transpose of the design matrix X. The predictor consists of an estimated mean value for location s_0, plus a weighted mean of the residuals from the mean function, with weights $v'V^{-1}$, known as the *simple kriging weights*.

The predictor (8.6) has prediction error variance

$$\sigma^2(s_0) = \sigma_0^2 - v'V^{-1}v + \delta(X'V^{-1}X)^{-1}\delta', \qquad (8.7)$$

where σ_0^2 is $\text{var}(Z(s_0))$, or the variance of the Z process, and where $\delta = x(s_0) - v'V^{-1}X$. The term $v'V^{-1}v$ is zero if v is zero, that is if all observations are uncorrelated with $Z(s_0)$, and equals σ_0^2 when s_0 is identical to an observation location. The third term of (8.7) is the contribution of the estimation error $\text{var}(\hat{\beta} - \beta) = (X'V^{-1}X)^{-1}$ to the prediction (8.6): it is zero if s_0 is an observation location, and increases, for example when $x(s_0)$ is more distant from X, as when we extrapolate in the space of X.

8.5.1 Universal, Ordinary, and Simple Kriging

The instances of this best linear unbiased prediction method with the number of predictors $p > 0$ are usually called *universal kriging*. Sometimes the term *kriging with external drift* is used for the case where $p = 1$ and X does not include coordinates.

A special case is that of (8.2), for which $p = 0$ and $X_0 \equiv 1$. The corresponding prediction is called *ordinary kriging*.

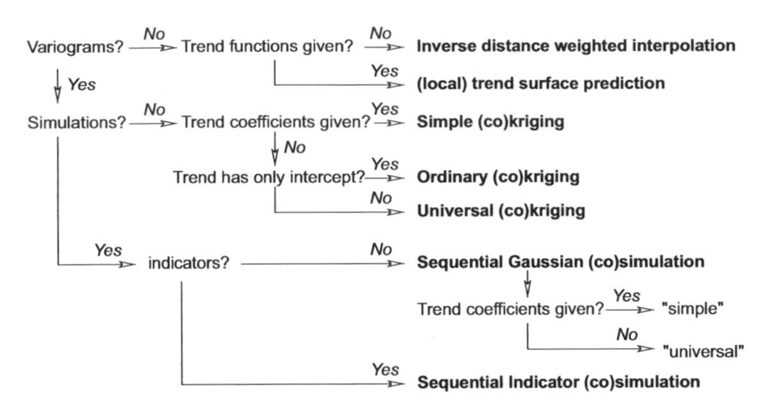

Fig. 8.9. Decision tree for the gstat `predict` method

Simple kriging is obtained when, for whatever reason, β is a priori assumed to be known. The known β can then be substituted for $\hat{\beta}$ in (8.6). The simple kriging variance is obtained by omitting the third term, which is associated with the estimation error of $\hat{\beta}$ in (8.7).

Applying these techniques is much more straightforward than this complicated jargon suggests, as an example will show:

```
> lz.sk <- krige(log(zinc) ~ 1, meuse, meuse.grid, v.fit,
+     beta = 5.9)

[using simple kriging]

> lz.ok <- krige(log(zinc) ~ 1, meuse, meuse.grid, v.fit)

[using ordinary kriging]

> lz.uk <- krige(log(zinc) ~ sqrt(dist), meuse, meuse.grid,
+     vt.fit)

[using universal kriging]
```

Clearly, the `krige` command chooses the kriging method itself, depending on the information it is provided with: are trend coefficients given? is the trend constant or more complex? How this is done is shown in the decision tree of Fig. 8.9.

8.5.2 Multivariable Prediction: Cokriging

The kriging predictions equations can be simply extended to obtain multivariable prediction equations, see, for example Hoef and Cressie (1993); Pebesma (2004). The general idea is that multiple variables may be cross correlated, meaning that they exhibit not only autocorrelation but that the spatial variability of variable A is correlated with variable B, and can therefore be used for its prediction, and vice versa. Typically, both variables are assumed to be measured on a limited set of locations, and the interpolation addresses unmeasured locations.

The technique is not limited to two variables. For each prediction location, multivariable prediction for q variables yields a $q \times 1$ *vector* with a prediction for each variable, and a $q \times q$ matrix with prediction error variances and covariances from which we can obtain the error correlations:

```
> cok.maps <- predict(vm.fit, meuse.grid)

Linear Model of Coregionalization found. Good.
[using ordinary cokriging]

> names(cok.maps)

 [1] "logCd.pred"      "logCd.var"       "logCu.pred"
 [4] "logCu.var"       "logPb.pred"      "logPb.var"
 [7] "logZn.pred"      "logZn.var"       "cov.logCd.logCu"
[10] "cov.logCd.logPb" "cov.logCu.logPb" "cov.logCd.logZn"
[13] "cov.logCu.logZn" "cov.logPb.logZn"
```

Clearly, only the unique matrix elements are stored; to get an overview of the prediction error variance and covariances, a utility function wrapping `spplot` is available; the output of

```
> spplot.vcov(cok.maps)
```

is given in Fig. 8.10.

Before the cokriging starts, **gstat** reports success in finding a Linear Model of Coregionalization (LMC). This is good, as it will assure non-negative cokriging variances. If this is not the case, for example because the ranges differ,

```
> vm2.fit <- vm.fit
> vm2.fit$model[[3]]$range = c(0, 900)
> predict(vm2.fit, meuse.grid)
```

will stop with an error message. Stopping on this check can be avoided by

```
> vm2.fit$set <- list(nocheck = 1)
> x <- predict(vm2.fit, meuse.grid)

Now checking for Cauchy-Schwartz inequalities:
variogram(var0,var1) passed Cauchy-Schwartz
variogram(var0,var2) passed Cauchy-Schwartz
variogram(var1,var2) passed Cauchy-Schwartz
variogram(var0,var3) passed Cauchy-Schwartz
variogram(var1,var3) passed Cauchy-Schwartz
variogram(var2,var3) passed Cauchy-Schwartz
[using ordinary cokriging]

> names(as.data.frame(x))

 [1] "logCd.pred"      "logCd.var"       "logCu.pred"
 [4] "logCu.var"       "logPb.pred"      "logPb.var"
 [7] "logZn.pred"      "logZn.var"       "cov.logCd.logCu"
[10] "cov.logCd.logPb" "cov.logCu.logPb" "cov.logCd.logZn"
[13] "cov.logCu.logZn" "cov.logPb.logZn" "x"
[16] "y"
```

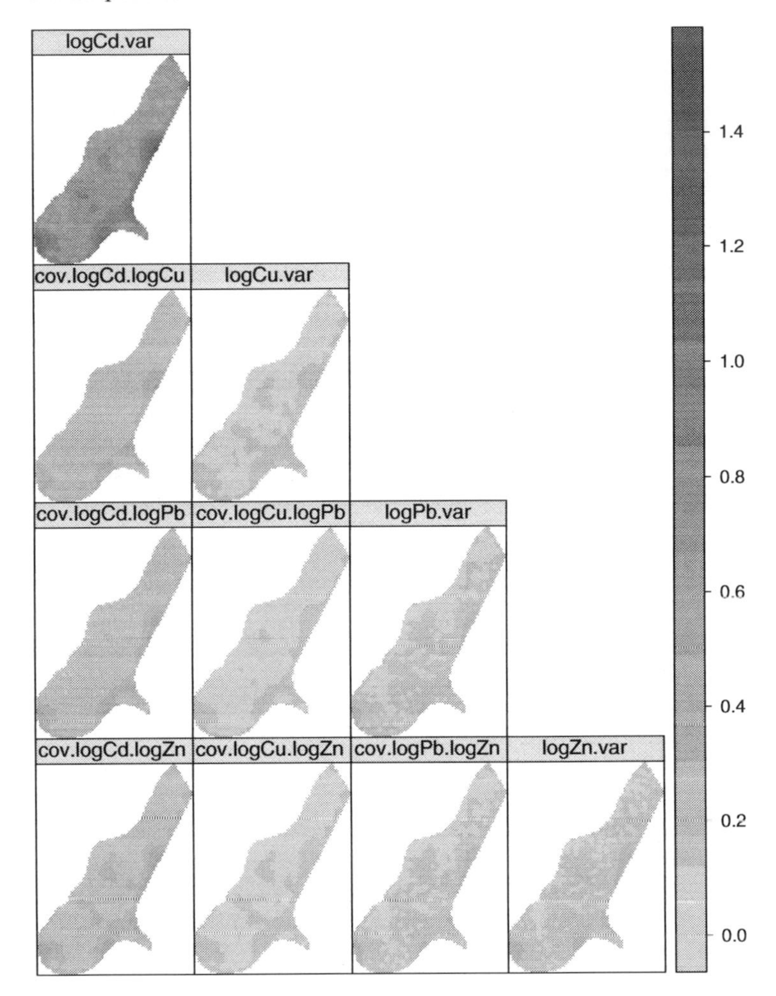

Fig. 8.10. Cokriging variances (*diagonal*) and covariances (*off-diagonal*)

```
> any(as.data.frame(x)[c(2, 4, 6, 8)] < 0)
[1] FALSE
```

which does check for pairwise Cauchy-Schwartz inequalities, that is $|\gamma_{ij}(h)| \leq \sqrt{\gamma_i(h)\gamma_j(h)}$, but will not stop on violations. Note that this latter check is not sufficient to guarantee positive variances. The final check confirms that we actually did not obtain any negative variances, for this particular case.

8.5.3 Collocated Cokriging

Collocated cokriging is a special case of cokriging, where a secondary variable is available at all prediction locations, and instead of choosing all observations of the secondary variable or those in a local neighbourhood, we restrict the secondary variable search neighbourhood to this single value on the prediction location. For instance, consider `log(zinc)` as primary and `dist` as secondary variable:

```
> g.cc <- gstat(NULL, "log.zinc", log(zinc) ~ 1, meuse,
+     model = v.fit)
> meuse.grid$distn <- meuse.grid$dist - mean(meuse.grid$dist) +
+     mean(log(meuse$zinc))
> vd.fit <- v.fit
> vov <- var(meuse.grid$distn)/var(log(meuse$zinc))
> vd.fit$psill <- v.fit$psill * vov
> g.cc <- gstat(g.cc, "distn", distn ~ 1, meuse.grid, nmax = 1,
+     model = vd.fit, merge = c("log.zinc", "distn"))
> vx.fit <- v.fit
> vx.fit$psill <- sqrt(v.fit$psill * vd.fit$psill) * cor(meuse$dist,
+     log(meuse$zinc))
> g.cc <- gstat(g.cc, c("log.zinc", "distn"), model = vx.fit)
> x <- predict(g.cc, meuse.grid)
```

```
Linear Model of Coregionalization found. Good.
[using ordinary cokriging]
```

Figure 8.11 shows the predicted maps using ordinary kriging, collocated cokriging and universal cokriging, using `log(zinc)` $\sim$ `sqrt(dist)` as trend.

8.5.4 Cokriging Contrasts

Cokriging error covariances can be of value when we want to compute functions of multiple predictions. Suppose Z_1 is measured on time 1, and Z_2 on time 2, and both are non-collocated. When we want to estimate the change $Z_2 - Z_1$, we can use the estimates for both moments, but for the error in the change we need in addition the prediction error covariance. The function `get.contr` helps computing the expected value and error variance for *any* linear combination (contrast) in a set of predictors, obtained by cokriging. A demo in **gstat**,

```
> demo(pcb)
```

gives a full space-time cokriging example that shows how time trends can be estimated for PCB-138 concentration in sea floor sediment, from four consecutive five-yearly rounds of monitoring, using universal cokriging.

8.5.5 Kriging in a Local Neighbourhood

By default, all spatial predictions method provided by **gstat** use all available observations for each prediction. In many cases, it is more convenient to use only the data in the neighbourhood of the prediction location. The reasons for this may be statistical or computational. Statistical reasons include that the

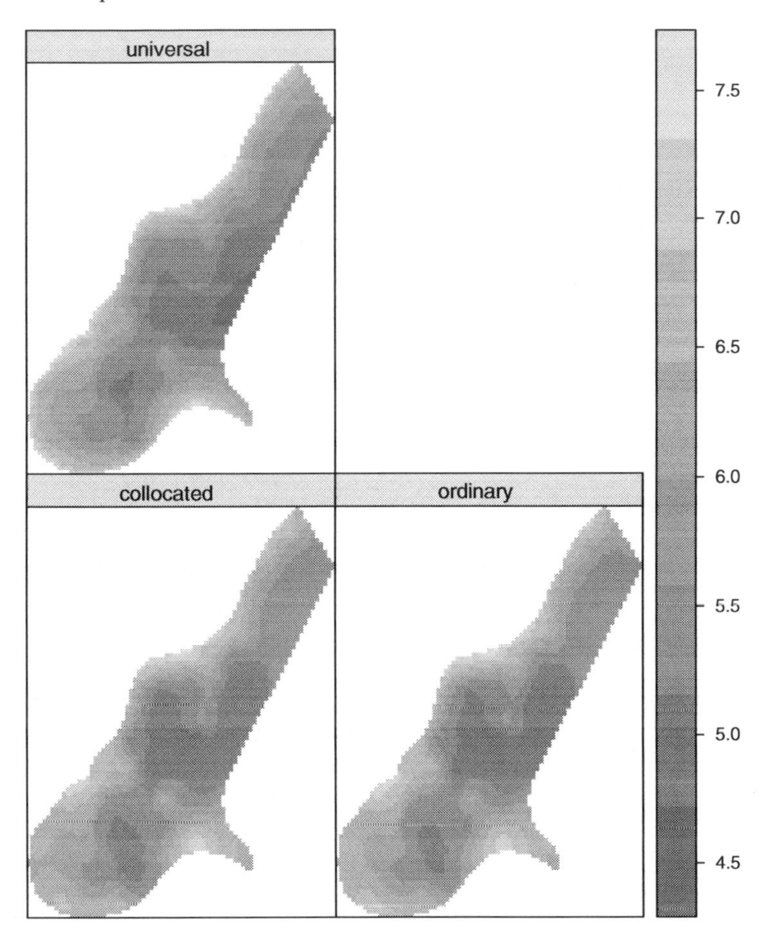

Fig. 8.11. Predictions for collocated cokriging, ordinary kriging, and universal kriging

hypothesis of constant mean or mean function should apply locally, or that the assumed knowledge of the variogram is only valid up to a small distance. Computational issues may involve both memory and speed: kriging for n data requires solving an $n \times n$ system. For large n (say more than 1,000) this may be too slow, and discarding anything but the closest say 100 observations may not result in notable different predictions.

It should be noted that for global kriging the matrix V needs to be decomposed only once, after which the result is re-used for each prediction location to obtain $V^{-1}v$. Decomposing a linear system of equations is an $O(n^2)$ operation, solving another system $O(n)$. Therefore, if a neighbourhood size is chosen slightly smaller than the global neighbourhood, the computation time may even increase, compared to using a global neighbourhood.

Neighbourhoods in **gstat** are defined by passing the arguments `nmax`, `nmin`, and/or `maxdist` to functions like `predict`, `krige`, or `gstat`. Arguments `nmax` and `nmin` define a neighbourhood size in terms of number of nearest points, `maxdist` specifies a circular search range to select point. They may be combined: when less than `nmin` points are found in a search radius, a missing value is generated.

For finding neighbourhood selections fast, **gstat** first builds a PR bucket quadtree, or for three-dimensional data octree search index (Hjaltason and Samet, 1995). With this index it finds any neighbourhood selection with only a small number of distance evaluations.

8.5.6 Change of Support: Block Kriging

Despite the fact that no measurement can ever be taken on something that has size zero, in geostatistics, by default observations $Z(s_i)$ are treated as being observed on point location. Kriging a value with a physical size equal to that of the observations is called *point kriging*. In contrast, *block kriging* predicts averages of larger areas or volumes. The term block kriging originates from mining, where early geostatistics was developed (Journel and Huijbregts, 1978). In mines, predictions based on bore hole data had to be made for *mineable units*, which tended to have a block shape. *Change of support* occurs when predictions are made for a larger physical support based on small physical support observations. There is no particular reason why the larger support needs to have a rectangular shape, but it is common

Besides the practical relevance to the mining industry, a consideration in many environmental applications has been that point kriging usually exhibits large prediction errors. This is due to the larger variability in the observations. When predicting averages over larger areas, much of the variability (i.e. that *within* the blocks) averages out and block mean values have lower prediction errors, while still revealing spatial patterns if the blocks are not too large. In environmental problems, legislation may be related to means or medians over larger areas, rather than to point values.

Block kriging (or other forms of prediction for blocks) can be obtained by **gstat** in three ways:

1. For regular blocks, by specifying a block size
2. For irregular but constant 'blocks', by specifying points that discretise the irregular form
3. For blocks or areas of varying size, by passing an object of class `SpatialPolygons` to the `newdata` argument (i.e. replacing `meuse.grid`)

Ordinary block kriging for blocks of size 40 × 40 is simply obtained by

```
> lz.ok <- krige(log(zinc) ~ 1, meuse, meuse.grid, v.fit,
+     block = c(40, 40))

[using ordinary kriging]
```

For a circular shape with radius 20, centred on the points of `meuse.grid`, one could select points on a regular grid within a circle:

```
> xy <- expand.grid(x = seq(-20, 20, 4), y = seq(-20, 20,
+     4))
> xy <- xy[(xy$x^2 + xy$y^2) <= 20^2, ]
> lz.ok <- krige(log(zinc) ~ 1, meuse, meuse.grid, v.fit,
+     block = xy)
```

```
[using ordinary kriging]
```

For block averages over varying regions, the `newdata` argument, usually a grid, can be replaced by a polygons object. Suppose `meuse.polygons` contains polygons for which we want to predict block averages, then this is done by

```
> lz.pols <- krige(log(zinc) ~ 1, meuse, meuse.polygons,
+     v.fit)
```

To discretise each (top level) Polygons object, coordinates that discretize the polygon are obtained by

```
> spsample(polygon, n = 500, type = "regular", offset = c(0.5,
+     0.5))
```

meaning that a regular discretisation is sought with approximately 500 points. These default arguments to `spsample` can be modified by altering the `sps.args` argument to `predict.gstat`; `spsample` is described on p. 118.

A much less efficient way to obtain block kriging predictions and prediction errors is to use Gaussian conditional simulation (Sect. 8.7) over a fine grid, calculate block means from each realisation, and obtain the mean and variance from a large number of realisations. In the limit, this should equal the analytical block kriging prediction.

When instead of a block *average* a non-linear spatial aggregation is required, such as a quantile value of points within a block, or the area fraction of a block where points exceed a threshold (Sect. 6.6), the simulation path is the more natural approach.

8.5.7 Stratifying the Domain

When a categorical variable is available that splits the area of interest in a number of disjoint areas, for example based on geology, soil type, land use or some other factor, we might want to apply separate krigings to the different units. This is called stratified kriging. The reason for doing kriging per-class may be that the covariance structure (semivariogram) is different for the different classes. In contrast to universal kriging with a categorical predictor, no correlation is assumed between residuals from different classes.

The example assumes there is a variable `part.a`, which partitions the area in two sub-areas, where `part.a` is 0 and where it is 1. First we can try to find out in which grid cells the observations lie:

```
> meuse$part.a <- idw(part.a ~ 1, meuse.grid,
+     meuse, nmax = 1)$var1.pred

[inverse distance weighted interpolation]
```

here, any interpolation may do, as we basically use the first nearest neighbour as predictor. A more robust approach may be to use the `overlay` method,

```
> meuse$part.a <- meuse.grid$part.a[overlay(meuse.grid,
+     meuse)]
```

because it will ignore observations not covered by the grid, when present.

Next, we can perform kriging for each of the sub-domains, store them in `x1` and `x2`, and merge the result using `rbind` in their non-spatial representation:

```
> x1 <- krige(log(zinc) ~ 1, meuse[meuse$part.a == 0, ],
+     meuse.grid[meuse.grid$part.a == 0, ], model = vgm(0.548,
+         "Sph", 900, 0.0654), nmin = 20, nmax = 40, maxdist = 1000)

[using ordinary kriging]

> x2 <- krige(log(zinc) ~ 1, meuse[meuse$part.a == 1, ],
+     meuse.grid[meuse.grid$part.a == 1, ], model = vgm(0.716,
+         "Sph", 900), nmin = 20, nmax = 40, maxdist = 1000)

[using ordinary kriging]

> lz.stk <- rbind(as.data.frame(x1), as.data.frame(x2))
> coordinates(lz.stk) <- c("x", "y")
> lz.stk <- as(x, "SpatialPixelsDataFrame")

> spplot(lz.stk["var1.pred"], main = "stratified kriging predictions")
```

8.5.8 Trend Functions and their Coefficients

For cases of exploratory data analysis or analysis of specific regression diagnostics, it may be of interest to limit prediction to the trend component $x(s_0)\hat{\beta}$, ignoring the prediction of the residual, that is ignoring the second term in the right-hand side of (8.6). This can be accomplished by setting argument `BLUE = TRUE` to `predict.gstat`:

```
> g.tr <- gstat(formula = log(zinc) ~ sqrt(dist), data = meuse,
+     model = v.fit)
> predict(g.tr, meuse[1, ])

[using universal kriging]
        coordinates var1.pred     var1.var
1 (181072, 333611)  6.929517 4.964793e-33

> predict(g.tr, meuse[1, ], BLUE = TRUE)

[generalized least squares trend estimation]
        coordinates var1.pred   var1.var
1 (181072, 333611)  6.862085 0.06123864
```

The first output yields the observed value (with zero variance), the second yields the generalised least squares trend component.

If we want to do significance testing of regression coefficients under a full model with spatially correlated residuals, we need to find out what the estimated regression coefficients and their standard errors are. For this, we can use **gstat** in a debug mode, in which case it will print a lot of information about intermediate calculations to the screen; just try

```
> predict(g, meuse[1, ], BLUE = TRUE, debug = 32)
```

but this does not allow saving the actual coefficients as data in R. Another way is to 'fool' the prediction mode with a specific contrast on the regression coefficients, for example the vector $x(s_0) = (0, 1)$, such that $x(s_0)\hat{\beta} = 0\hat{\beta}_0 + 1\hat{\beta}_1 = \hat{\beta}_1$. Both regression coefficient estimates are obtained by

```
> meuse$Int <- rep(1, 155)
> g.tr <- gstat(formula = log(zinc) ~ -1 + Int + sqrt(dist),
+     data = meuse, model = v.fit)
> rn <- c("Intercept", "beta1")
> df <- data.frame(Int = c(0, 1), dist = c(1, 0), row.names = rn)
> spdf <- SpatialPointsDataFrame(SpatialPoints(matrix(0,
+     2, 2)), df)
> spdf

          coordinates Int dist
Intercept      (0, 0)   0    1
beta1          (0, 0)   1    0

> predict(g.tr, spdf, BLUE = TRUE)

[generalized least squares trend estimation]
  coordinates var1.pred   var1.var
1      (0, 0) -2.471753 0.20018883
2      (0, 0)  6.953173 0.06633691
```

The `Int` variable is a 'manual' intercept to replace the automatic intercept, and the `-1` in the formula removes the automatic intercept. This way, we can control it and give it the zero value. The predictions now contain the generalised least squares estimates of the regression model.

8.5.9 Non-Linear Transforms of the Response Variable

For predictor variables, a non-linear transform simply yields a new variable and one can proceed as if nothing has happened. Searching for a good transform, such as using `sqrt(dist)` instead of direct `dist` values, may help in approaching the relationship with a straight line. For dependent variables this is not the case: because statistical expectation ('averaging') is a linear operation, $E(g(X)) = g(E(X))$ only holds if $g(\cdot)$ is a linear operator. This means that if we compute kriging predictors for zinc on the log scale, we do not obtain the expected zinc concentration by taking the exponent of the kriging predictor.

A large class of monotonous transformations is provided by the Box–Cox family of transformations, which take a value λ:

$$f(y, \lambda) = \begin{cases} (y^\lambda - 1)/\lambda & \text{if } \lambda \neq 0, \\ \ln(y) & \text{if } \lambda = 0. \end{cases}$$

A likelihood profile plot for lambda is obtained by the `boxcox` method in the package bundle **MASS**. For example, the plot resulting from

```
> library(MASS)
> boxcox(zinc ~ sqrt(dist), data = as.data.frame(meuse))
```

suggests that a Box–Cox transform with a slightly negative value for λ, for example $\lambda = -0.2$, might be somewhat better than log-transforming for those who like their data normal.

Yet another transformation is the normal score transform (Goovaerts, 1997) computed by the function `qqnorm`, defined as

```
> meuse$zinc.ns <- qqnorm(meuse$zinc, plot.it = FALSE)$x
```

Indeed, the resulting variable has mean zero, variance close to 1 (exactly one if n is large), and plots a straight line on a normal probability plot. So simple as this transform is, so complex can the back-transform be: it requires linear interpolation between the sample points, and for the extrapolation of values outside the data range the cited reference proposes several different models for tail distributions, all with different coefficients. There seems to be little guidance as how to choose between them based on sample data. It should be noted that back-transforming values outside the data range is not so much a problem with interpolation, as it is with simulation (Sect. 8.7).

Indicator kriging is obtained by indicator transforming a continuous variable, or reducing a categorical variable to a binary variable. An example for the indicator whether zinc is below 500 ppm is

```
> ind.f <- I(zinc < 500) ~ 1
> ind.fit <- fit.variogram(variogram(ind.f, meuse), vgm(1,
+      "Sph", 800, 1))
> ind.kr <- krige(ind.f, meuse, meuse.grid, ind.fit)

[using ordinary kriging]

> summary(ind.kr$var1.pred)

    Min.  1st Qu.   Median     Mean  3rd Qu.     Max.
-0.03472  0.47490  0.80730  0.70390  0.94540  1.08800
```

Clearly, this demonstrates the difficulty of interpreting the resulting estimates of ones and zeros as probabilities, as one has to deal with negative values and values larger than one.

When it comes to non-linear transformations such as the log transform, the question whether to transform or not to transform is often a hard one. On the one hand, it introduces the difficulties mentioned; on the other hand, transformation solves problems like negative predictions for non-negative variables, and, for example heteroscedasticity: for non-negative variables the variability is larger for areas with larger values, which opposes the stationarity assumption where variability is independent from location.

When a continuous transform is taken, such as the log-transform or the Box–Cox transform, it is possible to back-transform quantiles using the inverse transform. So, under log-normal assumptions the exponent of the kriging mean on the log scale is an estimate of the median on the working scale. From back-transforming a large number of quantiles, the mean value and variance may be worked out.

8.5.10 Singular Matrix Errors

Kriging cannot deal with duplicate observations, or observations that share the same location, because they are perfectly correlated, and lead to singular covariance matrices V, meaning that $V^{-1}v$ has no unique solution. Obtaining errors due to a singular matrix is a common case.

```
> meuse.dup <- rbind(as.data.frame(meuse)[1, ], as.data.frame(meuse))
> coordinates(meuse.dup) = ~x + y
> krige(log(zinc) ~ 1, meuse.dup, meuse[1, ], v.fit)
```

will result in the output:

```
[using ordinary kriging]

"chfactor.c", line 130: singular matrix in function LDLfactor()
Error in predict.gstat(g, newdata=newdata, block=block, nsim=nsim:
        LDLfactor
```

which points to the C function where the actual error occurred (`LDLfactor`). The most common case where this happens is when duplicate observations are present in the data set. Duplicate observations can be found by

```
> zd <- zerodist(meuse.dup)
> zd

     [,1] [,2]
[1,]    1    2

> krige(log(zinc) ~ 1, meuse.dup[-zd[, 1], ], meuse[1,
+     ], v.fit)

[using ordinary kriging]
       coordinates var1.pred     var1.var
1 (181072, 333611)  6.929517 1.963167e-33
```

which tells that observations 1 and 2 have identical location; the third command removes the first of the pair. Near-duplicate observations are found by increasing the `zero` argument of function `zerodist` to a very small threshold.

Other common causes for singular matrices are the following:

- The use of variogram models that cause observations to have nearly perfect correlation, despite the fact that they do not share the same location, for example from `vgm(0, "Nug", 0)` or `vgm(1, "Gau", 1e20)`. The Gaussian model is *always* a suspect if errors occur when it is used; adding a small nugget often helps.
- Using a regression model with perfectly correlated variables; note that, for example a global regression model may lead to singularity in a local neighbourhood where a predictor may be constant and correlate perfectly with the intercept, or otherwise perfect correlation occurs.

Stopping execution on singular matrices is usually best: the cause needs to be found and somehow resolved. An alternative is to skip those locations and continue. For instance,

```
> setL <- list(cn_max = 1e+10)
> krige(log(zinc) ~ 1, meuse.dup, meuse[1, ], v.fit, set = setL)

[using ordinary kriging]
       coordinates var1.pred var1.var
1 (181072, 333611)        NA       NA
```

checks whether the estimated condition number for V and $X'V^{-1}X$ exceeds 10^{10}, in which case `NA` values are generated for prediction. Larger condition numbers indicate that a matrix is closer to singular. This is by no means a solution. It will also report whether V or $X'V^{-1}X$ *are* singular; in the latter case the cause is more likely related to collinear regressors, which reside in X.

Near-singularity may not be picked up, and can potentially lead to dramatically bad results: predictions that are orders of magnitude different from the observation data. The causes should be sought in the same direction as real singularity. Setting the `cn_max` value may help finding where this occurs.

8.6 Model Diagnostics

The model diagnostics we have seen so far are fitted and residual plots (for linear regression models), spatial identification of groups of points in the variogram cloud, visual and numerical inspection of variogram models, and visual and numerical inspection of kriging results. Along the way, we have seen many model decisions that needed to be made; the major ones being the following:

- Possible transformation of the dependent variable
- The form of the trend function
- The cutoff, lag width, and possibly directional dependence for the sample variogram
- The variogram model type
- The variogram model coefficient values, or fitting method
- The size and criterion to define a local neighbourhood

and we have seen fairly little *statistical* guidance as to which choices are better. To some extent we can 'ask' the data what a good decision is, and for this we may use cross validation. We see that there are some model choices that do not seem very important, and others that cross validation simply cannot inform us about.

8.6.1 Cross Validation Residuals

Cross validation splits the data set into two sets: a modelling set and a validation set. The modelling set is used for variogram modelling and kriging on the locations of the validation set, and then the validation measurements can be compared to their predictions. If all went well, cross validation residuals are small, have zero mean, and no apparent structure.

How should we choose or isolate a set for validation? A possibility is to randomly partition the data in a model and test set. Let us try this for the `meuse` data set, splitting it in 100 observations for modelling and 55 for testing:

```
> sel100 <- sample(1:155, 100)
> m.model <- meuse[sel100, ]
> m.valid <- meuse[-sel100, ]
> v100.fit <- fit.variogram(variogram(log(zinc) ~ 1, m.model),
+     vgm(1, "Sph", 800, 1))
> m.valid.pr <- krige(log(zinc) ~ 1, m.model, m.valid,
+     v100.fit)
```

```
[using ordinary kriging]
```

```
> resid.kr <- log(m.valid$zinc) - m.valid.pr$var1.pred
> summary(resid.kr)
```

```
    Min.  1st Qu.   Median     Mean  3rd Qu.     Max.
-0.79990 -0.18240 -0.03922 -0.01881  0.19210  1.06900
```

```
> resid.mean <- log(m.valid$zinc) - mean(log(m.valid$zinc))
> R2 <- 1 - sum(resid.kr^2)/sum(resid.mean^2)
> R2
```

```
[1] 0.717017
```

which indicates that kriging prediction is a better predictor than the mean, with an indicative R^2 of 0.72. Running this analysis again will result in different values, as another random sample is chosen. Also note that no visual verification that the variogram model fit is sensible has been applied. A map with cross validation residuals can be obtained by

```
> m.valid.pr$res <- resid.kr
```

```
> bubble(m.valid.pr, "res")
```

A similar map is shown for 155 residuals in Fig. 8.12. Here, symbol size denotes residual size, with symbol area proportional to absolute value.

To use the data to a fuller extent, we would like to use all observations to create a residual once; this may be used to find influential observations. It can be done by replacing the first few lines in the example above with

```
> nfold <- 3
> part <- sample(1:nfold, 155, replace = TRUE)
> sel <- (part != 1)
> m.model <- meuse[sel, ]
> m.valid <- meuse[-sel, ]
```

and next define `sel = (part != 2)`, etc. Again, the random splitting brings in a random component to the outcomes. This procedure is threefold cross validation, and it can be easily extended to n-fold cross validation. When n equals the number of observations, the procedure is called leave-one-out cross validation.

A more automated way to do this is provided by the **gstat** functions `krige.cv` for univariate cross validation and `gstat.cv` for multivariable cross validation:

```
> v.fit <- vgm(0.59, "Sph", 874, 0.04)
> cv155 <- krige.cv(log(zinc) ~ 1, meuse, v.fit, nfold = 5)

[using ordinary kriging]
[using ordinary kriging]
[using ordinary kriging]
[using ordinary kriging]
[using ordinary kriging]

> bubble(cv155, "residual", main = "log(zinc): 5-fold CV residuals")
```

the result of which is shown in Fig. 8.12. It should be noted that these functions do not re-fit variograms for each fold; usually a variogram is fit on the complete data set, and in that case validation residuals are not completely independent from modelling data, as they already did contribute to the variogram model fitting.

8.6.2 Cross Validation z-Scores

The `krige.cv` object returns more than residuals alone:

```
> summary(cv155)

Object of class SpatialPointsDataFrame
Coordinates:
     min    max
x 178605 181390
y 329714 333611
Is projected: NA
proj4string : [NA]
Number of points: 155
Data attributes:
   var1.pred         var1.var          observed         residual
 Min.   :4.808   Min.    :0.1102   Min.   :4.727   Min.   :-0.955422
 1st Qu.:5.380   1st Qu.:0.1519   1st Qu.:5.288   1st Qu.:-0.218794
```

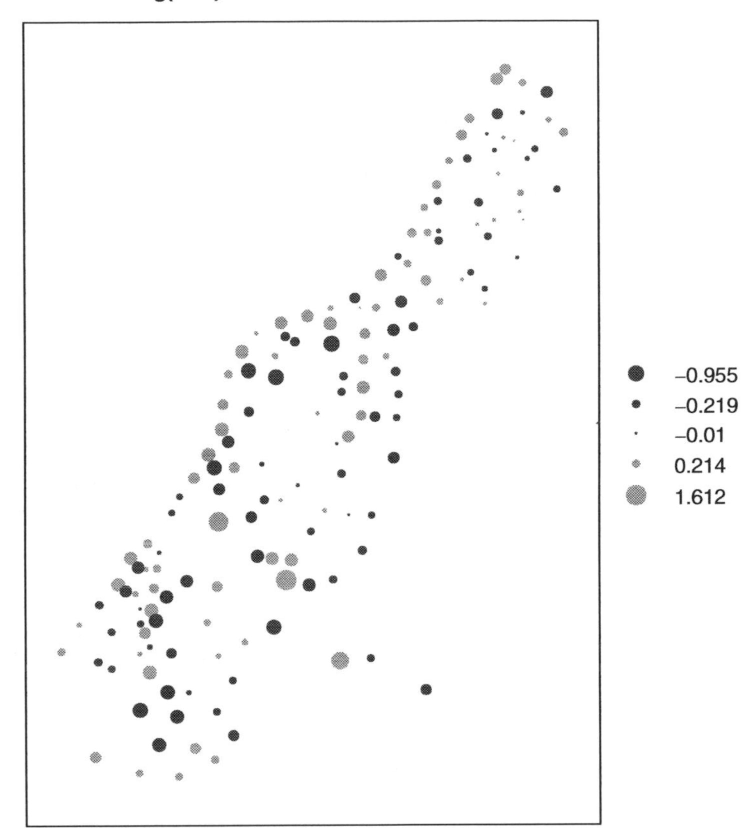

Fig. 8.12. Cross validation residuals for fivefold cross validation; symbol size denotes residual magnitude, positive residuals indicate under-prediction

```
 Median :5.881   Median :0.1779   Median :5.787   Median :-0.010007
 Mean   :5.887   Mean   :0.1914   Mean   :5.886   Mean   :-0.001047
 3rd Qu.:6.333   3rd Qu.:0.2145   3rd Qu.:6.514   3rd Qu.: 0.213743
 Max.   :7.257   Max.   :0.5370   Max.   :7.517   Max.   : 1.611694
     zscore              fold
 Min.   :-2.270139   Min.   :1.000
 1st Qu.:-0.508470   1st Qu.:2.000
 Median :-0.023781   Median :3.000
 Mean   : 0.001421   Mean   :2.987
 3rd Qu.: 0.498811   3rd Qu.:4.000
 Max.   : 3.537729   Max.   :5.000
```

the variable `fold` shows to which fold each record belonged, and the variable `zscore` is the z-score, computed as

$$z_i = \frac{Z(s_i) - \hat{Z}_{[i]}(s_i)}{\sigma_{[i]}(s_i)},$$

with $\hat{Z}_{[i]}(s_i)$ the cross validation prediction for s_i, and $\sigma_{[i]}(s_i)$ the corresponding kriging standard error. In contrast to standard residuals the z-score takes the kriging variance into account: it is a standardised residual, and if the variogram model is correct, the z-score should have mean and variance values close to 0 and 1. If, in addition, $Z(s)$ follows a normal distribution, so should the z-score do.

8.6.3 Multivariable Cross Validation

Multivariable cross validation is obtained using the `gstat.cv` function:

```
> g.cv <- gstat.cv(g, nmax = 40)
```

Here, the neighbourhood size is set to the nearest 40 observations for computational reasons. With multivariable cross validation, two additional parameters need be considered:

- `remove.all = FALSE` By default only the first variable is cross-validated, and all other variables are used to their full extent for prediction on the validation locations; if set to `TRUE`, also secondary data at the validation locations are removed.
- `all.residuals = FALSE` By default only residuals are computed and returned for the primary variable; if set to `TRUE`, residuals are computed and returned for all variables.

In a truly multivariable setting, where there is no hierarchy between the different variables to be considered, both should be set to `TRUE`.

8.6.4 Limitations to Cross Validation

Cross validation can be useful to find artefacts in data, but it should be used with caution for confirmatory purposes: one needs to be careful not to conclude that our (variogram and regression) model is correct if cross validation does not lead to unexpected findings. It is for instance not good at finding what is not in the data.

As an example, the Meuse data set does not contain point pairs with a separation distance closer than 40. Therefore, the two variogram models

```
> v1.fit <- vgm(0.591, "Sph", 897, 0.0507)
> v2.fit <- vgm(0.591, "Sph", 897, add.to = vgm(0.0507,
+     "Sph", 40))
```

that only differ with respect to the spatial correlation at distances smaller than 40 m yield identical cross validation results:

```
> set.seed(13331)
> cv155.1 <- krige.cv(log(zinc) ~ 1, meuse, v1.fit, nfold = 5)
```

```
[using ordinary kriging]
[using ordinary kriging]
[using ordinary kriging]
[using ordinary kriging]
[using ordinary kriging]
```

```
> set.seed(13331)
> cv155.2 <- krige.cv(log(zinc) ~ 1, meuse, v2.fit, nfold = 5)
```

```
[using ordinary kriging]
[using ordinary kriging]
[using ordinary kriging]
[using ordinary kriging]
[using ordinary kriging]
```

```
> summary(cv155.1$residual - cv155.2$residual)
```

```
   Min. 1st Qu.  Median    Mean 3rd Qu.    Max.
      0       0       0       0       0       0
```

Note that the `set.seed(13331)` was used here to force identical assignments of the otherwise random folding. When used for block kriging, the behaviour of the variogram at the origin is of utmost importance, and the two models yield strongly differing results. As an example, consider block kriging predictions at the `meuse.grid` cells:

```
> b1 <- krige(log(zinc) ~ 1, meuse, meuse.grid, v1.fit,
+     block = c(40, 40))$var1.var
```

```
[using ordinary kriging]
```

```
> b2 <- krige(log(zinc) ~ 1, meuse, meuse.grid, v2.fit,
+     block = c(40, 40))$var1.var
```

```
[using ordinary kriging]
```

```
> summary((b1 - b2)/b1)
```

```
   Min. 1st Qu.  Median    Mean 3rd Qu.    Max.
-0.4313 -0.2195 -0.1684 -0.1584 -0.1071  0.4374
```

where some kriging variances drop, but most increase, up to 30% when using the variogram without nugget instead of the one with a nugget. The decision which variogram to choose for distances shorter than those available in the data is up to the analyst, and matters.

8.7 Geostatistical Simulation

Geostatistical simulation refers to the simulation of possible realisations of a random field, given the specifications for that random field (e.g. mean structure, residual variogram, intrinsic stationarity) and possibly observation data.

Conditional simulation produces realisations that exactly honour observed data at data locations, unconditional simulations ignore observations and only reproduce means and prescribed variability.

Geostatistical simulation is fun to do, to see how much (or little) realisations can vary given the model description and data, but it is a poor means of quantitatively communicating uncertainty: many realisations are needed and there is no obvious ordering in which they can be viewed. They are, however, often needed when the uncertainty of kriging predictions is, for example input to a next level of analysis, and spatial correlation plays a role. An example could be the use of rainfall fields as input to spatially distributed rainfall-runoff models: interpolated values and their variances are of little value, but running the rainfall-runoff model with a large number of simulated rainfall fields may give a realistic assertion of the uncertainty in runoff, resulting from uncertainty in the rainfall field.

Calculating runoff given rainfall and catchment characteristic can be seen as a non-linear spatial aggregation process. Simpler non-linear aggregations are, for example for a given area or block the fraction of the variable that exceeds a critical limit, the 90th percentile within that area, or the actual area where (or its size distribution for which) a measured concentration exceeds a threshold. Simulation can give answers in terms of predictions as well as predictive distributions for all these cases. Of course, the outcomes can never be better than the degree to which the simulation model reflects reality.

8.7.1 Sequential Simulation

For simulating random fields, package **gstat** only provides the sequential simulation algorithm (see, e.g. Goovaerts (1997) for an explanation), and provides this for Gaussian simulation and indicator simulation, possibly multivariable, optionally with simulation of trend components, and optionally for block mean values. Package **RandomFields** provides a large number of other simulation algorithms.

Sequential simulation proceeds as follows; following a random path through the simulation locations, it repeats the following steps:

1. Compute the conditional distribution given data and previously simulated values, using simple kriging
2. Draw a value from this conditional distribution
3. Add this value to the data set
4. Go to the next unvisited location, and go back to 1

until all locations have been visited. In step 2, either the Gaussian distribution is used or the indicator kriging estimates are used to approximate a conditional distribution, interpreting kriging estimates as probabilities (after some fudging!).

Step 1 of this algorithm will become the most computationally expensive when all data (observed and previously simulated) are used. Also, as the number of simulation nodes is usually much larger than the number of observations, the simulation process will slow down more and more when global neighbourhoods are used. To obtain simulations with a reasonable speed, we need to set a maximum to the neighbourhood. This is best done with the `nmax` argument, as spatial data density increases when more and more simulated values are added. For simulation we again use the functions `krige` or `predict.gstat`; the argument `nsim` indicates how many realisations are requested:

```
> lzn.sim <- krige(log(zinc) ~ 1, meuse, meuse.grid, v.fit,
+     nsim = 6, nmax = 40)

drawing 6 GLS realisations of beta...
[using conditional Gaussian simulation]

> spplot(lzn.sim)
```

the result of which is shown in Fig. 8.13. It should be noted that these realisations are created following a single random path, in which case the expensive results ($V^{-1}v$ and the neighbourhood selection) can be re-used. Alternatively, one could use six function calls, each with `nsim = 1`.

The simulation procedure above also gave the output line `drawing 6 GLS realisations of beta...`, which confirms that prior to simulation of the field *for each realisation* a trend vector (in this case a mean value only) is drawn

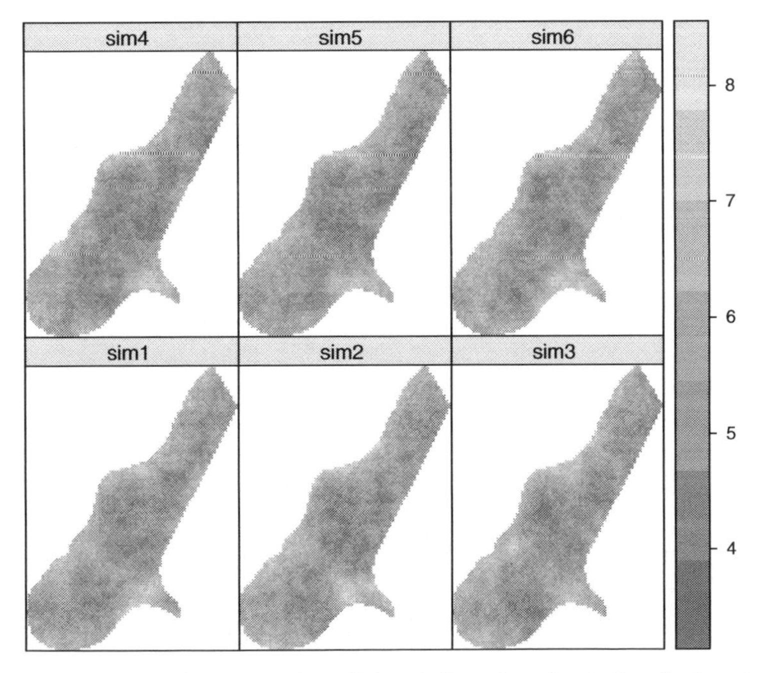

Fig. 8.13. Six realisations of conditional Gaussian simulation for log-zinc

from the normal distribution with mean $(X'V^{-1}X)^{-1}X'V^{-1}Z(s)$ and variance $(X'V^{-1}X)^{-1}$, that is the generalised least squares estimate and estimation variance. This procedure leads to simulations that have mean and variance equal to the ordinary or universal kriging mean and variance, and that have residual spatial correlation according to the variogram prescribed (Abrahamsen and Benth, 2001). For simulations that omit the simulation of the trend coefficients, the vector β should be passed, for example as `beta = 5.9` to the `krige` function, as with the simple kriging example. In that case, the simulated fields will follow the simple kriging mean and variance.

8.7.2 Non-Linear Spatial Aggregation and Block Averages

Suppose the area shown in Fig. 8.14 is the target area for which we want to know the fraction above a threshold; the area being available as a `SpatialPolygons` object `area`. We can now compute the distribution of the areal fraction above a cutoff of 500 ppm by simulation:

```
> nsim <- 1000
> cutoff <- 500
> grd <- overlay(meuse.grid, area.sp)
> sel.grid <- meuse.grid[!is.na(grd), ]
> lzn.sim <- krige(log(zinc) ~ 1, meuse, sel.grid, v.fit,
+     nsim = nsim, nmax = 40)

drawing 1000 GLS realisations of beta...
[using conditional Gaussian simulation]

> res <- apply(as.data.frame(lzn.sim)[1:nsim], 2, function(x) mean(x >
+     log(cutoff)))

> hist(res, main = paste("fraction above", cutoff), xlab = NULL,
+     ylab = NULL)
```

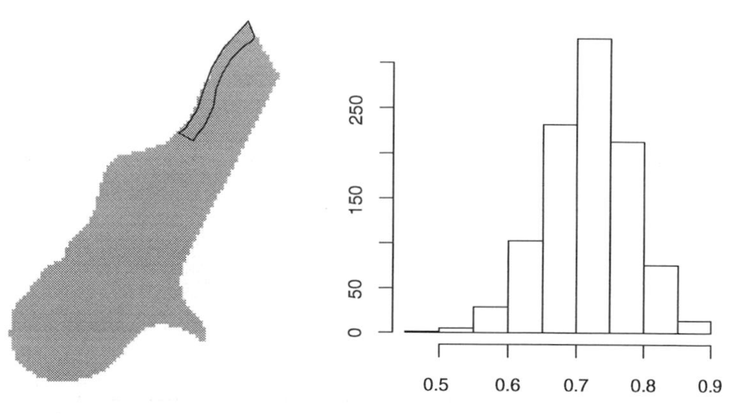

Fig. 8.14. A non-rectangular area for which a non-linear aggregation is required (*left*) and distribution of the areal fraction with zinc concentration above 500 ppm

shown in the right-hand side of Fig. 8.14. Note that if we had been interested in the probability of `mean(x) > log(cutoff)`, which is a rather different issue, then block kriging would have been sufficient:

```
> bkr <- krige(log(zinc) ~ 1, meuse, area.sp, v.fit)

[using ordinary kriging]

> 1 - pnorm(log(cutoff), bkr$var1.pred, sqrt(bkr$var1.var))

[1] 0.999879
```

Block averages can be simulated directly by supplying the block argument to `krige`; simulating points and aggregating these to block means *may* be more efficient because simulating blocks calls for the calculation of many block–block covariances, which involves the computation of quadruple integrals.

8.7.3 Multivariable and Indicator Simulation

Multivariable simulation is as easy as cokriging, try

```
> cok.sims <- predict(vm.fit, meuse.grid, nsim = 1000)
```

after passing the `nmax = 40`, or something similar to the `gstat` calls used to build up `vm.fit` (Sect. 8.4.5).

Simulation of indicators is done along the same lines. Suppose we want to simulate soil class 1, available in the Meuse data set:

```
> table(meuse$soil)

 1  2  3
97 46 12

> s1.fit <- fit.variogram(variogram(I(soil == 1) ~ 1, meuse),
+     vgm(1, "Sph", 800, 1))
> s1.sim <- krige(I(soil == 1) ~ 1, meuse, meuse.grid,
+     s1.fit, nsim = 6, indicators = TRUE, nmax = 40)

drawing six GLS realisations of beta... [using conditional
indicator simulation]

> spplot(s1.sim)
```

which is shown in Fig. 8.15.

8.8 Model-Based Geostatistics and Bayesian Approaches

Up to now, we have only seen kriging approaches where it was assumed that the variogram model, fitted from sample data, is assumed to be *known* when we do the kriging or simulation: any uncertainty about it is ignored. Diggle et al. (1998) give an approach, based on linear mixed and generalized linear

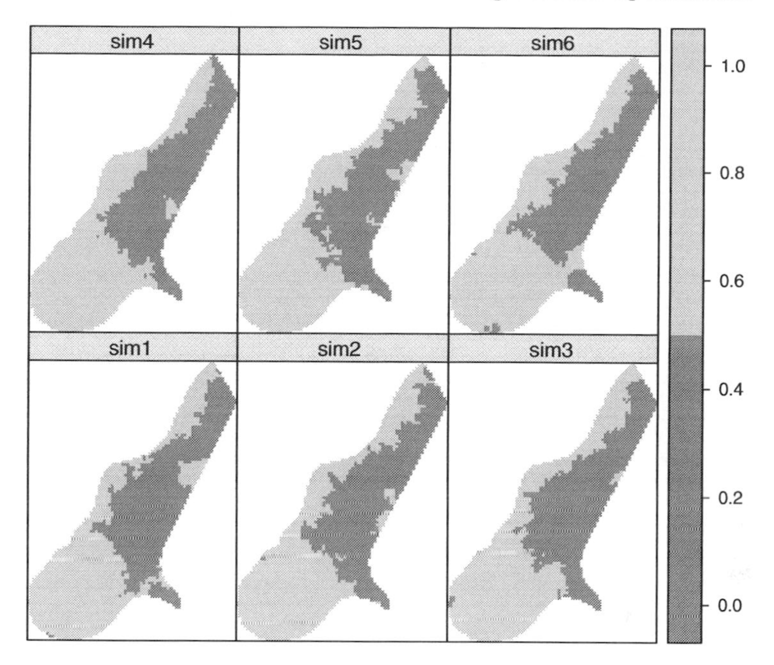

Fig. 8.15. Six realisations of conditional indicator simulation for soil type 1

mixed models, to provide what they call *model-based* geostatistical predictions. It incorporates the estimation error of the variogram coefficients.

When is uncertainty of the variogram an important issue? Obviously, when the sample is small, or, for example when variogram modelling is problematic due to the presence of extreme observations or data that come from a strongly skewed distribution.

8.9 Monitoring Network Optimization

Monitoring costs money. Monitoring programs have to be designed, started, stopped, evaluated, and sometimes need to be enlarged or shrunken. The difficulty of finding optimal network designs is that a quantitative criterion is often a priori not present. For example, should one focus on mean kriging variances, or variance of some global mean estimator, or rather on the ability to delineate a particular contour?

A very simple approach towards monitoring network optimization is to find the point whose removal leads to the smallest increase in mean kriging variance:

```
> m1 <- sapply(1:155, function(x) mean(krige(log(zinc) ~
+     1, meuse[-x, ], meuse.grid, v.fit)$var1.var))
> which(m1 == min(m1))
```

which will point to observation 72 as the first candidate for removal. Looking at the sorted magnitudes of change in mean kriging variance by

```
> plot(sort(m1))
```

will reveal that for several other candidate points their removal will have an almost identical effect on the mean variance.

Another approach could be, for example to delineate say the 500 ppm contour. We could, for example express the doubt about whether a location is below or above 500 as the closeness of $G((\hat{Z}(s_0) - 500)/\sigma(s_0))$ to 0.5, with $G(\cdot)$ the Gaussian distribution function.

```
> cutoff <- 1000
> f <- function(x) {
+     kr = krige(log(zinc) ~ 1, meuse[-x, ], meuse.grid,
+         v.fit)
+     mean(abs(pnorm((kr$var1.pred - log(cutoff))/sqrt(kr$var1.var)) -
+         0.5))
+ }
> m2 <- sapply(1:155, f)
> which(m2 == max(m2))
```

Figure 8.16 shows that different objectives lead to different candidate points. Also, deciding based on the kriging variance alone results in an outcome that is highly predictable from the points configuration alone: points in the densest areas are candidate for removal.

For *adding* observation points, one could loop over a fixed grid and find the point that increases the objective most; this is more work as the number

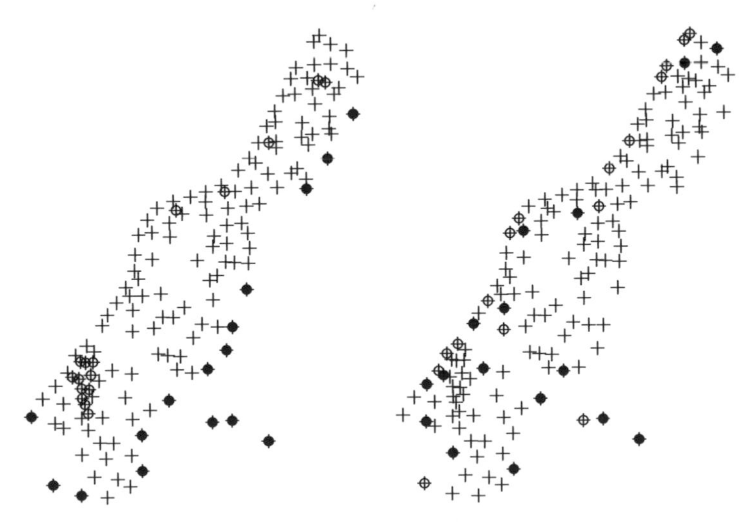

Fig. 8.16. Candidate points for removal. (*Left*) For mean kriging variance, (*right*) for delineating the 1,000 ppm contour. (*Open circles*) 10% most favourite points, (*closed circles*) 10% least favourite points

of options for addition are much larger than that for removal. Evaluating on the kriging variance is not a problem, as the observation value does not affect the kriging variance. For the second criterion, it does.

The problem when two or more points have to be added or removed *jointly* becomes computationally very intensive, as the sequential solution (first the best point, then the second best) is not necessarily equal to the joint solution, for example which configuration of n points is best. Instead of an exhaustive search a more clever optimisation strategy such as simulated annealing or a genetic algorithm should be used.

Package **fields** has a function `cover.design` that finds a set of points on a finite grid that minimises a geometric space-filling criterion

8.10 Other R Packages for Interpolation and Geostatistics

8.10.1 Non-Geostatistical Interpolation

Other interpolation methods may be used, for example based on generalised additive models or smoothers, such as given in package **mgcv**. Additive models in coordinates without interaction will not yield rotation-invariant solutions, but two-dimensional smoothing splines will. The interested reader is referred to Wood (2006).

Package **fields** also provides a function `Tps`, for thin plate (smoothing) splines. Package **akima** provides interpolation methods based on bilinear or bicubic splines (Akima, 1978).

Package **stinepack** provides a 'consistently well behaved method of interpolation based on piecewise rational functions using Stineman's algorithm' (Stineman, 1980).

An interpolation method that also has this property but that does take observation configuration into account is *natural neighbour* interpolation (Sibson, 1981), but this is not available in R.

None of the packages mentioned in this sub-section accept or return data in one of the `Spatial` classes of package **sp**.

8.10.2 spatial

Package **spatial** is part of the **VR** bundle that comes with Venables and Ripley (2002) and is probably one of the oldest spatial packages available in R. It provides calculation of spatial correlation using functions `correlogram` or `variogram`. It allows ordinary and universal point kriging over a grid for spherical, exponential, and Gaussian covariance models. For universal kriging predictors it only allows polynomials in the spatial coordinates. Function `surf.gls` fits a trend surface (i.e. a polynomial in the coordinates) by generalised least squares, and it has a `predict` method.

8.10.3 RandomFields

RandomFields (version 1.3.30) offers sample variogram computation, variogram fitting by least squares or maximum likelihood or restricted maximum likelihood. Simple and ordinary point kriging are provided, and unconditional and conditional simulation using a large variety of modern simulation methods different from sequential simulation, many of them based on spectral methods and Fourier transforms; their abbreviated code is shown as column labels in the table below; an explanation of them is available in the help for `PrintMethodList`.

A large class of covariance functions was proved, shown as row labels in the table below. Their functional form and literature references are found in the help for `PrintModelList`:

```
> PrintModelList()

                        List of models
                        ==============
                 [See also PrintMethodList()]

             Circ cut intr TBM2 TBM3 spec drct nugg add hyp oth
      bessel  X    -    -    -    -    X    X    -    -   -   -
      cauchy  X    X    X    X    X    -    X    -    -   -   -
   cauchytbm  X    -    -    +    X    -    X    -    -   -   -
    circular  X    -    -    +    -    -    X    -    X   -   -
        cone  -    -    -    -    -    -    -    -    X   -   -
    constant  X    -    -    X    X    -    X    -    -   -   -
       cubic  X    -    -    +    X    -    X    -    -   -   -
      cutoff  X    h    -    -    -    -    X    -    -   -   -
dampedcosine  X    -    -    +    X    -    X    -    -   -   -
 exponential  X    X    X    X    X    X    X    -    -   X   -
          FD  X    -    -    -    -    -    X    -    -   -   -
    fractalB  -    -    X    -    -    -    -    -    -   -   -
  fractgauss  X    -    -    -    -    -    X    -    -   -   -
       gauss  X    -    -    +    X    X    X    -    X   -   -
  gencauchy  X    X    X    +    X    -    X    -    -   -   -
 gengneiting  X    -    -    +    X    -    X    -    -   -   -
     gneiting  X    -    -    +    X    -    X    -    -   -   -
  hyperbolic  X    -    -    +    X    -    X    -    -   -   -
  iacocesare  X    -    -    -    -    -    X    -    -   -   -
        lgd1  X    -    -    +    -    -    X    -    -   -   -
     mastein  X    -    -    -    -    -    X    -    -   -   -
        nsst  X    -    -    +    X    -    X    -    -   -   -
       nsst2  X    -    -    +    X    -    X    -    -   -   -
      nugget  X    o    o    o    o    o    X    X    o   o   o
       penta  X    -    -    +    X    -    X    -    -   -   -
       power  X    -    -    +    X    -    X    -    -   -   -
qexponential  X    -    -    +    X    -    X    -    -   -   -
   spherical  X    -    -    X    X    -    X    -    X   -   -
```

```
         stable  X  X  X  +  X  -  X  -  -  -  -
          Stein  X  -  h  -  -  -  X  -  -  -  -
       steinst1  X  -  -  -  -  -  X  -  -  -  -
           wave  X  -  -  -  -  X  X  -  -  -  -
 whittlematern  X  X  X  +  X  X  X  -  -  -  -
               Circ cut intr TBM2 TBM3 spec drct nugg add hyp oth
Legend:'-': method not available
'X': method available for at least some parameter values
'+': parts are evaluated only approximatively
'o': given method is ignored and an alternative one is used
'h': available only as internal model within the method
```

8.10.4 geoR and geoRglm

In addition to variogram estimation, variogram model function fitting using least squares or (restricted) maximum likelihood (`likfit`), and ordinary and universal point kriging, package **geoR** allows for Bayesian kriging (function `krige.bayes` of (transformed) Gaussian variables). This requires the user to specify priors for each of the variogram model parameters (but not for the trend coefficients); `krige.bayes` will then compute the posterior kriging distribution. The function documentation points to a long list of documents that describe the implementation details. The book by Diggle and Ribeiro Jr. (2007) describes more details and gives worked examples.

Package **geoR** uses its own class for spatial data, called `geodata`. It contains coercion method for point data in **sp** format, try, for example

```
> library(geoR)
> plot(variog(as.geodata(meuse["zinc"]), max.dist = 1500))
```

Package **geoR** also has an `xvalid` function for leave-one-out cross validation that (optionally) allows for re-estimating model parameters (trend and variogram coefficients) when leaving out each observation. It also provides the `eyefit`, for interactive visual fitting of functions to sample variograms (see Sect. 8.4.3).

Package **geoRglm** extends package **geoR** for binomial and Poisson processes, and includes Bayesian and conventional kriging methods for trans-Gaussian processes. It mostly uses MCMC approaches, and may be slow for larger data sets.

8.10.5 fields

Package **fields** is an R package for curve and function fitting with an emphasis on spatial data. The main spatial prediction methods are thin plate splines (function `Tps`) and kriging (function `Krig` and `krig.image`). The kriging functions allow you to supply a covariance function that is written in native code. Functions that are positive definite on a sphere (i.e. for unprojected data) are available. Function `cover.design` is written for monitoring network optimisation.

9

Areal Data and Spatial Autocorrelation

9.1 Introduction

Spatial data are often observed on polygon entities with defined boundaries. The polygon boundaries are defined by the researcher in some fields of study, may be arbitrary in others and may be administrative boundaries created for very different purposes in others again. The observed data are frequently aggregations within the boundaries, such as population counts. The areal entities may themselves constitute the units of observation, for example when studying local government behaviour where decisions are taken at the level of the entity, for example setting local tax rates. By and large, though, areal entities are aggregates, bins, used to tally measurements, like voting results at polling stations. Very often, the areal entities are an exhaustive tessellation of the study area, leaving no part of the total area unassigned to an entity. Of course, areal entities may be made up of multiple geometrical entities, such as islands belonging to the same county; they may also surround other areal entities completely, and may contain holes, like lakes.

The boundaries of areal entities may be defined for some other purpose than their use in data analysis. Postal code areas can be useful for analysis, but were created to facilitate postal deliveries. It is only recently that national census organisations have accepted that frequent, apparently justified, changes to boundaries are a major problem for longitudinal analyses. In Sect. 5.1, we discussed the concept of spatial support, which here takes the particular form of the *modifiable areal unit problem* (Waller and Gotway, 2004, pp. 104–108). Arbitrary areal unit boundaries are a problem if their modification could lead to different results, with the case of political gerrymandering being a sobering reminder of how changes in aggregation may change outcomes.[1] They may also

[1] The CRAN **BARD** package for automated redistricting and heuristic exploration of redistricter revealed preference is an example of the use of R for studying this problem.

get in the way of the analysis if the spatial scale or footprint of an underlying data generating process is not matched by the chosen boundaries.

If data collection can be designed to match the areal entities to the data, the influence of the choice of aggregates will be reduced. An example could be the matching of labour market data to local labour markets, perhaps defined by journeys to work. On the other hand, if we are obliged to use arbitrary boundaries, often because there are no other feasible sources of secondary data, we should be aware of potential difficulties. Such mismatches are among the reasons for finding spatial autocorrelation in analysing areal aggregates; other reasons include substantive spatial processes in which entities influence each other by contagion, such as the adoption of similar policies by neighbours, and model misspecification leaving spatially patterned information in the model residuals. These causes of observed spatial autocorrelation can occur in combination, making the correct identification of the actual spatial processes an interesting undertaking.

A wide range of scientific disciplines have encountered spatial autocorrelation among areal entities, with the term 'Galton's problem' used in several. The problem is to establish how many effectively independent observations are present, when arbitrary boundaries have been used to divide up a study area. In his exchange with Tyler in 1889, Galton questioned whether observations of marriage laws across areal entities constituted independent observations, since they could just reflect a general pattern from which they had all descended. So positive spatial dependence tends to reduce the amount of information contained in the observations, because proximate observations can in part be used to predict each other.

In Chap. 8, we have seen how distances on a continuous surface can be used to structure spatial autocorrelation, for example with the variogram. Here we will be concerned with areal entities that are defined as neighbours, for chosen definitions of neighbours. On a continuous surface, all points are neighbours of each other, though some may carry very little weight, because they are very distant. On a tessellated surface, we can choose neighbour definitions that partition the set of all entities (excluding observation i) into members or non-members of the neighbour set of observation i. We can also decide to give each neighbour relationship an equal weight, or vary the weights on the arcs of the directed graph describing the spatial dependence.

The next two sections will cover the construction of neighbours and of weights that can be applied to neighbourhoods. Once this important and often demanding prerequisite is in place, we go on to look at ways of measuring spatial autocorrelation, bearing in mind that the spatial patterning we find may only indicate that our current model of the data is not appropriate. This applies to areal units not fitting the data generation process, to missing variables including variables with the wrong functional form, and differences between our assumptions about the data and their actual distributions, often shown as over-dispersion in count data. The modelling of areal data will be dealt with in the next chapter, with extensions in Chap. 11.

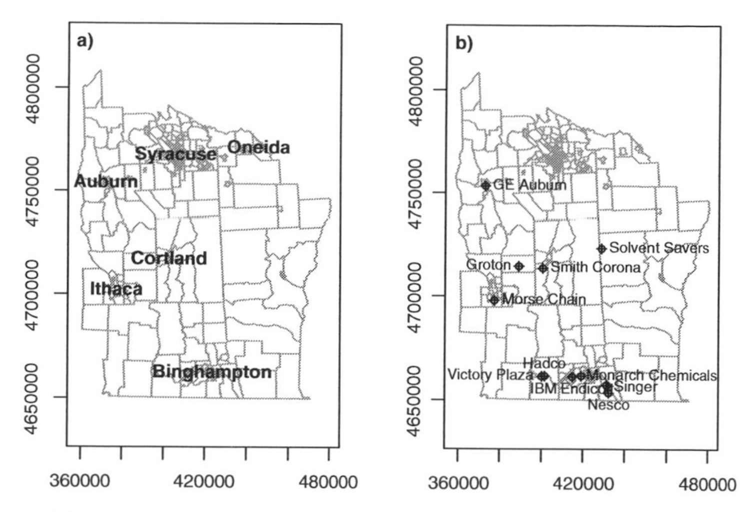

Fig. 9.1. (**a**) Major cities in the eight-county upper New York State study area; (**b**) locations of 11 inactive hazardous waste sites in the study area

While the tests build on models of spatial processes, we look at tests first, and only subsequently move on to modelling. We will also be interested to show how spatial autocorrelation can be introduced into independent data, so that simulations can be undertaken. The 281 census tract data set for eight central New York State counties featured prominently in Waller and Gotway (2004) will be used in many of the examples,[2] supplemented with tract boundaries derived from TIGER 1992 and distributed by SEDAC/CIESIN. This file is not identical with the boundaries used in the original source, but is very close and may be re-distributed, unlike the version used in the book. The area has an extent of about 160 km north–south and 120 km east–west; Fig. 9.1 shows the major cities in the study area and the location of 11 hazardous waste sites. The figures in Waller and Gotway (2004) include water bodies, which are not present in this version of the tract boundaries, in which tract boundaries follow the centre lines of lakes, rather than their shores.

9.2 Spatial Neighbours

Creating spatial weights is a necessary step in using areal data, perhaps just to check that there is no remaining spatial patterning in residuals. The first step is to define which relationships between observations are to be given a non-zero weight, that is to choose the neighbour criterion to be used; the second is to assign weights to the identified neighbour links. Trying to detect pattern in maps of residuals visually is not an acceptable choice, although one sometimes

[2] The boundaries have been projected from geographical coordinates to UTM zone 18.

hears comments explaining the lack of formal analysis such as 'they looked random', or alternatively 'I can see the clusters'. Making the neighbours and weights is, however, not easy to do, and so a number of functions are included in the **spdep** package to help. Further functions are found in some ecology packages, such as the **ade4** package – this package also provides `nb2neig` and `neig2nb` converters for inter-operability. The construction of spatial weights is touched on by Cressie (1993, pp. 384–385), Schabenberger and Gotway (2005, p. 18), Waller and Gotway (2004, pp. 223–225), Fortin and Dale (2005, pp. 113–118), O'Sullivan and Unwin (2003, pp. 193–194) and Banerjee et al. (2004, pp. 70–71). The paucity of treatments in the literature contrasts with the strength of the prior information being introduced by the analyst at this stage, and is why we have chosen to devote a more than proportionally large part of the book to this topic, since analysing areal data is crucially dependent on the choices made in constructing the spatial weights.

9.2.1 Neighbour Objects

In the **spdep** package, neighbour relationships between n observations are represented by an object of class `nb`; the class is an old-style class as presented on p. 24. It is a list of length n with the index numbers of neighbours of each component recorded as an integer vector. If any observation has no neighbours, the component contains an integer zero. It also contains attributes, typically a vector of character region identifiers, and a logical value indicating whether the relationships are symmetric. The region identifiers can be used to check for integrity between the data themselves and the neighbour object. The helper function `card` returns the cardinality of the neighbour set for each object, that is, the number of neighbours; it differs from the application of `length` to the list components because no-neighbour entries are coded as a single element integer vector with the value of zero.

```
> library(spdep)

> library(rgdal)
> NY8 <- readOGR(".", "NY8_utm18")
> NY_nb <- read.gal("NY_nb.gal", region.id = row.names(as(NY8,
+     "data.frame")))

> summary(NY_nb)

Neighbour list object:
Number of regions: 281
Number of nonzero links: 1522
Percentage nonzero weights: 1.927534
Average number of links: 5.41637
Link number distribution:

 1  2  3  4  5  6  7  8  9 10 11
 6 11 28 45 59 49 45 23 10  3  2
```

```
6 least connected regions:
55 97 100 101 244 245 with 1 link
2 most connected regions:
34 82 with 11 links
```

```
> isTRUE(all.equal(attr(NY_nb, "region.id"), row.names(as(NY8,
+     "data.frame"))))
```

```
[1] TRUE
```

```
> plot(NY8, border = "grey60")
> plot(NY_nb, coordinates(NY8), pch = 19, cex = 0.6, add = TRUE)
```

Starting from the census tract contiguities used in Waller and Gotway (2004) and provided as a DBF file on their website, a GAL format file has been created and read into R– we return to the import and export of neighbours on p. 255. Since we now have an `nb` object to examine, we can present the standard methods for these objects. There are `print`, `summary`, `plot`, and other methods; the `summary` method presents a table of the link number distribution, and both `print` and `summary` methods report asymmetry and the presence of no-neighbour observations; asymmetry is present when i is a neighbour of j but j is not a neighbour of i. Figure 9.2 shows the complete neighbour graph for the eight-county study area. For the sake of simplicity in showing how to create neighbour objects, we work on a subset of the map consisting of the census tracts within Syracuse, although the same principles apply to the full data set. We retrieve the part of the neighbour list in Syracuse using the `subset` method.

```
> Syracuse <- NY8[NY8$AREANAME == "Syracuse city", ]
> Sy0_nb <- subset(NY_nb, NY8$AREANAME == "Syracuse city")
> isTRUE(all.equal(attr(Sy0_nb, "region.id"), row.names(as(Syracuse,
+     "data.frame"))))
```

```
[1] TRUE
```

```
> summary(Sy0_nb)
```

```
Neighbour list object:
Number of regions: 63
Number of nonzero links: 346
Percentage nonzero weights: 8.717561
Average number of links: 5.492063
Link number distribution:

 1  2  3  4  5  6  7  8  9
 1  1  5  9 14 17  9  6  1
1 least connected region:
164 with 1 link
1 most connected region:
136 with 9 links
```

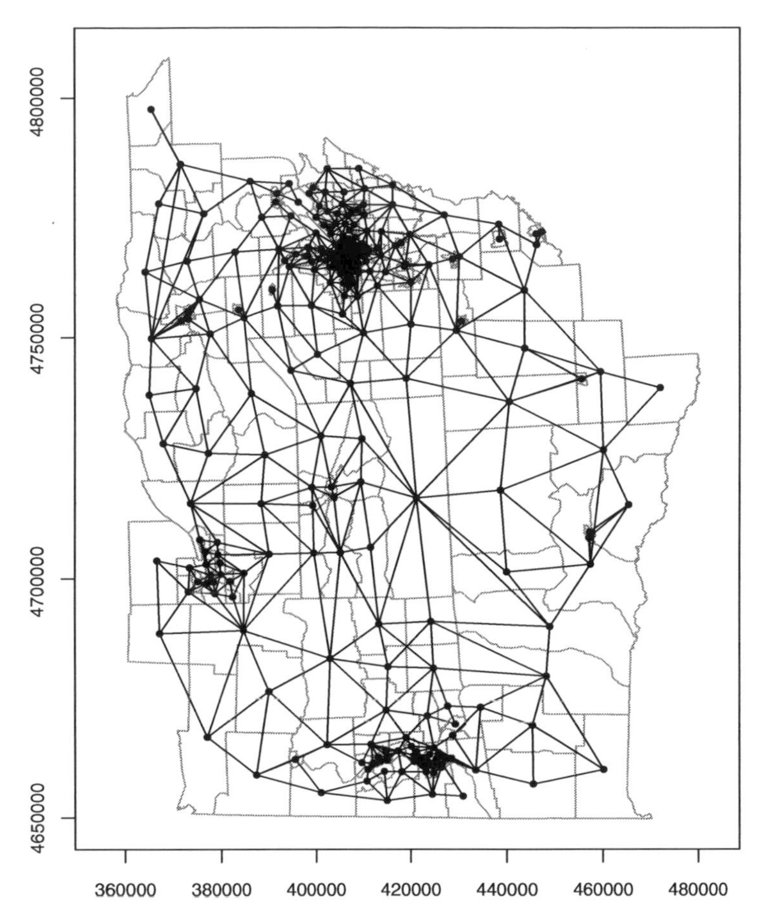

Fig. 9.2. Census tract contiguities, New York eight-county census tracts

9.2.2 Creating Contiguity Neighbours

We can create a copy of the same neighbours object for polygon contiguities using the `poly2nb` function in **spdep**. It takes an object extending the `SpatialPolygons` class as its first argument, and using heuristics identifies polygons sharing boundary points as neighbours. It also has a `snap` argument, to allow the shared boundary points to be a short distance from one another.

```
> class(Syracuse)

[1] "SpatialPolygonsDataFrame"
attr(,"package")
[1] "sp"

> Sy1_nb <- poly2nb(Syracuse)
> isTRUE(all.equal(Sy0_nb, Sy1_nb, check.attributes = FALSE))

[1] TRUE
```

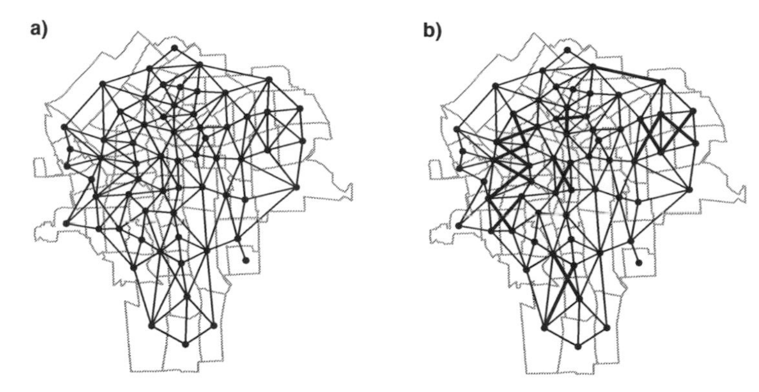

Fig. 9.3. (**a**) Queen-style census tract contiguities, Syracuse; (**b**) Rook-style contiguity differences shown as thicker lines

As we can see, creating the contiguity neighbours from the `Syracuse` object reproduces the neighbours from Waller and Gotway (2004). Careful examination of Fig. 9.2 shows, however, that the graph of neighbours is not planar, since some neighbour links cross each other. By default, the contiguity condition is met when at least one point on the boundary of one polygon is within the snap distance of at least one point of its neighbour. This relationship is given by the argument `queen=TRUE` by analogy with movements on a chessboard. So when three or more polygons meet at a single point, they all meet the contiguity condition, giving rise to crossed links. If `queen=FALSE`, at least two boundary points must be within the snap distance of each other, with the conventional name of a 'rook' relationship. Figure 9.3 shows the crossed line differences that arise when polygons touch only at a single point, compared to the stricter rook criterion.

```
> Sy2_nb <- poly2nb(Syracuse, queen = FALSE)
> isTRUE(all.equal(Sy0_nb, Sy2_nb, check.attributes = FALSE))

[1] FALSE
```

If we have access to a GIS such as GRASS or ArcGIS™, we can export the `SpatialPolygonsDataFrame` object and use the topology engine in the GIS to find contiguities in the graph of polygon edges – a shared edge will yield the same output as the rook relationship. Integration with GRASS was discussed in Sect. 4.4, and functions in **RArcInfo** and the equivalent `readOGR` function in **rgdal** for reading ArcGIS™ coverages in Sects. 4.2.2 and 4.2.1[3] can also be used for retrieving rook neighbours.

This procedure does, however, depend on the topology of the set of polygons being clean, which holds for this subset, but not for the full eight-county data set. Not infrequently, there are small artefacts, such as slivers where boundary lines intersect or diverge by distances that cannot be seen on plots,

[3] A script to access ArcGIS™ coverages using Python and R(D)COM using `readOGR` is on the book website.

but which require intervention to keep the geometries and data correctly associated. When these geometrical artefacts are present, the topology is not clean, because unambiguous shared polygon boundaries cannot be found in all cases; artefacts typically arise when data collected for one purpose are combined with other data or used for another purpose. Topologies are usually cleaned in a GIS by 'snapping' vertices closer than a threshold distance together, removing artefacts – for example, snapping across a river channel where the correct boundary is the median line but the input polygons stop at the channel banks on each side. The `poly2nb` function does have a `snap` argument, which may also be used when input data possess geometrical artefacts.

```
> library(spgrass6)
> writeVECT6(Syracuse, "SY0")
> contig <- vect2neigh("SY0")

> Sy3_nb <- sn2listw(contig)$neighbours
> isTRUE(all.equal(Sy3_nb, Sy2_nb, check.attributes = FALSE))

[1] TRUE
```

Similar approaches may also be used to read ArcGIS™ coverage data by tallying the left neighbour and right neighbour arc indices with the polygons in the data set, using either **RArcInfo** or **rgdal**.

In our Syracuse case, there are no exclaves or 'islands' belonging to the data set, but not sharing boundary points within the snap distance. If the number of polygons is moderate, the missing neighbour links may be added interactively using the `edit` method for `nb` objects, and displaying the polygon background. The same method may be used for removing links which, although contiguity exists, may be considered void, such as across a mountain range.

9.2.3 Creating Graph-Based Neighbours

Continuing with irregularly located areal entities, it is possible to choose a point to represent the polygon-support entities. This is often the polygon centroid, which is not the average of the coordinates in each dimension, but takes proper care to weight the component triangles of the polygon by area. It is also possible to use other points, or if data are available, construct, for example population-weighted centroids. Once representative points are available, the criteria for neighbourhood can be extended from just contiguity to include graph measures, distance thresholds, and k-nearest neighbours.

The most direct graph representation of neighbours is to make a Delaunay triangulation of the points, shown in the first panel in Fig. 9.4. The neighbour relationships are defined by the triangulation, which extends outwards to the convex hull of the points and which is planar. Note that graph-based representations construct the interpoint relationships based on Euclidean distance, with no option to use Great Circle distances for geographical coordinates. Because it joins distant points around the convex hull, it may be worthwhile

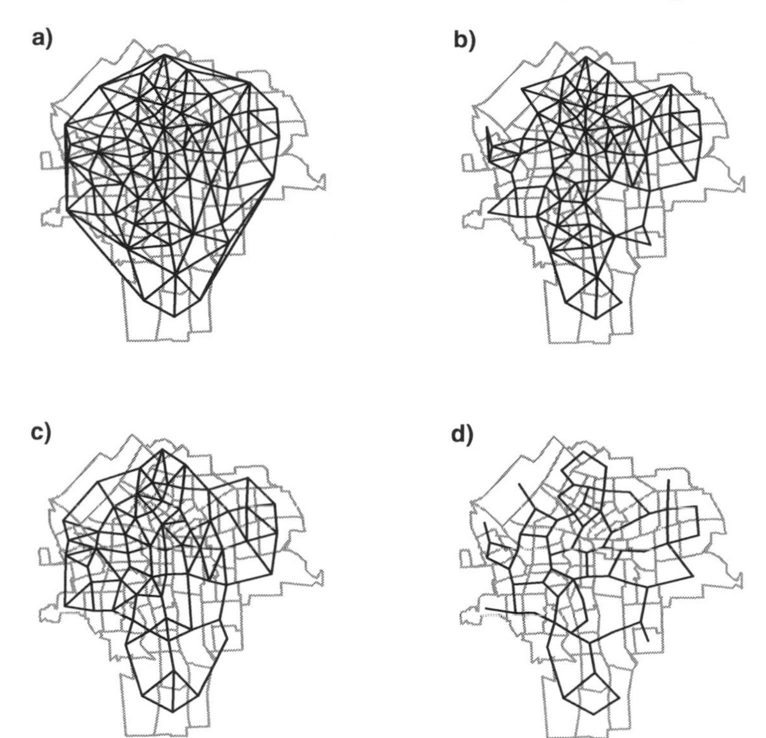

Fig. 9.4. (**a**) Delauney triangulation neighbours; (**b**) Sphere of influence neighbours; (**c**) Gabriel graph neighbours; (**d**) Relative graph neighbours

to thin the triangulation as a Sphere of Influence (SOI) graph, removing links that are relatively long. Points are SOI neighbours if circles centred on the points, of radius equal to the points' nearest neighbour distances, intersect in two places (Avis and Horton, 1985).[4]

```
> coords <- coordinates(Syracuse)
> IDs <- row.names(as(Syracuse, "data.frame"))
> library(tripack)
> Sy4_nb <- tri2nb(coords, row.names = IDs)
> Sy5_nb <- graph2nb(soi.graph(Sy4_nb, coords), row.names = IDs)
> Sy6_nb <- graph2nb(gabrielneigh(coords), row.names = IDs)
> Sy7_nb <- graph2nb(relativeneigh(coords), row.names = IDs)
```

Delaunay triangulation neighbours and SOI neighbours are symmetric by design – if i is a neighbour of j, then j is a neighbour of i. The Gabriel graph is also a subgraph of the Delaunay triangulation, retaining a different set of neighbours (Matula and Sokal, 1980). It does not, however, guarantee symmetry; the same applies to Relative graph neighbours (Toussaint, 1980). The `graph2nb` function takes a `sym` argument to insert links to restore symmetry,

[4] Functions for graph-based neighbours were kindly contributed by Nicholas Lewin-Koh.

but the graphs then no longer exactly fulfil their neighbour criteria. All the graph-based neighbour schemes always ensure that all the points will have at least one neighbour. Subgraphs of the full triangulation may also have more than one graph after trimming. The functions `is.symmetric.nb` can be used to check for symmetry, with argument `force=TRUE` if the symmetry attribute is to be overridden, and `n.comp.nb` reports the number of graph components and the components to which points belong (after enforcing symmetry, because the algorithm assumes that the graph is not directed). When there are more than one graph component, the matrix representation of the spatial weights can become block-diagonal if observations are appropriately sorted.

```
> nb_l <- list(Triangulation = Sy4_nb, SOI = Sy5_nb, Gabriel = Sy6_nb,
+     Relative = Sy7_nb)
> sapply(nb_l, function(x) is.symmetric.nb(x, verbose = FALSE,
+     force = TRUE))
```

```
Triangulation           SOI       Gabriel      Relative
         TRUE          TRUE         FALSE         FALSE
```

```
> sapply(nb_l, function(x) n.comp.nb(x)$nc)
```

```
Triangulation           SOI       Gabriel      Relative
            1             1             1             1
```

9.2.4 Distance-Based Neighbours

An alternative method is to choose the k nearest points as neighbours – this adapts across the study area, taking account of differences in the densities of areal entities. Naturally, in the overwhelming majority of cases, it leads to asymmetric neighbours, but will ensure that all areas have k neighbours. The `knearneigh` returns an intermediate form converted to an `nb` object by `knn2nb`; `knearneigh` can also take a `longlat` argument to handle geographical coordinates.

```
> Sy8_nb <- knn2nb(knearneigh(coords, k = 1), row.names = IDs)
> Sy9_nb <- knn2nb(knearneigh(coords, k = 2), row.names = IDs)
> Sy10_nb <- knn2nb(knearneigh(coords, k = 4), row.names = IDs)
> nb_l <- list(k1 = Sy8_nb, k2 = Sy9_nb, k4 = Sy10_nb)
> sapply(nb_l, function(x) is.symmetric.nb(x, verbose = FALSE,
+     force = TRUE))
```

```
   k1    k2    k4
FALSE FALSE FALSE
```

```
> sapply(nb_l, function(x) n.comp.nb(x)$nc)
```

```
k1 k2 k4
15  1  1
```

Figure 9.5 shows the neighbour relationships for $k = 1, 2, 4$, with many components for $k = 1$. If need be, k-nearest neighbour objects can be made symmetrical using the `make.sym.nb` function. The $k = 1$ object is also useful in

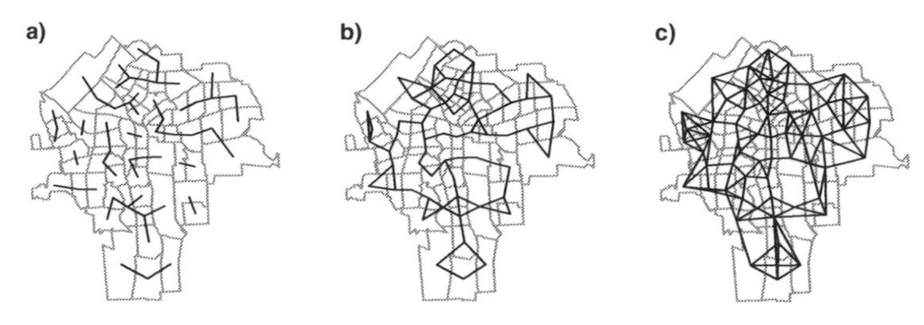

Fig. 9.5. (**a**) $k = 1$ neighbours; (**b**) $k = 2$ neighbours; (**c**) $k = 4$ neighbours

finding the minimum distance at which all areas have a distance-based neighbour. Using the `nbdists` function, we can calculate a list of vectors of distances corresponding to the neighbour object, here for first nearest neighbours. The greatest value will be the minimum distance needed to make sure that all the areas are linked to at least one neighbour. The `dnearneigh` function is used to find neighbours with an interpoint distance, with arguments `d1` and `d2` setting the lower and upper distance bounds; it can also take a `longlat` argument to handle geographical coordinates.

```
> dsts <- unlist(nbdists(Sy8_nb, coords))
> summary(dsts)

   Min. 1st Qu.  Median    Mean 3rd Qu.    Max.
  395.7   587.3   700.1   760.4   906.1  1545.0

> max_1nn <- max(dsts)
> max_1nn

[1] 1544.615

> Sy11_nb <- dnearneigh(coords, d1 = 0, d2 = 0.75 * max_1nn,
+     row.names = IDs)
> Sy12_nb <- dnearneigh(coords, d1 = 0, d2 = 1 * max_1nn,
+     row.names = IDs)
> Sy13_nb <- dnearneigh(coords, d1 = 0, d2 = 1.5 * max_1nn,
+     row.names = IDs)
> nb_l <- list(d1 = Sy11_nb, d2 = Sy12_nb, d3 = Sy13_nb)
> sapply(nb_l, function(x) is.symmetric.nb(x, verbose = FALSE,
+     force = TRUE))

  d1   d2   d3
TRUE TRUE TRUE

> sapply(nb_l, function(x) n.comp.nb(x)$nc)

d1 d2 d3
 4  1  1
```

Figure 9.6 shows how the numbers of distance-based neighbours increase with moderate increases in distance. Moving from 0.75 times the minimum

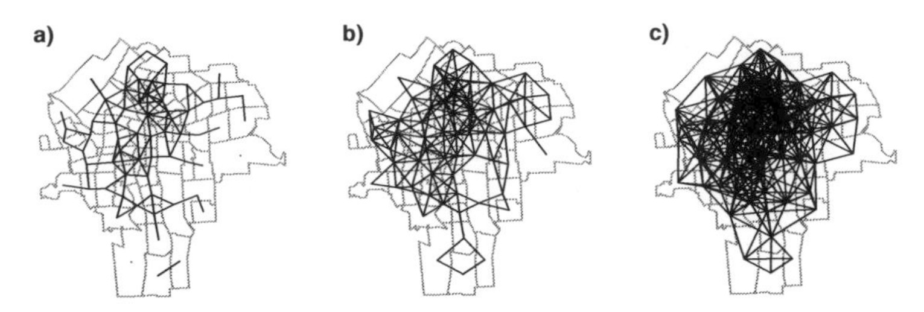

Fig. 9.6. (**a**) Neighbours within 1,158 m; (**b**) neighbours within 1,545 m; (**c**) neighbours within 2,317 m

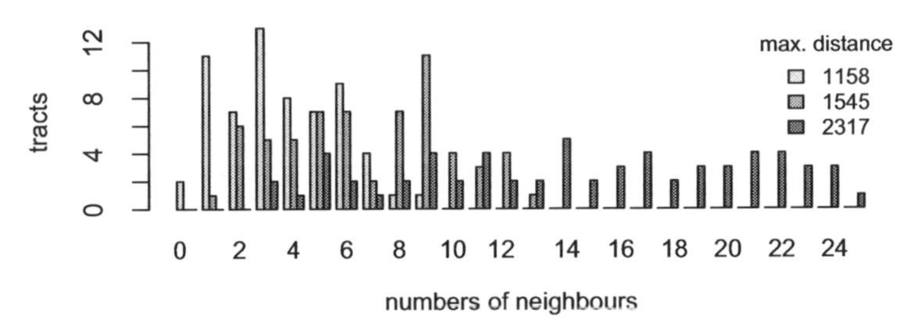

Fig. 9.7. Distance-based neighbours: frequencies of numbers of neighbours by census tract

all-included distance, to the all-included distance, and 1.5 times the minimum all-included distance, the numbers of links grow rapidly. This is a major problem when some of the first nearest neighbour distances in a study area are much larger than others, since to avoid no-neighbour areal entities, the distance criterion will need to be set such that many areas have many neighbours. Figure 9.7 shows the counts of sizes of sets of neighbours for the three different distance limits. In Syracuse, the census tracts are of similar areas, but were we to try to use the distance-based neighbour criterion on the eight-county study area, the smallest distance securing at least one neighbour for every areal entity is over 38 km.

```
> dsts0 <- unlist(nbdists(NY_nb, coordinates(NY8)))
> summary(dsts0)

   Min. 1st Qu.  Median    Mean 3rd Qu.    Max.
   82.7  1505.0  3379.0  5866.0  8954.0 38440.0
```

If the areal entities are approximately regularly spaced, using distance-based neighbours is not necessarily a problem. Provided that care is taken to handle the side effects of ‘weighting’ areas out of the analysis, using lists of neighbours with no-neighbour areas is not necessarily a problem either, but certainly ought to raise questions. Different disciplines handle the definition of neighbours in their own ways by convention; in particular, it seems that ecologists frequently use distance bands. If many distance bands are used, they approach the variogram, although the underlying understanding of spatial autocorrelation seems to be by contagion rather than continuous.

9.2.5 Higher-Order Neighbours

Distance bands can be generated by using a sequence of **d1** and **d2** argument values for the **dnearneigh** function if needed to construct a spatial autocorrelogram as understood in ecology. In other conventions, correlograms are constructed by taking an input list of neighbours as the first-order sets, and stepping out across the graph to second-, third-, and higher-order neighbours based on the number of links traversed, but not permitting cycles, which could risk making i a neighbour of i itself (O'Sullivan and Unwin, 2003, p. 203). The **nblag** function takes an existing neighbour list and returns a list of lists, from first to **maxlag** order neighbours.

```
> Sy0_nb_lags <- nblag(Sy0_nb, maxlag = 9)
```

Table 9.1 shows how the wave of connectedness in the graph spreads to the third order, receding to the eighth order, and dying away at the ninth

Table 9.1. Higher-order contiguities: frequencies of numbers of neighbours by order of neighbour list

	First	Second	Third	Fourth	Fifth	Sixth	Seventh	Eighth	Ninth
0	0	0	0	0	0	6	21	49	63
1	1	0	0	0	0	3	7	6	0
2	1	0	0	0	0	0	4	5	0
3	5	0	0	0	1	2	5	2	0
4	9	2	0	0	1	8	9	1	0
5	14	2	0	0	3	2	7	0	0
6	17	0	0	0	1	5	3	0	0
7	9	6	1	0	1	5	5	0	0
8	6	6	3	1	3	4	1	0	0
9	1	11	5	3	7	8	0	0	0
10	0	11	5	5	13	9	0	0	0
11	0	4	7	7	12	5	0	0	0
12	0	3	14	16	8	5	1	0	0
13	0	7	6	16	9	1	0	0	0
14	0	4	8	5	3	0	0	0	0
15	0	6	3	3	1	0	0	0	0
16	0	1	3	3	0	0	0	0	0
17	0	0	0	2	0	0	0	0	0
18	0	0	1	0	0	0	0	0	0
19	0	0	1	1	0	0	0	0	0
20	0	0	1	1	0	0	0	0	0
21	0	0	3	0	0	0	0	0	0
22	0	0	1	0	0	0	0	0	0
23	0	0	0	0	0	0	0	0	0
24	0	0	1	0	0	0	0	0	0

order – there are no tracts nine steps from each other in this graph. Both the distance bands and the graph step order approaches to spreading neighbourhoods can be used to examine the shape of relationship intensities in space, like the variogram, and can be used in attempting to look at the effects of scale.

9.2.6 Grid Neighbours

When the data are known to be arranged in a regular, rectangular grid, the `cell2nb` function can be used to construct neighbour lists, including those on a torus. These are useful for simulations, because, since all areal entities have equal numbers of neighbours, and there are no edges, the structure of the graph is as neutral as can be achieved. Neighbours can either be of type rook or queen.

```
> cell2nb(7, 7, type = "rook", torus = TRUE)
Neighbour list object:
Number of regions: 49
Number of nonzero links: 196
Percentage nonzero weights: 8.163265
Average number of links: 4
> cell2nb(7, 7, type = "rook", torus = FALSE)
Neighbour list object:
Number of regions: 49
Number of nonzero links: 168
Percentage nonzero weights: 6.997085
Average number of links: 3.428571
```

When a regular, rectangular grid is not complete, then we can use knowledge of the cell size stored in the grid topology to create an appropriate list of neighbours, using a tightly bounded distance criterion. Neighbour lists of this kind are commonly found in ecological assays, such as studies of species richness at a national or continental scale. It is also in these settings, with moderately large n, here $n = 3{,}103$, that the use of a sparse, list based representation shows its strength. Handling a 281×281 matrix for the eight-county census tracts is feasible, easy for a 63×63 matrix for Syracuse census tracts, but demanding for a $3{,}103 \times 3{,}103$ matrix.

```
> data(meuse.grid)
> coordinates(meuse.grid) <- c("x", "y")
> gridded(meuse.grid) <- TRUE
> dst <- max(slot(slot(meuse.grid, "grid"), "cellsize"))
> mg_nb <- dnearneigh(coordinates(meuse.grid), 0, dst)
> mg_nb
Neighbour list object:
Number of regions: 3103
Number of nonzero links: 12022
Percentage nonzero weights: 0.1248571
Average number of links: 3.874315
```

```
> table(card(mg_nb))

   1    2    3    4
   1  133  121 2848
```

9.3 Spatial Weights

The literature on spatial weights is surprisingly small, given their importance in measuring and modelling spatial dependence in areal data. Griffith (1995) provides sound practical advice, while Bavaud (1998) seeks to insert conceptual foundations under ad hoc spatial weights. Spatial weights can be seen as a list of weights indexed by a list of neighbours, where the weight of the link between i and j is the kth element of the ith weights list component, and k tells us which of the ith neighbour list component values is equal to j. If j is not present in the ith neighbour list component, j is not a neighbour of i. Consequently, some weights w_{ij} in the $\mathbf{W}$ weights matrix representation will set to zero, where j is not a neighbour of i. Here, we follow Tiefelsdorf et al. (1999) in our treatment, using their abstraction of spatial weights styles.

9.3.1 Spatial Weights Styles

Once the list of sets of neighbours for our study area is established, we proceed to assign spatial weights to each relationship. If we know little about the assumed spatial process, we try to avoid moving far from the binary representation of a weight of unity for neighbours (Bavaud, 1998), and zero otherwise. In this section, we review the ways that weights objects – `listw` objects – are constructed; the class is an old-style class as described on p. 24. Next, the conversion of these objects into dense and sparse matrix representations will be shown, concluding with functions for importing and exporting neighbour and weights objects.

The `nb2listw` function takes a neighbours list object and converts it into a weights object. The default conversion style is `W`, where the weights for each areal entity are standardised to sum to unity; this is also often called row standardisation. The `print` method for `listw` objects shows the characteristics of the underlying neighbours, the style of the spatial weights, and the spatial weights constants used in calculating tests of spatial autocorrelation. The `neighbours` component of the object is the underlying `nb` object, which gives the indexing of the `weights` component.

```
> Sy0_lw_W <- nb2listw(Sy0_nb)
> Sy0_lw_W

Characteristics of weights list object:
Neighbour list object:
Number of regions: 63
```

```
Number of nonzero links: 346
Percentage nonzero weights: 8.717561
Average number of links: 5.492063

Weights style: W
Weights constants summary:
   n   nn S0       S1      S2
W 63 3969 63 24.78291 258.564
```

```
> names(Sy0_lw_W)
```

```
[1] "style"      "neighbours" "weights"
```

```
> names(attributes(Sy0_lw_W))
```

```
[1] "names"     "class"     "region.id" "call"
```

For style="W", the weights vary between unity divided by the largest and smallest numbers of neighbours, and the sums of weights for each areal entity are unity. This spatial weights style can be interpreted as allowing the calculation of average values across neighbours. The weights for links originating at areas with few neighbours are larger than those originating at areas with many neighbours, perhaps boosting areal entities on the edge of the study area unintentionally. This representation is no longer symmetric, but is similar to symmetric – this matters as we see below in Sect. 10.2.1.

```
> 1/rev(range(card(Sy0_lw_W$neighbours)))
```

```
[1] 0.1111111 1.0000000
```

```
> summary(unlist(Sy0_lw_W$weights))
```

```
   Min. 1st Qu.  Median    Mean 3rd Qu.    Max.
 0.1111  0.1429  0.1667  0.1821  0.2000  1.0000
```

```
> summary(sapply(Sy0_lw_W$weights, sum))
```

```
   Min. 1st Qu.  Median    Mean 3rd Qu.    Max.
      1       1       1       1       1       1
```

Setting style="B" – 'binary' – retains a weight of unity for each neighbour relationship, but in this case, the sums of weights for areas differ according to the numbers of neighbour areas have.

```
> Sy0_lw_B <- nb2listw(Sy0_nb, style = "B")
> summary(unlist(Sy0_lw_B$weights))
```

```
   Min. 1st Qu.  Median    Mean 3rd Qu.    Max.
      1       1       1       1       1       1
```

```
> summary(sapply(Sy0_lw_B$weights, sum))
```

```
   Min. 1st Qu.  Median    Mean 3rd Qu.    Max.
  1.000   4.500   6.000   5.492   6.500   9.000
```

Two further styles with equal weights for all links are available: C and U, where the complete set of C weights sums to the number of areas, and U weights sum to unity.

```
> Sy0_lw_C <- nb2listw(Sy0_nb, style = "C")
> length(Sy0_lw_C$neighbours)/length(unlist(Sy0_lw_C$neighbours))

[1] 0.1820809

> summary(unlist(Sy0_lw_C$weights))

   Min. 1st Qu.  Median    Mean 3rd Qu.    Max.
 0.1821  0.1821  0.1821  0.1821  0.1821  0.1821

> summary(sapply(Sy0_lw_C$weights, sum))

   Min. 1st Qu.  Median    Mean 3rd Qu.    Max.
 0.1821  0.8194  1.0920  1.0000  1.1840  1.6390
```

Finally, the use of a variance-stabilising coding scheme has been proposed by Tiefelsdorf et al. (1999) and is provided as `style="S"`. The weights vary, less than for `style="W"`, but the row sums of weights by area vary more than for `style="W"` (where they are alway unity) and less than for styles B, C, or U. This style also makes asymmetric weights, but as with `style="W"`, they may be similar to symmetric if the neighbours list was itself symmetric. In the same way that the choice of the criteria to define neighbours may affect the results in testing or modelling of the use of weights constructed from those neighbours, results may also be changed by the choice of weights style. As indicated above, links coming from areal entities with many neighbours may be either weighted up or down, depending on the choice of style. The variance-stabilising coding scheme seeks to moderate these conflicting impacts.

```
> Sy0_lw_S <- nb2listw(Sy0_nb, style = "S")
> summary(unlist(Sy0_lw_S$weights))

   Min. 1st Qu.  Median    Mean 3rd Qu.    Max.
 0.1440  0.1633  0.1764  0.1821  0.1932  0.4321

> summary(sapply(Sy0_lw_S$weights, sum))

   Min. 1st Qu.  Median    Mean 3rd Qu.    Max.
 0.4321  0.9152  1.0580  1.0000  1.1010  1.2960
```

9.3.2 General Spatial Weights

The `glist` argument can be used to pass a list of vectors of general weights corresponding to the neighbour relationships to `nb2listw`. Say that we believe that the strength of neighbour relationships attenuates with distance, one of the cases considered by Cliff and Ord (1981, pp. 17–18); O'Sullivan and Unwin (2003, pp. 201–202) provide a similar discussion. We could set the weights to be proportional to the inverse distance between points representing the areas, using `nbdists` to calculate the distances for the given `nb` object. Using `lapply` to invert the distances, we can obtain a different structure of spatial weights from those above. If we have no reason to assume any more knowledge about neighbour relations than their existence or absence, this step is potentially misleading. If we do know, on the other hand, that migration or commuting flows describe the spatial weights' structure better than the binary alternative, it may be worth using them as general weights; there may, however, be symmetry problems, because such flows – unlike inverse distances – are only rarely symmetric.

```
> dsts <- nbdists(Sy0_nb, coordinates(Syracuse))
> idw <- lapply(dsts, function(x) 1/(x/1000))
> Sy0_lw_idwB <- nb2listw(Sy0_nb, glist = idw, style = "B")
> summary(unlist(Sy0_lw_idwB$weights))

   Min. 1st Qu.  Median    Mean 3rd Qu.    Max.
 0.3886  0.7374  0.9259  0.9963  1.1910  2.5270

> summary(sapply(Sy0_lw_idwB$weights, sum))

   Min. 1st Qu.  Median    Mean 3rd Qu.    Max.
  1.304   3.986   5.869   5.471   6.737   9.435
```

Figure 9.8 shows three representations of spatial weights for Syracuse displayed as matrices. The `style="W"` image on the left is evidently asymmetric, with darker greys showing larger weights for areas with few neighbours. The other two panels are symmetric, but express different assumptions about the strengths of neighbour relationships.

The final argument to `nb2listw` allows us to handle neighbour lists with no-neighbour areas. It is not obvious that the weight representation of the empty set is zero – perhaps it should be `NA`, which would lead to problems later.

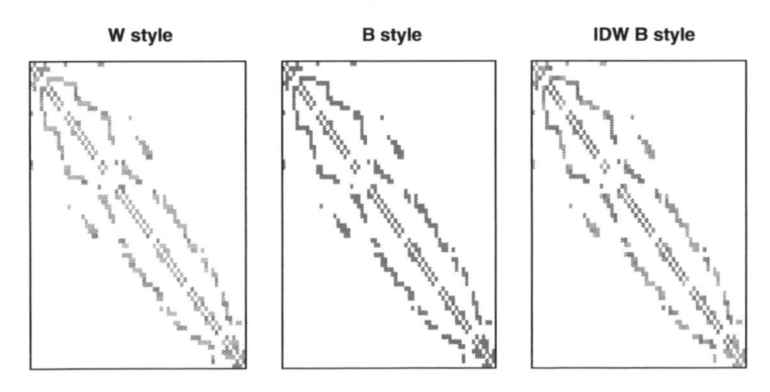

Fig. 9.8. Three spatial weights representations for Syracuse

For this reason, the default value of the argument is `zero.policy=FALSE`, leading to an error when given an `nb` argument with areas with no neighbours. Setting the argument to `TRUE` permits the creation of the spatial weights object, with zero weights. The `zero.policy` argument will subsequently need to be used in each function called, mainly to keep reminding the user that having areal entities with no neighbours is seen as unfortunate. The contrast between the set-based understanding of neighbours and conversion to a matrix representation is discussed by Bivand and Portnov (2004), and boils down to whether the product of a no-neighbour area's weights and an arbitrary n-vector should be a missing value or numeric zero. As we see later (p. 262), keeping the no-neighbour areal entities raises questions about the relevant size of n when testing for autocorrelation, among other issues.

```
> Sy0_lw_D1 <- nb2listw(Sy11_nb, style = "B")

Error in nb2listw(Sy11_nb, style = "B") : Empty neighbour sets found

> Sy0_lw_D1 <- nb2listw(Sy11_nb, style = "B", zero.policy = TRUE)
> print(Sy0_lw_D1, zero.policy = TRUE)

Characteristics of weights list object:
Neighbour list object:
Number of regions: 63
Number of nonzero links: 230
Percentage nonzero weights: 5.79491
Average number of links: 3.650794
2 regions with no links:
154 168

Weights style: B
Weights constants summary:
   n   nn  S0  S1   S2
B 61 3721 230 460 4496
```

The parallel problem of data sets with missing values in variables but with with fully specified spatial weights is approached through the `subset.listw` method, which re-generates the weights for the given subset of areas, for example given by `complete.cases`. Knowing which observations are incomplete, the underlying neighbours and weights can be subsetted in some cases, with the aim of avoiding the propagation of `NA` values when calculating spatially lagged values. Many tests and model fitting functions can carry this out internally if the appropriate argument flag is set, although the careful analyst will prefer to subset the input data and the weights before testing or modelling.

9.3.3 Importing, Converting, and Exporting Spatial Neighbours and Weights

Neighbour and weights objects produced in other software can be imported into R without difficulty, and such objects can be exported to other software too. As examples, some files have been generated in GeoDa[5] from the Syracuse census tracts written out as a shapefile, with the centroid used here stored in the data frame. The first two are for contiguity neighbours, using the queen and rook criteria, respectively. These so-called GAL-format files contain only neighbour information, and are described in detail in the help file accompanying the function `read.gal`.

```
> Sy14_nb <- read.gal("Sy_GeoDa1.GAL")
> isTRUE(all.equal(Sy0_nb, Sy14_nb, check.attributes = FALSE))
```

```
[1] TRUE
```

```
> Sy15_nb <- read.gal("Sy_GeoDa2.GAL")
> isTRUE(all.equal(Sy2_nb, Sy15_nb, check.attributes = FALSE))
```

```
[1] TRUE
```

The `write.nb.gal` function is used to write GAL-format files from `nb` objects. GeoDa also makes GWT-format files, described in the GeoDa documentation and the help file, which also contain distance information for the link between the areas, and are stored in a three-column sparse representation. They can be read using `read.gwt2nb`, here for a four-nearest-neighbour scheme, and only using the neighbour links. In general, **spdep** and GeoDa neighbours and weights are easy to exchange, not least because of generous contributions of code to **spdep** and time for testing by Luc Anselin, who created and administers GeoDa.

```
> Sy16_nb <- read.gwt2nb("Sy_GeoDa4.GWT")
> isTRUE(all.equal(Sy10_nb, Sy16_nb, check.attributes = FALSE))
```

```
[1] TRUE
```

A similar set of functions is available for exchanging spatial weights with the Spatial Econometrics Library[6] created by James LeSage. The sparse representation of weights is similar to the GWT-format and can be imported using `read.dat2listw`. Export to three different formats goes through the `listw2sn` function, which converts a spatial weights object to a three-column sparse representation, similar to the 'spatial.neighbor' class in the `S-PLUS`™ SpatialStats module. The output data frame can be written with `write.table` to a file to be read into `S-PLUS`™, written out as a GWT-format file with `write.sn2gwt` or as a text representation of a sparse matrix for Matlab™ with `write.sn2dat`. There is a function called `listw2WB` for creating a list of spatial weights for WinBUGS, to be written to file using `dput`.

In addition, `listw2mat` can be used to export spatial weights to, among others, Stata for use with the contributed `spatwmat` command there. This is done by writing the matrix out as a Stata™ data file, here for the binary contiguity matrix for Syracuse:

---

[5] `http://www.geoda.uiuc.edu/`, Anselin et al. (2006).

[6] `http://www.spatial-econometrics.com`.

```
> library(foreign)
> df <- as.data.frame(listw2mat(Sy0_lw_B))
> write.dta(df, file = "Sy0_lw_B.dta", version = 7)
```

The `mat2listw` can be used to reverse the process, when a dense weights matrix has been read into R, and needs to be made into a neighbour and weights list object. Unfortunately, this function does not set the `style` of the `listw` object to a known value, using `M` to signal this lack of knowledge. It is then usual to rebuild the `listw` object, treating the `neighbours` component as an `nb` object, the `weights` component as a list of general weights and setting the style in the `nb2listw` function directly. It was used for the initial import of the eight-county contiguities, as shown in detail on the `NY_data` help page provided with **spdep**.

Finally, there is a function `nb2lines` to convert neighbour lists into SpatialLinesDataFrame objects, given point coordinates representing the areas. This allows neighbour objects to be plotted in an alternative way, and if need be, to be exported as shapefiles.

9.3.4 Using Weights to Simulate Spatial Autocorrelation

In Fig. 9.8, use was made of `listw2mat` to turn a spatial weights object into a dense matrix for display. The same function is used for constructing a dense representation of the $(\mathbf{I} - \rho\mathbf{W})$ matrix to simulate spatial autocorrelation within the `invIrW` function, where $\mathbf{W}$ is a weights matrix, ρ is a spatial autocorrelation coefficient, and $\mathbf{I}$ is the identity matrix. This approach was introduced by Cliff and Ord (1973, pp. 146–147), and does not impose strict conditions on the matrix to be inverted (only that it be non-singular), and only applies to simulations from a simultaneous autoregressive process. The underlying framework for the covariance representation used here – simultaneous autoregression – will be presented in Sect. 10.2.1.

Starting with a vector of random numbers corresponding to the number of census tracts in Syracuse, we use the row-standardised contiguity weights to introduce autocorrelation.

```
> set.seed(987654)
> n <- length(Sy0_nb)
> uncorr_x <- rnorm(n)
> rho <- 0.5
> autocorr_x <- invIrW(Sy0_lw_W, rho) %*% uncorr_x
```

The outcome is shown in Fig. 9.9, where the spatial lag plot of the original, uncorrelated variable contrasts with that of the autocorrelated variable, which now has a strong positive relationship between tract values and the spatial lag – here the average of values of neighbouring tracts.

The `lag` method for `listw` objects creates 'spatial lag' values: $\mathrm{lag}(y_i) = \sum_{j \in N_i} w_{ij} y_j$ for observed values y_i; N_i is the set of neighbours of i. If the

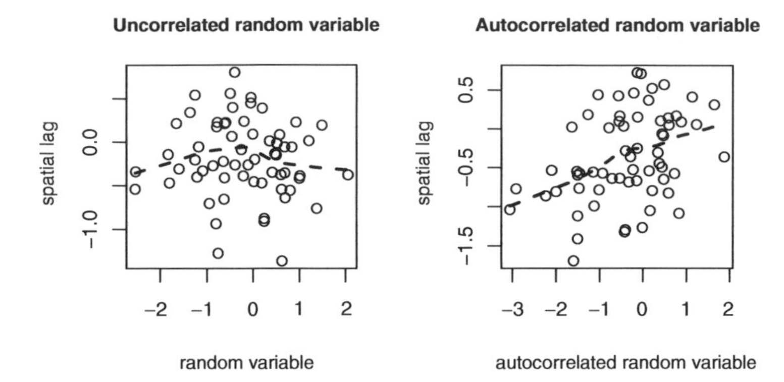

Fig. 9.9. Simulating spatial autocorrelation: spatial lag plots, showing a locally weighted smoother line

weights object style is row-standardisation, the lag(y_i) values will be averages over the sets of neighbours for each i, rather like a moving window defined by N_i and including values weighted by w_{ij}.

9.3.5 Manipulating Spatial Weights

There are three contributed packages providing support for sparse matrices, **SparseM**, **Matrix**, and **spam**. The **spdep** package began by using compiled code shipped with the package for sparse matrix handling, but changed first to **SparseM**, next adding **Matrix** wrappers, and more recently introducing the use of **spam** and deprecating the interface to **SparseM**. The `as.spam.listw` wrapper to the **spam** package `spam` class is used internally in spatial regression functions among others. The `as_dgRMatrix_listw` wrapper provides the same conversion to the **Matrix** `dgRMatrix` class.

A function that is used a good deal within testing and model fitting functions is `listw2U`, which returns a symmetric `listw` object representing the $\frac{1}{2}(W + W^{\mathrm{T}})$ spatial weights matrix.

Analysing areal data is crucially dependent on the construction of the spatial weights, which is why it has taken some time to describe the breadth of choices facing the researcher. We can now go on to test for spatial autocorrelation, and to model using assumptions about underlying spatial processes.

9.4 Spatial Autocorrelation: Tests

Now that we have a range of ways of constructing spatial weights, we can start using them to test for the presence of spatial autocorrelation. Before doing anything serious, it would be very helpful to review the assumptions being made in the tests; we will be using Moran's I as an example, but the consequences apply to other tests too. As Schabenberger and Gotway (2005, pp. 19–23) explain clearly, tests assume that the mean model of the data removes systematic spatial patterning from the data. If we are examining ecological data, but neglect environmental drivers such as temperature, precipitation, or elevation, we should not be surprised if the data seem to display spatial autocorrelation (for a discussion, see Bivand, 2008, pp. 9–15). Such misspecification of the mean model is not at all uncommon, and may be unavoidable where observations on variables needed to specify it correctly are not available. In fact, Cressie (1993, p. 442) only discusses the testing of residual autocorrelation, and then very briefly, preferring to approach autocorrelation through modelling.

Another issue that can arise is that the spatial weights we use for testing are not those that generated the autocorrelation – our chosen weights may, for example not suit the actual scales of interaction between areal entities. This is a reflection of misspecification of the model of the variance of the residuals from the mean model, which can also include making distributional assumptions that are not appropriate for the data, for example assuming homoskedasticity or regular shape parameters (for example, skewness and kurtosis). Some of these can be addressed by transforming the data and by using weighted estimation, but in any case, care is needed in interpreting apparent spatial autocorrelation that may actually stem from misspecification.

The use of global tests for spatial autocorrelation is covered in much more detail that the construction of spatial weights in the spatial data analysis texts that we are tracking. Waller and Gotway (2004, pp. 223–236) follow up the problem of mistaking the misspecification of the mean model for spatial autocorrelation. This is less evident in Fortin and Dale (2005, pp. 122–132) and O'Sullivan and Unwin (2003, pp. 180–203), but they devote more space to join count statistics for categorical data. Banerjee et al. (2004, pp. 71–73) are, like Cressie (1993), more concerned with modelling than testing.

We begin with the simulated variable for the Syracuse census tracts (see Sect. 9.3.4). Since the input variable is known to be drawn at random from the Normal distribution, we can manipulate it to see what happens to test results under different conditions. The test to be used in this introductory discussion is Moran's I, which is calculated as a ratio of the product of the variable of interest and its spatial lag, with the cross-product of the variable of interest, and adjusted for the spatial weights used:

$$I = \frac{n}{\sum_{i=1}^{n}\sum_{j=1}^{n} w_{ij}} \frac{\sum_{i=1}^{n}\sum_{j=1}^{n} w_{ij}(y_i - \bar{y})(y_j - \bar{y})}{\sum_{i=1}^{n}(y_i - \bar{y})^2},$$

where y_i is the ith observation, $\bar{y}$ is the mean of the variable of interest, and w_{ij} is the spatial weight of the link between i and j. Centring on the mean is equivalent to asserting that the correct model has a constant mean, and that any remaining patterning after centring is caused by the spatial relationships encoded in the spatial weights.

Table 9.2. Moran's I test results for five different data generating processes

	I	$E(I)$	var(I)	St. deviate	p-value
uncorr_x	−0.03329	−0.01613	0.00571	−0.227	0.59
autocorr_x	0.2182	−0.0161	0.0057	3.1	0.00096
autocorr_x k=1	0.1921	−0.0161	0.0125	1.86	0.031
trend_x	0.23747	−0.01613	0.00575	3.34	0.00041
lm(trend_x ∼ et)	−0.0538	−0.0309	0.0054	−0.312	0.62

The results for Moran's I are collated in Table 9.2 for five settings. The first column contains the observed value of I, the second is the expectation, which is $-1/(n-1)$ for the mean-centred cases, the third the variance of the statistic under randomisation, next the standard deviate $(I - E(I))/\sqrt{\text{var}(I)}$, and finally the p-value of the test for the alternative that $I > E(I)$. The test results are for the uncorrelated case first (`uncorr_x`) – there is no trace of spatial dependence with these weights. Even though a random drawing could show spatial autocorrelation, we would be unfortunate to find a pattern corresponding to our spatial weights by chance for just one draw. When the spatially autocorrelated variable is tested (`autocorr_x`), it shows, as one would expect, a significant result for these spatial weights. If we use spatial weights that differ from those used to generate the spatial autocorrelation (`autocorr_x k=1`), the value of I falls, and although it is marginally significant, it is worth remembering that, had the generating process been less strong, we might have come to the wrong conclusion based on the choice of spatial weights not matching the actual generating process.

```
> moran_u <- moran.test(uncorr_x, listw = Sy0_lw_W)
> moran_a <- moran.test(autocorr_x, listw = Sy0_lw_W)
> moran_a1 <- moran.test(autocorr_x, listw = nb2listw(Sy9_nb,
+     style = "W"))
```

The final two rows of Table 9.2 show what can happen when our assumption of a constant mean is erroneous (Schabenberger and Gotway, 2005, pp. 22–23). Introducing a gentle trend rising from west to east into the uncorrelated random variable, we have a situation in which there is no underlying spatial autocorrelation, just a simple linear trend. If we assume a constant mean, we reach the wrong conclusion shown in the fourth row of the table (`trend_x`). The final row shows how we get back to the uncorrelated residuals by including the trend in the mean, and again have uncorrelated residuals (`lm(trend_x ∼ et)`).

```
> et <- coords[, 1] - min(coords[, 1])
> trend_x <- uncorr_x + 0.00025 * et
> moran_t <- moran.test(trend_x, listw = Sy0_lw_W)
> moran_t1 <- lm.morantest(lm(trend_x ~ et), listw = Sy0_lw_W)
```

This shows how important it can be to understand that tests for spatial autocorrelation can also react to a misspecified model of the mean, and that the omission of a spatially patterned variable from the mean function will 'look like' spatial autocorrelation to the tests.

9.4.1 Global Tests

Moran's I – `moran.test` – is perhaps the most common global test, and for this reason we continue to use it here. Other global tests implemented in the **spdep** package include Geary's C (`geary.test()`), the global Getis-Ord G (`globalG.test()`), and the spatial general cross product Mantel test, which includes Moran's I, Geary's C, and the Sokal variant of Geary's C as alternative forms (`sp.mantel.mc()`). All these are for continuous variables, with `moran.test()` having an argument to use an adjustment for a ranked continuous variable, that is where the metric of the variable is by the ranks of its values rather than the values themselves. There are also join count tests for categorical variables, with the variable of interest represented as a factor (`joincount.test()` for same colour joins, `joincount.multi()` for same-colour and different colour joins).

The values of these statistics may be of some interest in themselves, but are not directly interpretable. The approach taken most generally is to standardise the observed value by subtracting the analytical expected value, and dividing the difference by the square root of the analytical variance for the spatial weights used, for a set of assumptions. The result is a standard deviate, and is compared with the Normal distribution to find the probability value of the observed statistic under the null hypothesis of no spatial dependence for the chosen spatial weights – most often the test is one-sided, with an alternative hypothesis of the observed statistic being significantly greater than its expected value.

As we see, outcomes can depend on the choices made, for example the style of the weights and to what extent the assumptions made are satisfied. It might seem that Monte Carlo or equivalently bootstrap permutation-based tests, in which the values of the variable of interest are randomly assigned to spatial entities, would provide protection against errors of inference. In fact, because tests for spatial autocorrelation are sensitive to spatial patterning in the variable of interest from any source, they are not necessarily – as we saw above – good guides to decide what is going on in the data generation process. Parametric bootstrapping or tests specifically tuned to the setting – or better specification of the variable of interest – are sometimes needed.

A further problem for which there is no current best advice is how to proceed if some areal entities have no neighbours. By default, test functions in **spdep** do not accept spatial weights with no-neighbour entities unless the `zero.policy` argument is set to `TRUE`. But even if the analyst accepts the presence of rows and columns with only zero entries in the spatial weights matrix, the correct size of n can be taken as the number of observations, or may be reduced to reflect the fact that some of the observations are effectively being ignored. By default, n is adjusted, but the `adjust.n` argument may be set to

FALSE. If n is not adjusted, for example for Moran's I, the absolute value of the statistic will increase, and the absolute value of its expectation and variance will decrease. When measures of autocorrelation were developed, it was generally assumed that all entities would have neighbours, so what one should do when some do not, is not obvious. The problem is not dissimilar to the choice of variogram bin widths and weights in geostatistics (Sect. 8.4.3).

We have already used the New York state eight-county census tract data set for examining the construction of neighbour lists and spatial weights. Now we introduce the data themselves, based on Waller and Gotway (2004, pp. 98, 345–353). There are 281 census tract observations, including as we have seen sparsely populated rural areas contrasting with dense, small, urban tracts. The numbers of incident leukaemia cases are recorded by tract, aggregated from census block groups, but because some cases could not be placed, they were added proportionally to other block groups, leading to non-integer counts. The counts are for the five years 1978–1982, while census variables, such as the tract population, are just for 1980. Other census variables are the percentage aged over 65, and the percentage of the population owning their own home. Exposure to TCE waste sites is represented as the logarithm of 100 times the inverse of the distance from the tract centroid to the nearest site. We return to these covariates in the next chapter.

The first example is of testing the number of cases by census tract (following Waller and Gotway (2004, p. 231)) for autocorrelation using the default spatial weights style of row standardisation, and using the analytical randomisation assumption in computing the variance of the statistic. The outcome, as we see, is that the spatial patterning of the variable of interest is significant, with neighbouring tracts very likely to have similar values for whatever reason.

```
> moran.test(NY8$Cases, listw = nb2listw(NY_nb))

	Moran's I test under randomisation

data:  NY8$Cases
weights: nb2listw(NY_nb)

Moran I statistic standard deviate = 3.978, p-value = 3.477e-05
alternative hypothesis: greater
sample estimates:
Moran I statistic       Expectation          Variance
         0.146883         -0.003571          0.001431
```

Changing the style of the spatial weights to make all weights equal and summing to the number of observations, we see that the resulting probability value is reduced about 20 times – we recall that row-standardisation favours observations with few neighbours, and that styles `'B'`, `'C'`, and `'U'` 'weight up' observations with many neighbours. In this case, style `'S'` comes down between `'C'` and `'W'`.

```
> lw_B <- nb2listw(NY_nb, style = "B")
> moran.test(NY8$Cases, listw = lw_B)

	Moran's I test under randomisation

data:  NY8$Cases
weights: lw_B

Moran I statistic standard deviate = 3.186, p-value = 0.0007207
alternative hypothesis: greater
sample estimates:
Moran I statistic       Expectation          Variance
         0.110387         -0.003571          0.001279
```

By default, `moran.test` uses the randomisation assumption, which differs from the simpler normality assumption by introducing a correction term based on the kurtosis of the variable of interest (here 3.63). When the kurtosis value corresponds to that of a normally distributed variable, the two assumptions yield the same variance, but as the variable departs from normality, the randomisation assumption compensates by increasing the variance and decreasing the standard deviate. In this case, there is little difference and the two return similar outcomes.

```
> moran.test(NY8$Cases, listw = lw_B, randomisation = FALSE)

	Moran's I test under normality

data:  NY8$Cases
weights: lw_B

Moran I statistic standard deviate = 3.183, p-value = 0.0007301
alternative hypothesis: greater
sample estimates:
Moran I statistic       Expectation          Variance
         0.110387         -0.003571          0.001282
```

It is useful to show here that the standard test under normality is in fact the same test as the Moran test for regression residuals for the model, including only the intercept. Making this connection here shows that we could introduce additional variables on the right-hand side of our model, over and above the intercept, and potentially other ways of handling misspecification.

```
> lm.morantest(lm(Cases ~ 1, NY8), listw = lw_B)

	Global Moran's I for regression residuals

data:
model: lm(formula = Cases ~ 1, data = NY8)
weights: lw_B
```

```
Moran I statistic standard deviate = 3.183, p-value = 0.0007301
alternative hypothesis: greater
sample estimates:
Observed Moran's I        Expectation          Variance
          0.110387          -0.003571          0.001282
```

Using the same construction, we can also use a Saddlepoint approximation rather than the analytical normal assumption (Tiefelsdorf, 2002), and an exact test (Tiefelsdorf, 1998, 2000; Hepple, 1998; Bivand et al., 2008). These methods are substantially more demanding computationally, and were originally regarded as impractical. For moderately sized data sets such as the one we are using, however, need less than double the time required for reaching a result. In general, exact and Saddlepoint methods make little difference to outcomes for global tests when the number of spatial entities is not small, as here, with the probability value only changing by a factor of two. We see later that the impact of differences between the normality assumption and the Saddlepoint approximation and exact test is stronger for local indicators of spatial association.

```
> lm.morantest.sad(lm(Cases ~ 1, NY8), listw = lw_B)

	Saddlepoint approximation for global Moran's I
	(Barndorff-Nielsen formula)

data:
model:lm(formula = Cases ~ 1, data = NY8)
weights: lw_B

Saddlepoint approximation = 2.993, p-value = 0.001382
alternative hypothesis: greater
sample estimates:
Observed Moran's I
            0.1104

> lm.morantest.exact(lm(Cases ~ 1, NY8), listw = lw_B)

	Global Moran's I statistic with exact p-value

data:
model:lm(formula = Cases ~ 1, data = NY8)
weights: lw_B

Exact standard deviate = 2.992, p-value = 0.001384
alternative hypothesis: greater
sample estimates:
[1] 0.1104
```

We can also use a Monte Carlo test, a permutation bootstrap test, in which the observed values are randomly assigned to tracts, and the statistic of interest computed nsim times. Since we have enough observations in the global case, we can repeat this permutation potentially very many times without repetition.

```
> set.seed(1234)
> bperm <- moran.mc(NY8$Cases, listw = lw_B, nsim = 999)
> bperm

	Monte-Carlo simulation of Moran's I

data:  NY8$Cases
weights: lw_B
number of simulations + 1: 1000

statistic = 0.1104, observed rank = 998, p-value = 0.002
alternative hypothesis: greater
```

Waller and Gotway (2004, p. 231) also include a Poisson constant risk parametric bootstrap assessment of the significance of autocorrelation in the case counts. The constant global rate r is calculated first, and used to create expected counts for each census tract by multiplying by the population.

```
> r <- sum(NY8$Cases)/sum(NY8$POP8)
> rni <- r * NY8$POP8
> CR <- function(var, mle) rpois(length(var), lambda = mle)
> MoranI.pboot <- function(var, i, listw, n, S0, ...) {
+     return(moran(x = var, listw = listw, n = n, S0 = S0)$I)
+ }
> set.seed(1234)

> boot2 <- boot(NY8$Cases, statistic = MoranI.pboot,
+     R = 999, sim = "parametric", ran.gen = CR,
+     listw = lw_B, n = length(NY8$Cases), S0 = Szero(lw_B),
+     mle = rni)

> pnorm((boot2$t0 - mean(boot2$t))/sd(boot2$t), lower.tail = FALSE)

[1] 0.1472
```

The expected counts can also be expressed as the fitted values of a null Poisson regression with an offset set to the logarithm of tract population – with a log-link, this shows the relationship to generalised linear models (because Cases are not all integer, warnings are generated):

```
> rni <- fitted(glm(Cases ~ 1 + offset(log(POP8)), data = NY8,
+     family = "poisson"))
```

These expected counts rni are fed through to the lambda argument to rpois to generate the synthetic data sets by sampling from the Poisson distribution. The output probability value is calculated from the same observed Moran's I

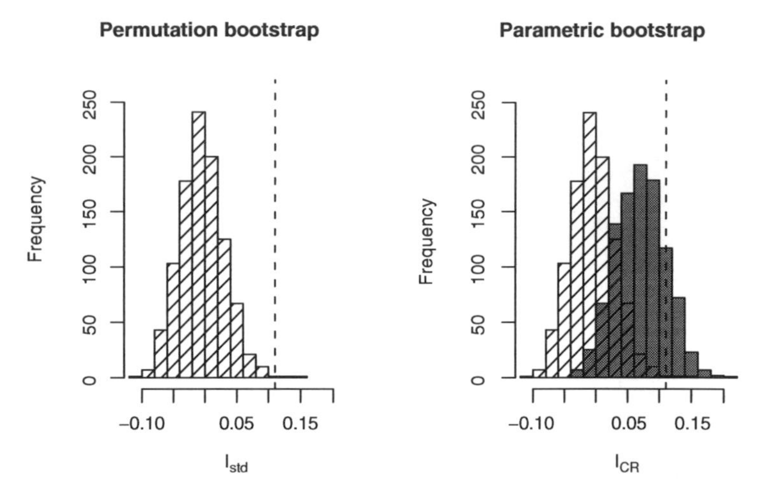

Fig. 9.10. Histograms of simulated values of Moran's I under random permutations of the data and parametric samples from constant risk expected values; the observed values of Moran's I are marked by vertical lines

minus the mean of the simulated I values, and divided by their standard deviation. Figure 9.10 corresponds to Waller and Gotway (2004, p. 232, Fig. 7.8), with the parametric simulations shifting the distribution of Moran's I rightwards, because it is taking the impact of the heterogeneous tract populations into account.

There is a version of Moran's I adapted to use an Empirical Bayes rate by Assunção and Reis (1999) that, unlike the rate results above, shrinks extreme rates for tracts with small populations at risk towards the rate for the area as a whole – it also uses Monte Carlo methods for inference:

```
> set.seed(1234)
> EBImoran.mc(n = NY8$Cases, x = NY8$POP8, listw = nb2listw(NY_nb,
+     style = "B"), nsim = 999)

	Monte-Carlo simulation of Empirical Bayes Index

data:  cases: NY8$Cases, risk population: NY8$POP8
weights: nb2listw(NY_nb, style = "B")
number of simulations + 1: 1000

statistic = 0.0735, observed rank = 980, p-value = 0.02
alternative hypothesis: greater
```

The results for the Empirical Bayes rates suggest that one reason for the lack of significance of the parametric bootstrapping of the constant risk observed and expected values could be that unusual and extreme values were observed in tracts with small populations. Once the rates have been smoothed, some global autocorrelation is found.

```
> cor8 <- sp.correlogram(neighbours = NY_nb, var = NY8$Cases,
+     order = 8, method = "I", style = "C")
```

```
> print(cor8, p.adj.method = "holm")
```

```
Spatial correlogram for NY8$Cases
method: Moran's I
   estimate expectation  variance standard deviate Pr(I) two sided
1  0.110387   -0.003571  0.001279             3.19         0.01009 *
2  0.095113   -0.003571  0.000564             4.16         0.00026 ***
3  0.016711   -0.003571  0.000348             1.09         0.83111
4  0.037506   -0.003571  0.000255             2.57         0.06104 .
5  0.026920   -0.003571  0.000203             2.14         0.12960
6  0.026428   -0.003571  0.000175             2.27         0.11668
7  0.009341   -0.003571  0.000172             0.98         0.83111
8  0.002119   -0.003571  0.000197             0.41         0.83111
---
Signif. codes:  0 '***' 0.001 '**' 0.01 '*' 0.05 '.' 0.1 ' ' 1
```

Another approach is to plot and tabulate values of a measure of spatial autocorrelation for higher orders of neighbours or bands of more distant neighbours where the spatial entities are points. The **spdep** package provides the first type as a wrapper to `nblag` and `moran.test`, so that here the first-order contiguous neighbours we have used until now are 'stepped out' to the required number of orders. Figure 9.11 shows the output plot in the left panel, and suggests that second-order neighbours are also positively autocorrelated (although the probability values should be adjusted for multiple comparisons).

The right panel in Fig. 9.11 presents the output of the `correlog` function in the **pgirmess** package by Patrick Giraudoux; the function is a wrapper for `dnearneigh` and `moran.test`. The function automatically selects distance bands of almost 10 km, spanning the whole study area. In this case, the first two bands of 0–10 and 10–20 km have significant values.

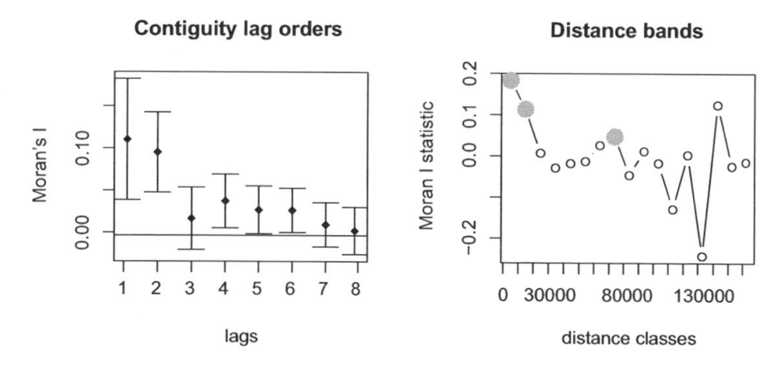

Fig. 9.11. Correlograms: (*left*) values of Moran's I for eight successive lag orders of contiguous neighbours; (*right*) values of Moran's I for a sequence of distance band neighbour pairs

```
> library(pgirmess)
> corD <- correlog(coordinates(NY8), NY8$Cases, method = "Moran")

> corD

Moran I statistic
      dist.class      coef   p.value     n
 [1,]       4996  0.183373 3.781e-09 10728
 [2,]      14822  0.113940 1.128e-08  9248
 [3,]      24649  0.007028 3.025e-01  5718
 [4,]      34475 -0.028733 8.968e-01  5376
 [5,]      44302 -0.017871 7.686e-01  5342
 [6,]      54128 -0.012729 6.940e-01  5578
 [7,]      63954  0.026370 5.040e-02  5524
 [8,]      73781  0.047751 3.620e-03  5976
 [9,]      83607 -0.046052 9.730e-01  4334
[10,]      93434  0.011773 2.632e-01  3862
[11,]     103260 -0.017014 7.074e-01  8756
[12,]     113086 -0.128775 1.000e+00  5816
[13,]     122913  0.003007 4.246e-01  1958
[14,]     132739 -0.243455 9.987e-01   320
[15,]     142566  0.124519 8.951e-02    92
[16,]     152392 -0.024368 4.435e-01    44
[17,]     162218 -0.014310 1.495e-01     6
```

9.4.2 Local Tests

Global tests for spatial autocorrelation are calculated from the local relationships between the values observed at a spatial entity and its neighbours, for the neighbour definition chosen. Because of this, we can break global measures down into their components, and by extension, construct localised tests intended to detect 'clusters' – observations with very similar neighbours – and 'hotspots' – observations with very different neighbours. These are discussed briefly by Schabenberger and Gotway (2005, pp. 23–25) and O'Sullivan and Unwin (2003, pp. 203–205), and at greater length by Waller and Gotway (2004, pp. 236–242) and Fortin and Dale (2005, pp. 153–159). They are covered in some detail by Lloyd (2007, pp. 65–70) in a book concentrating on local models.

First, let us examine a Moran scatterplot of the leukaemia case count variable. The plot (shown in Fig. 9.12) by convention places the variable of interest on the x-axis, and the spatially weighted sum of values of neighbours – the spatially lagged values – on the y-axis. Global Moran's I is a linear relationship between these and is drawn as a slope. The plot is further partitioned into quadrants at the mean values of the variable and its lagged values: low–low, low–high, high–low, and high–high.

```
> moran.plot(NY8$Cases, listw = nb2listw(NY_nb, style = "C"))
```

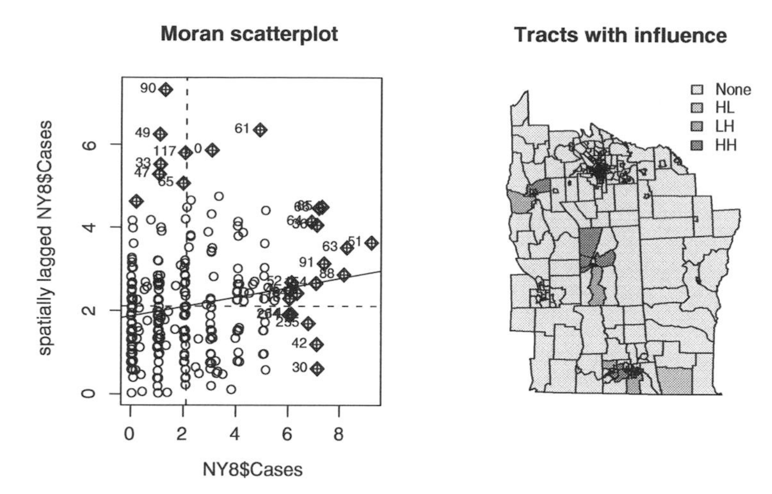

Fig. 9.12. (*Left*) Moran scatterplot of leukaemia incidence; (*right*) tracts with influence by Moran scatterplot quadrant

Since global Moran's I is, like similar correlation coefficients, a linear relationship, we can also apply standard techniques for detecting observations with unusually strong influence on the slope. Specifically, `moran.plot` calls `influence.measures` on the linear model of `lm(wx ~ x)` providing the slope coefficient, where `wx` is the spatially lagged value of `x`. This means that we can see whether particular local relationships are able to influence the slope more than proportionally. The map in the right panel of Fig. 9.12 shows tracts with significant influence (using standard criteria) coded by their quadrant in the Moran scatterplot.

Local Moran's I_i values are constructed as the n components summed to reach global Moran's I:

$$I_i = \frac{(y_i - \bar{y}) \sum_{j=1}^{n} w_{ij}(y_j - \bar{y})}{\frac{\sum_{i=1}^{n}(y_i - \bar{y})^2}{n}},$$

where once again we assume that the global mean $\bar{y}$ is an adequate representation of the variable of interest y. The two components in the numerator, $(y_i - \bar{y})$ and $\sum_{j=1}^{n} w_{ij}(y_j - \bar{y})$, appear without centring in the Moran scatterplot.

As with the global statistic, the local statistics can be tested for divergence from expected values, under assumptions of normality, and randomisation analytically, and using Saddlepoint approximations and exact methods. The two latter methods can be of importance because the number of neighbours of each observation is very small, and this in turn may make the adoption of the normality assumption problematic. Using numerical methods, which would previously have been considered demanding, the Saddlepoint approximation or exact local probability values can be found in well under 10 s, about

20 times slower than probability values based on normality or randomisation assumptions, for this moderately sized data set.

Trying to detect residual local patterning in the presence of global spatial autocorrelation is difficult. For this reason, results for local dependence are not to be seen as 'absolute', but are conditioned at least by global spatial autocorrelation, and more generally by the possible influence of spatial data generating processes at a range of scales from global through local to dependence not detected at the scale of the observations.

```
> lm1 <- localmoran(NY8$Cases, listw = nb2listw(NY_nb,
+     style = "C"))
> lm2 <- as.data.frame(localmoran.sad(lm(Cases ~ 1, NY8),
+     nb = NY_nb, style = "C"))
> lm3 <- as.data.frame(localmoran.exact(lm(Cases ~ 1, NY8),
+     nb = NY_nb, style = "C"))
```

Waller and Gotway (2004, p. 239) extend their constant risk hypothesis treatment to local Moran's I_i, and we can follow their lead:

```
> r <- sum(NY8$Cases)/sum(NY8$POP8)
> rni <- r * NY8$POP8
> lw <- nb2listw(NY_nb, style = "C")
> sdCR <- (NY8$Cases - rni)/sqrt(rni)
> wsdCR <- lag(lw, sdCR)
> I_CR <- sdCR * wsdCR
```

Figure 9.13 shows the two sets of values of local Moran's I_i, calculated in the standard way and using the Poisson assumption for the constant risk hypothesis. We already know that global Moran's I can vary in value and in

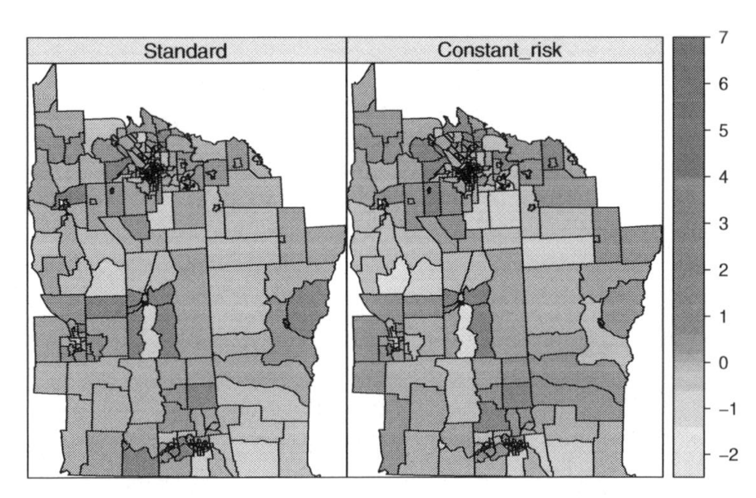

Fig. 9.13. Local Moran's I_i values calculated directly and using the constant risk hypothesis

inference depending on our assumptions – for example that inference should take deviations from our distributional assumptions into account. The same applies here to the assumption for the Poisson distribution that its mean and standard deviation are equal, whereas over-dispersion seems to be a problem in data also displaying autocorrelation. There are some sign changes between the maps, with the constant risk hypothesis values somewhat farther from zero.

We can also construct a simple Monte Carlo test of the constant risk hypothesis local Moran's I_i values, simulating very much as in the global case, but now retaining all of the local results. Once the simulation is completed, we extract the rank of the observed constant risk local Moran's I_i value for each tract, and calculate its probability value for the number of simulations made. We use a parametric approach to simulating the local counts using the local expected count as the parameter to `rpois`, because the neighbour counts are very low and make permutation unwise. Carrying out permutation testing using the whole data set also seems unwise, because we would then be comparing like with unlike.

```
> set.seed(1234)
> nsim <- 999
> N <- length(rni)
> sims <- matrix(0, ncol = nsim, nrow = N)
> for (i in 1:nsim) {
+     y <- rpois(N, lambda = rni)
+     sdCRi <- (y - rni)/sqrt(rni)
+     wsdCRi <- lag(lw, sdCRi)
+     sims[, i] <- sdCRi * wsdCRi
+ }
> xrank <- apply(cbind(I_CR, sims), 1, function(x) rank(x)[1])
> diff <- nsim - xrank
> diff <- ifelse(diff > 0, diff, 0)
> pval <- punif((diff + 1)/(nsim + 1))
```

The probability values shown in Fig. 9.14 are in general very similar to each other. We follow Waller and Gotway (2004) in not adjusting for multiple comparisons, and will consequently not interpret the probability values as more than indications. Values close to zero are said to indicate clusters in the data where tracts with similar values neighbour each other (positive local autocorrelation and a one-sided test). Values close to unity indicate hotspots where the values of contiguous tracts differ more than might be expected (negative local autocorrelation and a one-sided test). Of course, finding clusters or hotspots also needs to be qualified by concerns about misspecification in the underlying model of the data generation process.

Finally, we zoom in to examine the local Moran's I_i probability values for three calculation methods for the tracts in and near the city of Binghampton (Fig. 9.15). It appears that the use of the constant risk approach handles the heterogeneity in the counts better than the alternatives. These results broadly

Fig. 9.14. Probability values for all census tracts, local Moran's I_i: normality and randomisation assumptions, Saddlepoint approximation, exact values, and constant risk hypothesis

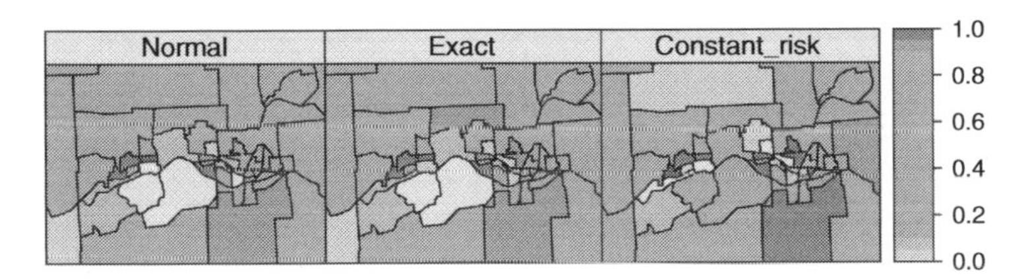

Fig. 9.15. Probability values for census tracts in and near the city of Binghampton, local Moran's I_i: normality assumption, exact values, and constant risk hypothesis

agree with those reached by Waller and Gotway (2004, p. 241), but we note that our underlying model is very simplistic. Finding spatial autocorrelation is not a goal in itself, be it local or global, but rather just one step in a process leading to a proper model. It is to this task that we now turn.

10

Modelling Areal Data

10.1 Introduction

We have seen in Chap. 9 that the lack of independence between observations in spatial data – spatial autocorrelation – is commonplace, and that tests are available. In an ideal world, one would prefer to gather data in which the observations were mutually independent, and so avoid problems in inference from analytical results. Most applied data analysts, however, do not have this option, and must work with the data that are available, or that can be collected with available technologies. It is quite often the case that observations on relevant covariates are not available at all, and that the detection of spatial autocorrelation in data or model residuals in fact constitutes the only way left to model the remaining variation.

In this chapter, we show how spatial structure in dependence between observations may be modelled, in particular for areal data, but where necessary also using alternative representations. We look at spatial econometrics approaches separately, because the terminology used in that domain differs somewhat from other areas of spatial statistics. We cover spatial filtering using Moran eigenvectors and geographically weighted regression in this chapter, but leave Bayesian hierarchical models until Chap. 11.

The problems we face when trying to fit models in the presence of spatial autocorrelation are challenging, not least because the spatial autocorrelation that we seem to have found may actually come from model misspecification (see Sect. 9.4). If this is the case, effort spent on modelling the spatial structure would be better used on improving the model itself, perhaps by handling heteroskedasticity, by adding a missing covariate, by revisiting the functional form of included covariates, or by reconsidering the distributional representation of the response variable.

10.2 Spatial Statistics Approaches

Spatial dependence can be modelled in different ways using statistical models. In many cases, it is common to assume that observations are independent and identically distributed, but this may not be the case when working with spatial data. Observations are not independent because there may exist some correlation between neighbouring areas. It may also be difficult to pick apart the impact of spatial autocorrelation and spatial differences in the distribution of the observation. Cressie (1993, pp. 402–448, 458–477, 548–568) provides a very wide discussion of these approaches, including reviews of the background for their development and comprehensive worked examples. Schabenberger and Gotway (2005, pp. 335–348) and Waller and Gotway (2004, pp. 362–380) concentrate on the spatial autoregressive models to be used in this section. Wall (2004) provides a useful comparative review of the ways in which spatial processes for areal data are modelled. Banerjee et al. (2004, pp. 79–87) also focus on these models, because the key features carry through to hierarchical models. Fortin and Dale (2005, pp. 229–233) indicate that spatial autoregressive models may play a different role in ecology, although reviews like Dormann et al. (2007) suggest that they may be of use.

In this section, we have followed Waller and Gotway (2004, Chap. 9) quite closely, as their examples highlight issues such as transforming the response variable and using weights to try to handle heteroskedasticity.

From a statistical point of view, it is possible to account for correlated observations by considering a structure of the following kind in the model. If the vector of response variables is multivariate normal, we can express the model as follows:

$$Y = \mu + e,$$

where μ is the vector of area means, which can be modelled in different ways and e is the vector of random errors, which we assume is normally distributed with zero mean and generic variance V. The mean is often supposed to depend on a linear term on some covariates X, so that we will substitute the mean by $X^{\mathrm{T}}\beta$ in the model. On the other hand, correlation between areas is taken into account by considering a specific form of the variance matrix V.

For the case of non-Normal variables, we could transform the original data to achieve the desired Normality. Hence, the techniques described below can still be applied on the transformed data. In principle, many correlation structures could be feasible in order to account for spatial correlation. However, we focus on two approaches that are commonly used in practise,such as SAR (Simultaneous Autoregressive) and CAR (Conditionally Autoregressive) models.

In Chap. 9, we took the mean of the counts of leukaemia cases by tract as our best understanding of the data generation process, supplementing this with the constant risk approach to try to handle heterogeneity coming from variations in tract populations. One of the alternatives examined by Waller and Gotway (2004, p. 348) is to take a log transformation of the rate:

$$Z_i = \log \frac{1000(Y_i + 1)}{n_i}.$$

The transformed incidence proportions are not yet normal, with three outliers, tracts with small populations but unexpectedly large case counts. They could be smoothed away, but may in fact be interesting, as the patterns they display may be related to substantive covariates, such as closeness to TCE locations. As covariates, we have used the inverse distance to the closest TCE (PEXPOSURE), the proportion of people aged 65 or higher (PCTAGE65P) and the proportion of people who own their own home (PCTOWNHOME).

To set the scene, let us start with a linear model of the relationship between the transformed incidence proportions and the covariates. Note that most model fitting functions accept `Spatial*DataFrame` objects as their `data` argument values, and simply treat them as regular `data.frame` objects. This is not by inheritance, but because the same access methods are provided (see p. 35).

```
> library(spdep)

> nylm <- lm(Z ~ PEXPOSURE + PCTAGE65P + PCTOWNHOME, data = NY8)
> summary(nylm)

Call:
lm(formula = Z ~ PEXPOSURE + PCTAGE65P + PCTOWNHOME, data = NY8)

Residuals:
    Min      1Q  Median      3Q     Max
-1.7417 -0.3957 -0.0326  0.3353  4.1398

Coefficients:
            Estimate Std. Error t value Pr(>|t|)
(Intercept)  -0.5173     0.1586   -3.26   0.0012 **
PEXPOSURE     0.0488     0.0351    1.39   0.1648
PCTAGE65P     3.9509     0.6055    6.53  3.2e-10 ***
PCTOWNHOME   -0.5600     0.1703   -3.29   0.0011 **
---
Signif. codes:  0 '***' 0.001 '**' 0.01 '*' 0.05 '.' 0.1 ' ' 1

Residual standard error: 0.657 on 277 degrees of freedom
Multiple R-squared: 0.193,  Adjusted R-squared: 0.184
F-statistic: 22.1 on 3 and 277 DF,  p-value: 7.3e-13

> NY8$lmresid <- residuals(nylm)
```

Figure 10.1 shows the spatial distribution of residual values for the study area census tracts. The two census variables appear to contribute for explaining the variance in the response variable, but exposure to TCE does not. Moreover, although there is less spatial autocorrelation in the residuals from

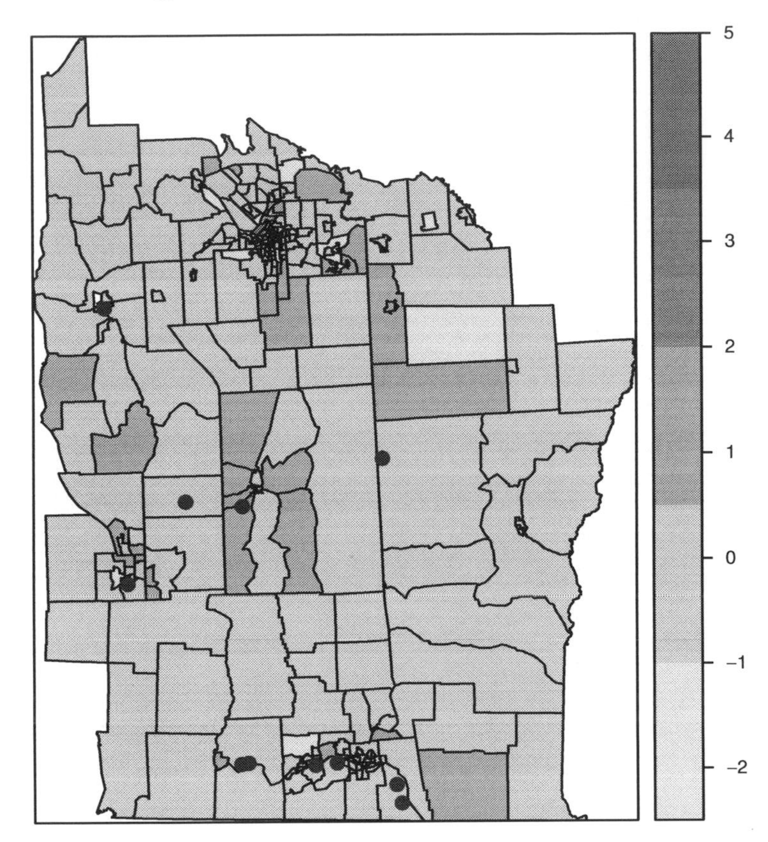

Fig. 10.1. Residuals from the linear model of transformed incidence proportions; TCE site locations shown for comparative purposes

the model with covariates than in the null model, it is clear that there is information in the residuals that we should try to use. An exact test for spatial autocorrelation in the residuals leads to similar conclusions.

Since the Moran test is intended to detect spatial autocorrelation, we can try to fit a model taking this into account. We should not, however, forget that the misspecifications detected by Moran's I can have a range of causes (see Sect. 9.4). It is also the case that if the fitted model exhibits multi-collinearity, the results of the test may be affected because of the numerical consequences of the model matrix not being of full rank for the expectation and variance of the statistic.

```
> NYlistw <- nb2listw(NY_nb, style = "B")
> lm.morantest(nylm, NYlistw)

	Global Moran's I for regression residuals

data:
```

```
model: lm(formula = Z ~ PEXPOSURE + PCTAGE65P + PCTOWNHOME,
data = NY8)
weights: NYlistw

Moran I statistic standard deviate = 2.638, p-value = 0.004169
alternative hypothesis: greater
sample estimates:
Observed Moran's I        Expectation            Variance
          0.083090          -0.009891            0.001242
```

10.2.1 Simultaneous Autoregressive Models

The SAR specification uses a regression on the values from the other areas to account for the spatial dependence. This means that the error terms ε are modelled so that they depend on each other in the following way:

$$e_i = \sum_{i=1}^{m} b_{ij} e_i + \varepsilon_i.$$

Here, ε_i are used to represent residual errors, which are assumed to be independently distributed according to a Normal distribution with zero mean and diagonal covariance matrix Σ_ε with elements $\sigma^2_{\varepsilon_i}, i = 1, \ldots, m$ (the same variance σ^2_ε is often considered though). The b_{ij} values are used to represent spatial dependence between areas. b_{ii} must be set to zero so that each area is not regressed on itself.

Note that if we express the error terms as $e = B(Y - X^{\mathrm{T}}\beta) + \varepsilon$, the model can also be expressed as

$$Y = X^{\mathrm{T}}\beta + B(Y - X^{\mathrm{T}}\beta) + \varepsilon.$$

Hence, this model can be formulated in a matrix form as follows:

$$(I - B)(Y - X^{\mathrm{T}}\beta) = \varepsilon,$$

where B is a matrix that contains the dependence parameters b_{ij} and I is the identity matrix of the required dimension. It is important to point out that in order for this SAR model to be well defined, the matrix $I - B$ must be non-singular.

Under this model, Y is distributed according to a multivariate normal with mean

$$E[Y] = X^{\mathrm{T}}\beta$$

and covariance matrix

$$\mathrm{Var}[Y] = (I - B)^{-1}\Sigma_\varepsilon(I - B^{\mathrm{T}})^{-1}.$$

Often Σ_ε is taken to depend on a single parameter σ^2, so that $\Sigma_\varepsilon = \sigma^2 I$ and then Var[Y] simplifies to

$$\text{Var}[Y] = \sigma^2 (I - B)^{-1}(I - B^{\text{T}})^{-1}.$$

It is also possible to specify Σ_ε as a diagonal matrix of weights associated with heterogeneity among the observations.

A useful re-parametrisation of this model can be obtained by writing $B = \lambda W$, where λ is a spatial autocorrelation parameter and W is a matrix that represents spatial dependence – it is often assumed to be symmetric. These structures can be chosen among those described in Chap. 9. With this specification, the variance of Y becomes

$$\text{Var}[Y] = \sigma^2 (I - \lambda W)^{-1}(I - \lambda W^{\text{T}})^{-1}.$$

These models can be estimated efficiently by maximum likelihood. In R this can be done by using function `spautolm` in package `spdep`. The model can be specified using a formula for the linear predictor, whilst matrix W must be passed as a `listw` object. To create this object from the list of neighbours we can use function `nb2listw`, which will take an object of class `nb`, as explained in Chap. 9.

The following code shows how to fit a simultaneous autoregression to the chosen model. We have fitted the standard model and the weighted model using the population size in 1980 (according to the US Census) in the areas as weights. This reproduces the example developed in Waller and Gotway (2004, Chap. 9, pp. 375–379), and the reader is referred to their discussion for more information. In the call to `nb2listw`, we specified `style = "B"` to construct W using a binary indicator of neighbourhood.

```
> nysar <- spautolm(Z ~ PEXPOSURE + PCTAGE65P + PCTOWNHOME,
+     data = NY8, listw = NYlistw)
> summary(nysar)

Call:
spautolm(formula = Z ~ PEXPOSURE + PCTAGE65P + PCTOWNHOME, data = NY8,
    listw = NYlistw)

Residuals:
      Min        1Q    Median        3Q       Max
-1.567536 -0.382389 -0.026430  0.331094  4.012191

Coefficients:
             Estimate Std. Error z value  Pr(>|z|)
(Intercept) -0.618193   0.176784 -3.4969 0.0004707
PEXPOSURE    0.071014   0.042051  1.6888 0.0912635
PCTAGE65P    3.754200   0.624722  6.0094 1.862e-09
PCTOWNHOME  -0.419890   0.191329 -2.1946 0.0281930
```

```
Lambda: 0.04049 LR test value: 5.244 p-value: 0.022026

Log likelihood: -276.1
ML residual variance (sigma squared): 0.4139, (sigma: 0.6433)
Number of observations: 281
Number of parameters estimated: 6
AIC: 564.2
```

According to the results obtained it seems that there is significant spatial correlation in the residuals because the estimated value of λ is 0.0405 and the p-value of the likelihood ratio test is 0.0220. In the likelihood ratio test we compare the model with no spatial autocorrelation (i.e. $\lambda = 0$) to the one which allows for it (i.e. the fitted model with non-zero autocorrelation parameter).

The proximity to a TCE seems not to be significant, although its p-value is close to being significant at the 95% level and it would be advisable not to discard a possible association and to conduct further research on this. The other two covariates are significant, suggesting that census tracts with larger percentages of older people and with lower percentages of house owners have higher transformed incidence rates.

However, this model does not account for the heterogeneous distribution of the population by tracts beyond the correction introduced in transforming incidence proportions. Weighted version of these models can be fitted so that tracts are weighted proportionally to the inverse of their population size. For this purpose, we include the parameter `weights=POP8` in the call to the function `lm`.

```
> nylmw <- lm(Z ~ PEXPOSURE + PCTAGE65P + PCTOWNHOME, data = NY8,
+     weights = POP8)
> summary(nylmw)

Call:
lm(formula = Z ~ PEXPOSURE + PCTAGE65P + PCTOWNHOME, data = NY8,
    weights = POP8)

Residuals:
    Min      1Q  Median      3Q     Max
-129.07  -14.71    5.82   25.62   70.72

Coefficients:
            Estimate Std. Error t value Pr(>|t|)
(Intercept)  -0.7784     0.1412   -5.51  8.0e-08 ***
PEXPOSURE     0.0763     0.0273    2.79   0.0056 **
PCTAGE65P     3.8566     0.5713    6.75  8.6e-11 ***
PCTOWNHOME   -0.3987     0.1531   -2.60   0.0097 **
---
Signif. codes:  0 '***' 0.001 '**' 0.01 '*' 0.05 '.' 0.1 ' ' 1
```

```
Residual standard error: 33.5 on 277 degrees of freedom
Multiple R-squared: 0.198,  Adjusted R-squared: 0.189
F-statistic: 22.8 on 3 and 277 DF,  p-value: 3.38e-13
```

```
> NY8$lmwresid <- residuals(nylmw)
```

Starting with the weighted linear model, we can see that the TCE exposure variable has become significant with the expected sign, indicating that tracts closer to the TCE sites have slightly higher transformed incidence proportions. The other two covariates now also have more significant coefficients. Figure 10.2 shows that information has been shifted from the model residuals to the model itself, with little remaining spatial structure visible on the map.

```
> lm.morantest(nylmw, NYlistw)

	Global Moran's I for regression residuals

data:
```

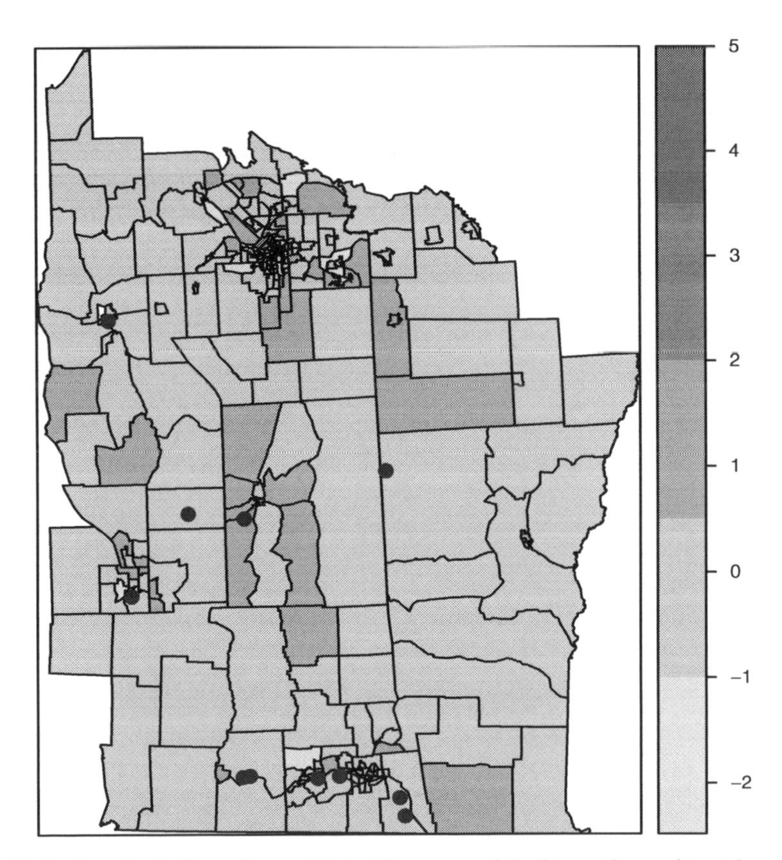

Fig. 10.2. Residuals from the weighted linear model of transformed incidence proportions; TCE site locations shown for comparative purposes

```
model: lm(formula = Z ~ PEXPOSURE + PCTAGE65P + PCTOWNHOME,
data = NY8, weights = POP8)
weights: NYlistw

Moran I statistic standard deviate = 0.4773, p-value = 0.3166
alternative hypothesis: greater
sample estimates:
Observed Moran's I        Expectation          Variance
          0.007533          -0.009310          0.001245
```

The Moran tests for regression residuals can also be used with a weighted linear model object. The results are interesting, suggesting that the misspecification detected by Moran's I is in fact related to heteroskedasticity more than to spatial autocorrelation. We can check this for the SAR model too, since `spautolm` also takes a `weights` argument:

```
> nysarw <- spautolm(Z ~ PEXPOSURE + PCTAGE65P + PCTOWNHOME,
+     data = NY8, listw = NYlistw, weights = POP8)
> summary(nysarw)

Call:
spautolm(formula = Z ~ PEXPOSURE + PCTAGE65P + PCTOWNHOME, data = NY8,
    listw = NYlistw, weights = POP8)

Residuals:
     Min       1Q   Median       3Q      Max
-1.48488 -0.26823  0.09489  0.46552  4.28343

Coefficients:
             Estimate Std. Error z value  Pr(>|z|)
(Intercept) -0.797063   0.144054 -5.5331 3.146e-08
PEXPOSURE    0.080545   0.028334  2.8428  0.004473
PCTAGE65P    3.816731   0.576037  6.6258 3.453e-11
PCTOWNHOME  -0.380778   0.156507 -2.4330  0.014975

Lambda: 0.009564 LR test value: 0.3266 p-value: 0.56764

Log likelihood: -251.6
ML residual variance (sigma squared): 1104, (sigma: 33.23)
Number of observations: 281
Number of parameters estimated: 6
AIC: 515.2
```

The coefficients of the covariates change slightly in the new model, and all the coefficient p-values drop substantially. In this weighted SAR fit, proximity to a TCE site becomes significant. However, there are no traces of spatial autocorrelation left after adjusting for the heterogeneous size of the population. This suggests that the spatial variation in population between tracts is responsible for the observed residual spatial correlation after adjusting for covariates.

To compare both models and choose the best one, we use Akaike's Information Criterion (AIC) reported in the model summaries. The AIC is a weighted sum of the log-likelihood of the model and the number of fitted coefficients; according to the criterion, better models are those with the lower values of the AIC. Hence, the weighted model provides a better fitting since its AIC is considerably lower. This indicates the importance of accounting for heterogeneous populations in the analysis of this type of lattice data.

10.2.2 Conditional Autoregressive Models

The CAR specification relies on the conditional distribution of the spatial error terms. In this case, the distribution of e_i conditioning on e_{-i} (the vector of all random error terms minus e_i itself) is given. Instead of the whole e_{-i} vector, only the neighbours of area i, defined in a chosen way, are used. We represent them by $e_{j\sim i}$. Then, a simple way of putting the conditional distribution of e_i is

$$e_i|e_{j\sim i} \sim N\Big(\sum_{j\sim i} \frac{c_{ij}e_j}{\sum_{j\sim i} c_{ij}}, \frac{\sigma^2_{e_i}}{\sum_{j\sim i} c_{ij}}\Big),$$

where c_{ij} are dependence parameters similar to b_{ij}. However, specifying the conditional distributions of the error terms does not imply that the joint distribution exists. To have a proper distribution some constraints must be set on the parameters of the model. The reader is referred to Schabenberger and Gotway (2005, pp. 338–339) for a detailed description of CAR specifications. For our modelling purposes, the previous guidelines will be enough to obtain a proper CAR specification in most cases.

To fit a CAR model, we can use function `spautolm` again. This time we set the argument `family="CAR"` to specify that we are fitting this type of models.

```
> nycar <- spautolm(Z ~ PEXPOSURE + PCTAGE65P + PCTOWNHOME,
+     data = NY8, family = "CAR", listw = NYlistw)
> summary(nycar)

Call:
spautolm(formula = Z ~ PEXPOSURE + PCTAGE65P + PCTOWNHOME, data = NY8,
    listw = NYlistw, family = "CAR")

Residuals:
      Min        1Q    Median        3Q       Max
-1.539732 -0.384311 -0.030646  0.335126  3.808848

Coefficients:
             Estimate Std. Error z value  Pr(>|z|)
(Intercept) -0.648362   0.181129 -3.5796 0.0003442
PEXPOSURE    0.077899   0.043692  1.7829 0.0745986
```

```
PCTAGE65P     3.703830   0.627185  5.9055 3.516e-09
PCTOWNHOME   -0.382789   0.195564 -1.9574 0.0503053

Lambda: 0.08412 LR test value: 5.801 p-value: 0.016018

Log likelihood: -275.8
ML residual variance (sigma squared): 0.4076, (sigma: 0.6384)
Number of observations: 281
Number of parameters estimated: 6
AIC: 563.7
```

The estimated coefficients of the covariates in the model are very similar to those obtained with the SAR models. Nevertheless, the p-values of two covariates, the distance to the nearest TCE and the percentage of people owning a home, are slightly above the 0.05 threshold. The likelihood ratio test indicates that there is significant spatial autocorrelation and the estimated value of λ is 0.0841.

Considering a weighted regression, using the population size as weights, for the same model to account for the heterogeneous distribution of the population completely removes the spatial autocorrelation in the data. The coefficients of the covariates do not change much and all of them become significant. Hence, modelling spatial autocorrelation by means of SAR or CAR specifications does not change the results obtained; Waller and Gotway (2004, pp. 375–379) give a complete discussion of these results.[1]

```
> nycarw <- spautolm(Z ~ PEXPOSURE + PCTAGE65P + PCTOWNHOME,
+     data = NY8, family = "CAR", listw = NYlistw, weights = POP8)
> summary(nycarw)

Call:
spautolm(formula = Z ~ PEXPOSURE + PCTAGE65P + PCTOWNHOME, data = NY8,
    listw = NYlistw, weights = POP8, family = "CAR")

Residuals:
      Min        1Q    Median        3Q       Max
-1.491042 -0.270906  0.081435  0.451556  4.198134

Coefficients:
             Estimate Std. Error z value  Pr(>|z|)
(Intercept) -0.790154   0.144862 -5.4545 4.910e-08
PEXPOSURE    0.081922   0.028593  2.8651  0.004169
PCTAGE65P    3.825858   0.577720  6.6223 3.536e-11
PCTOWNHOME  -0.386820   0.157436 -2.4570  0.014010
```

[1] The fitted coefficient values of the weighted CAR model do not exactly reproduce those of Waller and Gotway (2004, p. 379), although the spatial coefficient is reproduced. In addition, the model cannot be fit with S-PLUS™ SpatialStats module slm, as the product of the two components of the model covariance matrix is not symmetric, while the two components taken separately are. This suggests that caution in using current implementations of weighted CAR models is justified.

```
Lambda: 0.02242 LR test value: 0.3878 p-value: 0.53343

Log likelihood: -251.6
ML residual variance (sigma squared): 1103, (sigma: 33.21)
Number of observations: 281
Number of parameters estimated: 6
AIC: 515.1
```

10.2.3 Fitting Spatial Regression Models

The `spautolm` function fits spatial regression models by maximum likelihood, by first finding the value of the spatial autoregressive coefficient, which maximises the log likelihood function for the model family chosen, and then fitting the other coefficients by generalised least squares at that point. This means that the spatial autoregressive coefficient can be found by line search using `optimize`, rather than by optimising over all the model parameters at the same time.

The most demanding part of the functions called to optimise the spatial autoregressive coefficient is the computation of the Jacobian, the log determinant of the $n \times n$ matrix $|I - B|$, or $|I - \lambda W|$ in our parametrisation. As n increases, the use of the short-cut of

$$\log(|I - \lambda W|) = \log \left(\prod_{i=1}^{n} (1 - \lambda \zeta_i) \right),$$

where ζ_i are the eigenvalues of W, becomes more difficult. The default method of `method="full"` uses eigenvalues, and can thus also set the lower and upper bounds for the line search for λ accurately (as $[1/\min_i(\zeta_i), 1/\max_i(\zeta_i)]$), but is not feasible for large n. It should also be noted that although eigenvalues are computed for intrinsically asymmetric spatial weights matrices, their imaginary parts are discarded, so that even for `method="full"`, the consequences of using such asymmetric weights matrices are unknown.

Alternative approaches involve finding the log determinant of a Cholesky decomposition of the sparse matrix $(I - \lambda W)$ directly. Here it is not possible to pre-compute eigenvalues, so one log determinant is computed for each value of λ used, but the number needed is in general not excessive, and much larger n become feasible on ordinary computers. A number of different sparse matrix approaches have been tried, with the use of **Matrix** and `method="Matrix"`, the one suggested currently. All of the sparse matrix approaches to computing the Jacobian require that matrix W be symmetric or at least similar to symmetric, thus providing for weights with `"W"` and `"S"` styles based on symmetric neighbour lists and symmetric general spatial weights, such as inverse distance. Matrices that are similar to symmetric have the same eigenvalues, so that the eigenvalues of symmetric $W^* = D^{1/2} W D^{1/2}$ and row-standardised

$W = DB$ are the same, for symmetric binary or general weights matrix B, and D a diagonal matrix of inverse row sums of B, $d_{ii} = 1/\sum_{j=1}^{n} b_{ij}$ (Ord, 1975, p. 125).

```
> nysarwM <- spautolm(Z ~ PEXPOSURE + PCTAGE65P + PCTOWNHOME,
+     data = NY8, family = "SAR", listw = NYlistw, weights = POP8,
+     method = "Matrix")

> summary(nysarwM)

Call:
spautolm(formula = Z ~ PEXPOSURE + PCTAGE65P + PCTOWNHOME, data = NY8,
    listw = NYlistw, weights = POP8, family = "SAR", method = "Matrix")

Residuals:
     Min       1Q   Median       3Q      Max
-1.48488 -0.26823  0.09489  0.46552  4.28343

Coefficients:
             Estimate Std. Error z value  Pr(>|z|)
(Intercept) -0.797063   0.144054 -5.5331 3.146e-08
PEXPOSURE    0.080545   0.028334  2.8428  0.004473
PCTAGE65P    3.816731   0.576037  6.6258 3.453e-11
PCTOWNHOME  -0.380778   0.156507 -2.4330  0.014975

Lambda: 0.009564 LR test value: 0.3266 p-value: 0.56764

Log likelihood: -251.6
ML residual variance (sigma squared): 1104, (sigma: 33.23)
Number of observations: 281
Number of parameters estimated: 6
AIC: 515.2
```

The output from fitting the weighted SAR model using functions from the **Matrix** package is identical with that from using the eigenvalues of W. Thanks to help from the **Matrix** package authors, Douglas Bates and Martin Mächler; additional facilities have been made available allowing the Cholesky decomposition to be computed once and updated for new values of the spatial coefficient. An internal vectorised version of this update method has also been made available, making the look-up time for many coefficient values small.

If it is of interest to examine values of the log likelihood function for a range of values of λ, the `llprof` argument may be used to give the number of equally spaced λ values to be chosen between the inverse of the smallest and largest eigenvalues for `method="full"`, or a sequence of such values more generally.

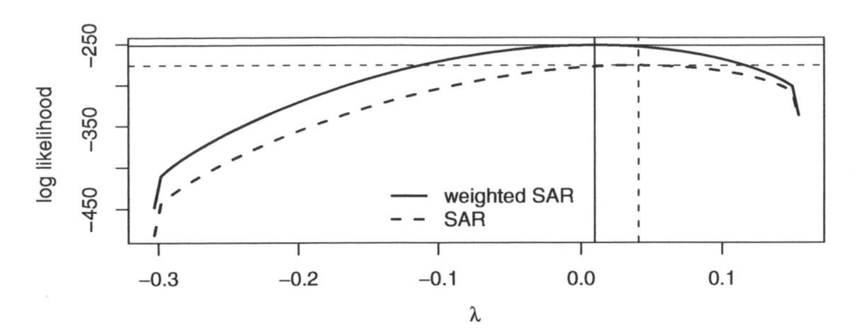

Fig. 10.3. Log likelihood values for a range of values of λ, weighted and unweighted SAR models; fitted spatial coefficient values and maxima shown

```
> 1/range(eigenw(NYlistw))
[1] -0.3029  0.1550
> nysar_ll <- spautolm(Z ~ PEXPOSURE + PCTAGE65P + PCTOWNHOME,
+     data = NY8, family = "SAR", listw = NYlistw, llprof = 100)
> nysarw_ll <- spautolm(Z ~ PEXPOSURE + PCTAGE65P + PCTOWNHOME,
+     data = NY8, family = "SAR", listw = NYlistw, weights = POP8,
+     llprof = 100)
```

Figure 10.3 shows the shape of the values of the log likelihood function along the feasible range of λ for the weighted and unweighted SAR models. We can see easily that the curves are very flat at the maxima, meaning that we could shift λ a good deal without impacting the function value much. The figure also shows the sharp fall-off in function values as the large negative values of the Jacobian kick in close to the ends of the feasible range.

Finally, `family="SMA"` for simultaneous moving average models is also available within the same general framework, but always involves handling dense matrices for fitting.

```
> nysmaw <- spautolm(Z ~ PEXPOSURE + PCTAGE65P + PCTOWNHOME,
+     data = NY8, family = "SMA", listw = NYlistw, weights = POP8)
> summary(nysmaw)

Call:
spautolm(formula = Z ~ PEXPOSURE + PCTAGE65P + PCTOWNHOME, data = NY8,
    listw = NYlistw, weights = POP8, family = "SMA")

Residuals:
      Min        1Q    Median        3Q       Max
-1.487080 -0.268990  0.093956  0.466055  4.284087

Coefficients:
             Estimate Std. Error z value  Pr(>|z|)
(Intercept) -0.795243   0.143749 -5.5321 3.163e-08
PEXPOSURE    0.080153   0.028237  2.8386  0.004531
PCTAGE65P    3.820316   0.575463  6.6387 3.165e-11
PCTOWNHOME  -0.382529   0.156160 -2.4496  0.014302
```

```
Lambda: 0.009184 LR test value: 0.3077 p-value: 0.57909

Log likelihood: -251.6
ML residual variance (sigma squared): 1105, (sigma: 33.24)
Number of observations: 281
Number of parameters estimated: 6
AIC: 515.2
```

Although there may be computing environments within which it seems easier to fit spatial regression models, arguably few give the analyst both reasonable defaults and the opportunity to examine in as much detail as is needed in the internal workings of the methods used, and of their implementations in software. Naturally, improvements will need to be made, perhaps including the fitting of more than one spatial autocorrelation parameter.

10.3 Mixed-Effects Models

The errors e_i which appear in the previous models are used to account for between-area variation, following a specified correlation structure. These terms are usually known as *random effects* because, contrary to what happens with fixed effects (the covariates), the value of the random effect can change from area to area. The range of application of random effects is quite wide, and they are often used to model different types of interaction between the observations. Although mixed-effects models belong to a different tradition from the spatial models discussed above, they are central to multi-level models and small area estimation, both of which can also be used in the analysis of spatial data. In the spatial context, Schabenberger and Gotway (2005, pp. 325–334) discuss linear mixed-effects models; other coverage is to be found in Pinheiro and Bates (2000, pp. 230–232, 237–238) for the implementation used here.

Using a similar notation as in previous sections, mixed-effect models (McCulloch and Searle, 2001) can be formulated as

$$Y = X\beta + Ze + \varepsilon.$$

Vector e represents the random effects, whilst Z is used to account for their structure. The distribution of e is assumed to be Normal with mean zero and generic covariance matrix Σ_e. This structure can reflect the influence of several elements of e on a single observation. Z is a design matrix that may be fixed or depend on any parameter. For example, Z can be set to a specific value to reproduce a SAR or CAR specification but, in this case, Z also depends on λ, which is another parameter to be estimated. Similar models may also be specified for areal data with point support using functions in the **spBayes** package.

Maximum Likelihood or Restricted Maximum Likelihood (McCulloch and Searle, 2001) are often employed to fit mixed-effects models. Packages `nlme` and `lme4` (Pinheiro and Bates, 2000) can fit these types of models. These packages allow the specification of different types of covariance matrices of the random effects, including spatial structure.

The following example illustrates how to fit a mixed-effects model using a correlation matrix, which depends on the distance between the centroids of the areas. First, we need to specify the correlation structure between the areas. This correlation structure is similar to those used in geostatistics and we have chosen a Gaussian variogram based on the Euclidean distances between the centroids of the regions.

```
> library(nlme)
> NY8$x <- coordinates(NY8)[, 1]/1000
> NY8$y <- coordinates(NY8)[, 2]/1000
> sp1 <- corSpatial(1, form = ~x + y, type = "gaussian")
> scor <- Initialize(sp1, as(NY8, "data.frame")[, c("x",
+     "y")], nugget = FALSE)
```

Once we have specified the correlation structure using `corSpatial`, we need to set up the model. The fixed part of the model is as in the previous SAR and CAR models. In the random part of the model we need to include a random effect per area. This is done by including `random=` $\sim$ `1|AREAKEY` in the call to `lme`. The fitting functions require that the `Spatial*DataFrame` object be coerced to a `data.frame` object in this case.

```
> spmodel <- lme(Z ~ PEXPOSURE + PCTAGE65P + PCTOWNHOME,
+     random = ~1 | AREAKEY, data = as(NY8, "data.frame"),
+     correlation = scor, method = "ML")

> summary(spmodel)

Linear mixed-effects model fit by maximum likelihood
 Data: as(NY8, "data.frame")
    AIC   BIC logLik
  571.5 596.9 -278.7


Random effects:
 Formula: ~1 | AREAKEY
        (Intercept) Residual
StdDev:      0.6508  0.04671

Correlation Structure: Gaussian spatial correlation
 Formula: ~x + y | AREAKEY
 Parameter estimate(s):
  range
0.01929
Fixed effects: Z ~ PEXPOSURE + PCTAGE65P + PCTOWNHOME
             Value Std.Error  DF t-value p-value
(Intercept) -0.517    0.1586 277  -3.262  0.0012
PEXPOSURE    0.049    0.0351 277   1.393  0.1648
PCTAGE65P    3.951    0.6055 277   6.525  0.0000
```

```
PCTOWNHOME  -0.560    0.1703 277  -3.288  0.0011
 Correlation:
           (Intr) PEXPOS PCTAGE
PEXPOSURE  -0.411
PCTAGE65P  -0.587 -0.075
PCTOWNHOME -0.741  0.082  0.147

Standardized Within-Group Residuals:
      Min         Q1        Med         Q3        Max
-0.191116 -0.043422 -0.003575  0.036788  0.454251

Number of Observations: 281
Number of Groups: 281
```

In this case for the un-weighted model, the coefficients of the fixed part are the same as for the linear model. The random effects can be arranged so that they follow a SAR or CAR specification, and it can be seen as a particular structure for Z. Note that when a SAR specification is added, Z, which models the structure of the random effects, may depend on further parameters.

10.4 Spatial Econometrics Approaches

One of the attractions of spatial data analysis is the wide range of scientific disciplines involved. Naturally, this leads to multiple approaches to many kinds of analysis, including accepted ways of applying tests and model fitting methods. It also leads to some sub-communities choosing their own sets of tools, not infrequently diverging from other sub-communities. During the 2003 Distributed Computational Statistics meeting, surprise and amusement was caused by the remark that the Internet domain `www.spatial-statistics.com` contains material chiefly relating to real estate research. But this connection is in fact quite reasonable, as real estate generates a lot of spatial data, and requires suitable methods. Indeed, good understanding of real estate markets and financing is arguably as important to society as a good understanding of the spatial dimensions of disease incidence.

Spatial econometrics is authoritatively described by Anselin (1988, 2002), with additional comments by Bivand (2002, 2006) with regard to doing spatial econometrics in R. While the use of weights, as we have seen above, has resolved a serious model mis-specification in public health data, it would be more typical for econometricians to test first for heteroskedasticity, and to try to relieve it by adjusting coefficient standard errors:

```
> library(lmtest)
> bptest(nylm)

	studentized Breusch-Pagan test

data:  nylm
BP = 9.214, df = 3, p-value = 0.02658
```

The Breusch–Pagan test (Johnston and DiNardo, 1997, pp. 198–200) results indicate the presence of heteroskedasticity when the residuals from the original linear model are regressed on the right-hand-side variables – the default test set. This might suggest the need to adjust the estimated coefficient standard errors using a variance–covariance matrix (Zeileis, 2004) taking heteroskedasticity into account:

```
> library(sandwich)
> coeftest(nylm)

t test of coefficients:

             Estimate Std. Error t value Pr(>|t|)
(Intercept)   -0.5173     0.1586   -3.26   0.0012 **
PEXPOSURE      0.0488     0.0351    1.39   0.1648
PCTAGE65P      3.9509     0.6055    6.53  3.2e-10 ***
PCTOWNHOME    -0.5600     0.1703   -3.29   0.0011 **
---
Signif. codes:  0 '***' 0.001 '**' 0.01 '*' 0.05 '.' 0.1 ' ' 1

> coeftest(nylm, vcov = vcovHC(nylm, type = "HC4"))

t test of coefficients:

             Estimate Std. Error t value Pr(>|t|)
(Intercept)   -0.5173     0.1617   -3.20  0.00154 **
PEXPOSURE      0.0488     0.0343    1.42  0.15622
PCTAGE65P      3.9509     0.9992    3.95  9.8e-05 ***
PCTOWNHOME    -0.5600     0.1672   -3.35  0.00092 ***
---
Signif. codes:  0 '***' 0.001 '**' 0.01 '*' 0.05 '.' 0.1 ' ' 1
```

There are only minor changes in the standard errors, and they do not affect our inferences.[2]

In spatial econometrics, Moran's I is supplemented by Lagrange Multiplier tests fully described in Anselin (1988, 2002) and Anselin et al. (1996). The development of these tests, as more generally in spatial econometrics, seems to assume the use of row-standardised spatial weights, so we move from symmetric binary weights used above to row-standardised similar to symmetric weights. A key concern is to try to see whether the data generating process is a spatial error SAR or a spatial lag SAR. The former is the SAR that we have already met, while the spatial lag model includes only the endogenous spatially lagged dependent variable in the model.

```
> NYlistwW <- nb2listw(NY_nb, style = "W")
> res <- lm.LMtests(nylm, listw = NYlistwW, test = "all")
> tres <- t(sapply(res, function(x) c(x$statistic, x$parameter,
```

[2] Full details of the test procedures can be found in the references to the function documentation in **lmtest** and **sandwich**.

```
+      x$p.value)))
> colnames(tres) <- c("Statistic", "df", "p-value")
> printCoefmat(tres)

       Statistic    df p-value
LMerr       5.17  1.00    0.02
LMlag       8.54  1.00  0.0035
RLMerr      1.68  1.00    0.20
RLMlag      5.05  1.00    0.02
SARMA      10.22  2.00    0.01
```

The robust LM tests take into account the alternative possibility, that is the `LMerr` test will respond to both an omitted spatially lagged dependent variable and spatially autocorrelated residuals, while the robust `RLMerr` is designed to test for spatially autocorrelated residuals in the possible presence of an omitted spatially lagged dependent variable. The `lm.LMtests` function here returns a list of five LM tests, which seem to point to a spatial lag specification. Further variants have been developed to take into account both spatial autocorrelation and heteroskedasticity, but are not yet available in R. Again, it is the case that if the fitted model exhibits multicollinearity, the results of the tests will be affected.

The spatial lag model takes the following form:

$$\mathbf{y} = \rho\mathbf{W}\mathbf{y} + \mathbf{X}\beta + \mathbf{e},$$

where $\mathbf{y}$ is the endogenous variable, $\mathbf{X}$ is a matrix of exogenous variables, and $\mathbf{W}$ is the spatial weights matrix. This contrasts with the spatial Durbin model, including the spatial lags of the covariates (independent variables) with coefficients γ:

$$\mathbf{y} = \rho\mathbf{W}\mathbf{y} + \mathbf{X}\beta + \mathbf{W}\mathbf{X}\gamma + \mathbf{e},$$

and the spatial error model:

$$\mathbf{y} - \lambda\mathbf{W}\mathbf{y} = \mathbf{X}\beta - \lambda\mathbf{W}\mathbf{X}\beta + \mathbf{e},$$

$$(\mathbf{I} - \lambda\mathbf{W})\mathbf{y} = (\mathbf{I} - \lambda\mathbf{W})\mathbf{X}\beta + \mathbf{e},$$

which can also be written as

$$\mathbf{y} = \mathbf{X}\beta + \mathbf{u},$$

$$\mathbf{u} = \lambda\mathbf{W}\mathbf{u} + \mathbf{e}.$$

First let us fit a spatial lag model by maximum likelihood, once again finding the spatial lag coefficient by line search, then the remaining coefficients by generalised least squares:

```
> nylag <- lagsarlm(Z ~ PEXPOSURE + PCTAGE65P + PCTOWNHOME,
+     data = NY8, listw = NYlistwW)
> summary(nylag)

Call:
lagsarlm(formula = Z ~ PEXPOSURE + PCTAGE65P + PCTOWNHOME, data = NY8,
    listw = NYlistwW)

Residuals:
      Min        1Q    Median        3Q       Max
-1.626029 -0.393321 -0.018767  0.326616  4.058315

Type: lag
Coefficients: (asymptotic standard errors)
             Estimate Std. Error z value  Pr(>|z|)
(Intercept) -0.505343   0.155850 -3.2425  0.001185
PEXPOSURE    0.045543   0.034433  1.3227  0.185943
PCTAGE65P    3.650055   0.599219  6.0914 1.120e-09
PCTOWNHOME  -0.411829   0.169095 -2.4355  0.014872

Rho: 0.2252 LR test value: 7.75 p-value: 0.0053703
Asymptotic standard error: 0.07954 z-value: 2.831 p-value: 0.0046378
Wald statistic: 8.015 p-value: 0.0046378

Log likelihood: -274.9 for lag model
ML residual variance (sigma squared): 0.41, (sigma: 0.6403)
Number of observations: 281
Number of parameters estimated: 6
AIC: 561.7, (AIC for lm: 567.5)
LM test for residual autocorrelation
test value: 0.6627 p-value: 0.41561

> bptest.sarlm(nylag)

	studentized Breusch-Pagan test

data:
BP = 7.701, df = 3, p-value = 0.05261
```

The spatial econometrics model fitting functions can also use sparse matrix techniques, but when the eigenvalue technique is used, asymptotic standard errors are calculated for the spatial coefficient. There is a numerical snag here, that if the variables in the model are scaled such that the other coefficients are scaled differently from the spatial autocorrelation coefficient, the inversion of the coefficient variance–covariance matrix may fail. The correct resolution is to re-scale the variables, but the tolerance of the inversion function called internally may be relaxed. In addition, an LM test on the residuals is carried out, suggesting that no spatial autocorrelation remains, and a spatial Breusch–Pagan test shows a lessening of heteroskedasticity.

Fitting a spatial Durbin model, a spatial lag model including the spatially lagged explanatory variables (but not the lagged intercept when the spatial weights are row standardised), we see that the fit is not improved significantly.

```
> nymix <- lagsarlm(Z ~ PEXPOSURE + PCTAGE65P + PCTOWNHOME,
+     data = NY8, listw = NYlistwW, type = "mixed")
> nymix

Call:
lagsarlm(formula = Z ~ PEXPOSURE + PCTAGE65P + PCTOWNHOME, data = NY8,
    listw = NYlistwW, type = "mixed")
Type: mixed

Coefficients:
           rho     (Intercept)       PEXPOSURE       PCTAGE65P
       0.17578        -0.32260         0.09039         3.61356
    PCTOWNHOME   lag.PEXPOSURE  lag.PCTAGE65P  lag.PCTOWNHOME
      -0.02687        -0.05188         0.13123        -0.69950

Log likelihood: -272.7

> anova(nymix, nylag)

      Model  df  AIC logLik Test L.Ratio p-value
nymix     1   9  563   -273    1
nylag     2   6  562   -275    2       4    0.22
```

If we impose the Common Factor constraint on the spatial Durbin model, that $\gamma = -\lambda\beta$, we fit the spatial error model:

```
> nyerr <- errorsarlm(Z ~ PEXPOSURE + PCTAGE65P + PCTOWNHOME,
+     data = NY8, listw = NYlistwW)
> summary(nyerr)

Call:errorsarlm(formula = Z ~ PEXPOSURE + PCTAGE65P + PCTOWNHOME,
    data = NY8, listw = NYlistwW)

Residuals:
      Min         1Q     Median         3Q        Max
-1.628589 -0.384745 -0.030234  0.324747  4.047906

Type: error
Coefficients: (asymptotic standard errors)
            Estimate Std. Error z value  Pr(>|z|)
(Intercept) -0.58662    0.17471 -3.3577  0.000786
PEXPOSURE    0.05933    0.04226  1.4039  0.160335
PCTAGE65P    3.83746    0.62345  6.1552 7.496e-10
PCTOWNHOME  -0.44428    0.18897 -2.3510  0.018721

Lambda: 0.2169 LR test value: 5.425 p-value: 0.019853
```

```
Asymptotic standard error: 0.08504 z-value: 2.551 p-value: 0.010749
Wald statistic: 6.506 p-value: 0.010749

Log likelihood: -276 for error model
ML residual variance (sigma squared): 0.4137, (sigma: 0.6432)
Number of observations: 281
Number of parameters estimated: 6
AIC: 564, (AIC for lm: 567.5)
```

Both the spatial lag and Durbin models appear to fit the data somewhat better than the spatial error model. However, in relation to our initial interest in the relationship between transformed incidence proportions and exposure to TCE sites, we are no further forward than we were with the linear model, and although we seem to have reduced the mis-specification found in the linear model by choosing the spatial lag model, the reduction in error variance is only moderate.

Spatial econometrics has also seen the development of alternatives to maximum likelihood methods for fitting models. Code for two of these has been contributed by Luc Anselin, and is available in **spdep**. For example, the spatial lag model may be fitted by analogy with two-stage least squares in a simultaneous system of equations, by using the spatial lags of the explanatory variables as instruments for the spatially lagged dependent variable.

```
> nystsls <- stsls(Z ~ PEXPOSURE + PCTAGE65P + PCTOWNHOME,
+     data = NY8, listw = NYlistwW)
> summary(nystsls)

Call:
stsls(formula = Z ~ PEXPOSURE + PCTAGE65P + PCTOWNHOME, data = NY8,
    listw = NYlistwW)

Residuals:
      Min        1Q    Median        3Q       Max
-1.593609 -0.368930 -0.029486  0.335873  3.991544

Coefficients:
             Estimate Std. Error t value  Pr(>|t|)
Rho          0.409651   0.171972  2.3821  0.017215
(Intercept) -0.495567   0.155743 -3.1820  0.001463
PEXPOSURE    0.042846   0.034474  1.2428  0.213924
PCTAGE65P    3.403617   0.636631  5.3463 8.977e-08
PCTOWNHOME  -0.290416   0.201743 -1.4395  0.149998

Residual variance (sigma squared): 0.4152, (sigma: 0.6444)
```

The implementation acknowledges that the estimate of the spatial coefficient will be biased, but because it can be used with very large data sets and does provide an alternative, it is worth mentioning. It is interesting that when the `robust` argument is chosen, adjusting not only standard errors but also coefficient values for heteroskedasticity over and above the spatial autocorrelation already taken into account, we see that the coefficient operationalising TCE exposure moves towards significance:

```
> nystslsR <- stsls(Z ~ PEXPOSURE + PCTAGE65P + PCTOWNHOME,
+     data = NY8, listw = NYlistwW, robust = TRUE)
> summary(nystslsR)

Call:
stsls(formula = Z ~ PEXPOSURE + PCTAGE65P + PCTOWNHOME, data = NY8,
    listw = NYlistwW, robust = TRUE)

Residuals:
      Min         1Q     Median        3Q       Max
-1.559044 -0.361838 -0.016518  0.353569  4.092810

Coefficients:
             Estimate Robust std. Error z value Pr(>|z|)
Rho          0.411452          0.184989  2.2242 0.026135
(Intercept) -0.499489          0.156801 -3.1855 0.001445
PEXPOSURE    0.056973          0.029993  1.8995 0.057494
PCTAGE65P    3.030160          0.955171  3.1724 0.001512
PCTOWNHOME  -0.267249          0.203269 -1.3148 0.188591

Asymptotic robust residual variance: 0.409, (sigma: 0.6395)
```

Finally, GMerrorsar is an implementation of the Kelejian and Prucha (1999) Generalised Moments (GM) estimator for the autoregressive parameter in a spatial model. It uses a GM approach to optimise λ and σ^2 jointly, and where the numerical search surface is not too flat, can be an alternative to maximum likelihood methods when n is large.

```
> nyGMerr <- GMerrorsar(Z ~ PEXPOSURE + PCTAGE65P + PCTOWNHOME,
+     data = NY8, listw = NYlistwW)
> summary(nyGMerr)

Call:GMerrorsar(formula = Z ~ PEXPOSURE + PCTAGE65P + PCTOWNHOME,
    data = NY8, listw = NYlistwW)

Residuals:
      Min         1Q     Median        3Q       Max
-1.640399 -0.384014 -0.031843  0.318732  4.057979

Type: GM SAR estimator
Coefficients: (GM standard errors)
             Estimate Std. Error z value  Pr(>|z|)
(Intercept) -0.577906   0.172566 -3.3489 0.0008114
PEXPOSURE    0.057984   0.041303  1.4039 0.1603604
PCTAGE65P    3.848771   0.621105  6.1967 5.768e-10
PCTOWNHOME  -0.458145   0.186666 -2.4544 0.0141138
```

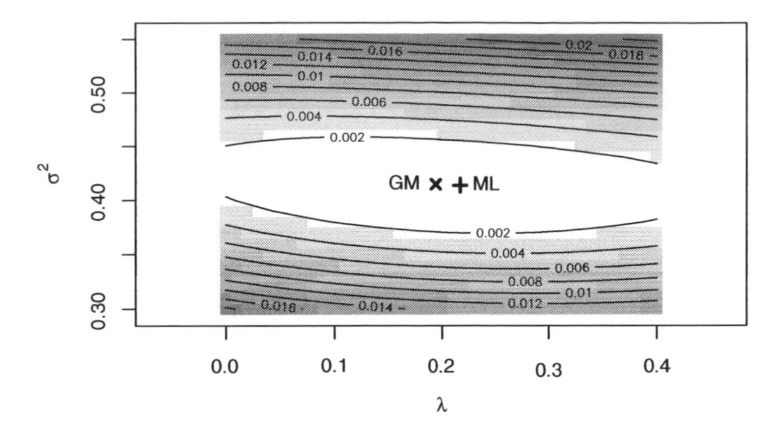

Fig. 10.4. Numerical search surface of the generalised moments estimator with GM and ML optima marked

```
Lambda: 0.1939 LR test value: 5.361 p-value: 0.020594

Log likelihood: -276.0 for GM model
ML residual variance (sigma squared): 0.4146, (sigma: 0.6439)
Number of observations: 281
Number of parameters estimated: 6
AIC: 564.1, (AIC for lm: 567.5)
```

Figure 10.4 shows, however, that there is much more variability in the surface on the σ^2 axis than on the λ axis, and so the optimiser may stop its search when the default joint criterion for termination is satisfied, rather than searching harder along λ. Of course, non-default settings may be passed to the optimiser to tune its performance, but this too requires care and insight.

10.5 Other Methods

Other methods can be used to model dependency between areas. In this section we introduce some of them, based in part on the recent applied survey reported by Dormann et al. (2007). A specific difficulty that we met above when considering mixed-effects models is that available functions for model fitting use point support rather than polygon support. This means that our prior description of the relationships between observations are distance-based, and so very similar to those described in detail in Chap. 8, where the focus was more on interpolation than modelling. These methods are discussed in the spatial context by Schabenberger and Gotway (2005, pp. 352–382) and Waller and Gotway (2004, pp. 380–409), and hierarchical methods are being employed with increasing frequency (Banerjee et al., 2004).

10.5.1 GAM, GEE, GLMM

Generalised Additive Models (GAM) are very similar to generalised linear models, but they also allow for including non-linear terms in the linear predictor term (Hastie and Tibshirani, 1990; Wood, 2006). It is worth noting that the `formula` argument to linear, generalised linear, spatial, and many other models may contain polynomial and spline terms if desired, but these need to be configured manually. Different types of non-linear functions are available, and may be chosen in the `s()` function in the formula. Here, an isotropic thin plate regression spline is used effectively as a semi-parametric trend surface to add smooth spatial structure from the residuals to the fit, as in Chap. 7 (p. 180).

```
> library(mgcv)

This is mgcv 1.3-29

> nyGAM1 <- gam(Z ~ PEXPOSURE + PCTAGE65P + PCTOWNHOME +
+     s(x, y), weights = POP8, data = NY8)
> anova(nylmw, nyGAM1, test = "Chisq")

Analysis of Variance Table

Model 1: Z ~ PEXPOSURE + PCTAGE65P + PCTOWNHOME
Model 2: Z ~ PEXPOSURE + PCTAGE65P + PCTOWNHOME + s(x, y)
  Res.Df    RSS     Df Sum of Sq P(>|Chi|)
1 277.00 310778
2 273.19 305229   3.81      5550      0.27
```

This does not add much to what we already knew from the weighted linear model, with the differences in the residual degrees of freedom showing that the thin plate regression spline term only takes 3.810 estimated degrees of freedom. This does not, however, exploit the real strengths of the technique. Because it can fit generalised models, we can step back from using the transformed incidence proportions to use the case counts (admittedly not integer because of the sharing-out of cases with unknown tract within blocks), offset by the logarithm of tract populations. Recall that we have said that distributional assumptions about the response variable matter – our response variable perhaps ought to be treated as discrete, so methods respecting this may be more appropriate.

Using the Poisson Generalised Linear Model (GLM) fitting approach, we fit first with `glm`; the Poisson model is introduced in Chap. 11. We can already see that this GLM approach yields interesting insights and that the effects of TCE exposure on the numbers of cases are significant.

```
> nyGLMp <- glm(Cases ~ PEXPOSURE + PCTAGE65P + PCTOWNHOME +
+     offset(log(POP8)), data = NY8, family = "poisson")

> summary(nyGLMp)
```

```
Call:
glm(formula = Cases ~ PEXPOSURE + PCTAGE65P + PCTOWNHOME +
    offset(log(POP8)), family = "poisson", data = NY8)

Deviance Residuals:
   Min       1Q   Median       3Q      Max
-2.678   -1.057   -0.198    0.633    3.266

Coefficients:
            Estimate Std. Error z value Pr(>|z|)
(Intercept)  -8.1344     0.1826  -44.54  < 2e-16 ***
PEXPOSURE     0.1489     0.0312    4.77  1.8e-06 ***
PCTAGE65P     3.9982     0.5978    6.69  2.3e-11 ***
PCTOWNHOME   -0.3571     0.1903   -1.88     0.06 .
---
Signif. codes:  0 '***' 0.001 '**' 0.01 '*' 0.05 '.' 0.1 ' ' 1

(Dispersion parameter for poisson family taken to be 1)

    Null deviance: 428.25  on 280  degrees of freedom
Residual deviance: 353.35  on 277  degrees of freedom
AIC: Inf

Number of Fisher Scoring iterations: 5
```

The use of the Moran's I test for regression residuals is speculative and provisional. Based on Lin and Zhang (2007), it takes the deviance residuals and the linear part of the GLM and provides an indication that the Poisson regression, like the weighted linear regression, does not have strong residual spatial autocorrelation (see Fig. 10.5). Much more work remains to be done, perhaps based on Jacqmin-Gadda et al. (1997), to reach a satisfactory spatial autocorrelation test for the residuals of GLM.

```
> NY8$lmpresid <- residuals(nyGLMp, type = "deviance")
> lm.morantest(nyGLMp, listw = NYlistwW)

    Global Moran's I for regression residuals

data:
model: glm(formula = Cases ~ PEXPOSURE + PCTAGE65P +
PCTOWNHOME + offset(log(POP8)), family = "poisson", data =
NY8)
weights: NYlistwW

Moran I statistic standard deviate = 0.7681, p-value = 0.2212
alternative hypothesis: greater
sample estimates:
Observed Moran's I        Expectation           Variance
          0.024654          -0.004487           0.001439
```

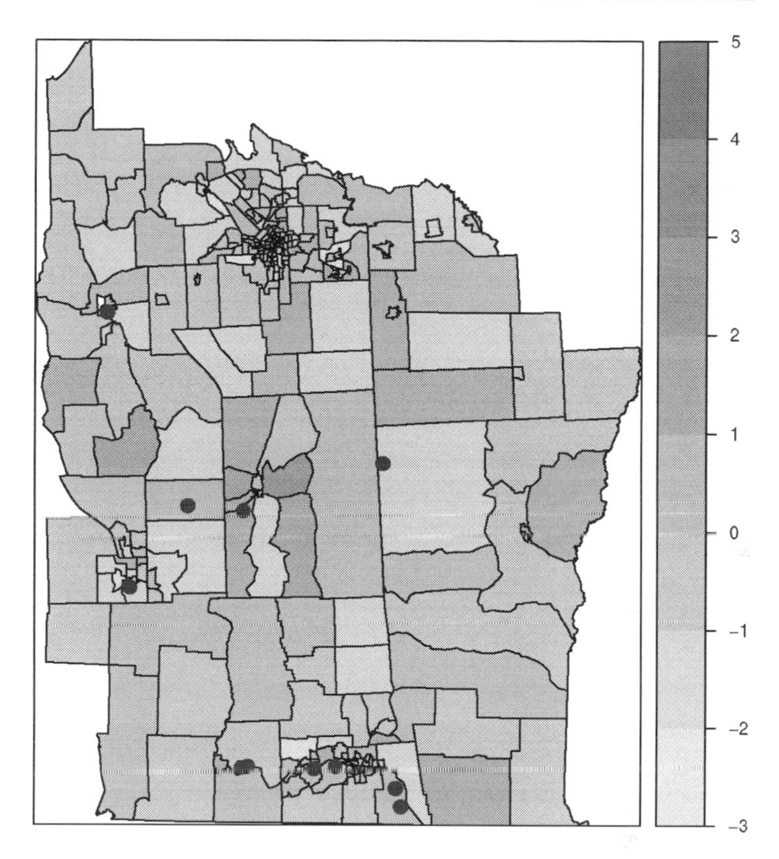

Fig. 10.5. Residuals from the Poisson regression model; TCE site locations shown for comparative purposes

With the GLM to start from, we again add an isotropic thin plate regression spline in `gam`. There is little over-dispersion present – fitting with `family=quasipoisson`, in which the dispersion parameter is not fixed at unity, so they can model over-dispersion that does not result in large changes. Model comparison shows that the presence of the spline term is now significant. While the coefficient values of the Poisson family fits are not directly comparable with the linear fits on the transformed incidence proportions, we can see that exposure to TCE sites is clearly more significant.

```
> nyGAMp <- gam(Cases ~ PEXPOSURE + PCTAGE65P + PCTOWNHOME +
+     offset(log(POP8)) + s(x, y), data = NY8, family = "poisson")
> summary(nyGAMp)

Family: poisson
Link function: log
```

```
Formula:
Cases ~ PEXPOSURE + PCTAGE65P + PCTOWNHOME + offset(log(POP8)) +
    s(x, y)

Parametric coefficients:
            Estimate Std. Error z value Pr(>|z|)
(Intercept)  -8.1366     0.2070  -39.31  < 2e-16 ***
PEXPOSURE     0.1681     0.0558    3.01   0.0026 **
PCTAGE65P     3.7199     0.6312    5.89  3.8e-09 ***
PCTOWNHOME   -0.3602     0.1951   -1.85   0.0649 .
---
Signif. codes:  0 '***' 0.001 '**' 0.01 '*' 0.05 '.' 0.1 ' ' 1

Approximate significance of smooth terms:
        edf Est.rank Chi.sq p-value
s(x,y) 7.71       16     24   0.089 .
---
Signif. codes:  0 '***' 0.001 '**' 0.01 '*' 0.05 '.' 0.1 ' ' 1

R-sq.(adj) =  0.394   Deviance explained = 21.4%
UBRE score = 0.2815  Scale est. = 1         n = 281

> anova(nyGLMp, nyGAMp, test = "Chisq")

Analysis of Deviance Table

Model 1: Cases ~ PEXPOSURE + PCTAGE65P + PCTOWNHOME +
    offset(log(POP8))

Model 2: Cases ~ PEXPOSURE + PCTAGE65P + PCTOWNHOME +
    offset(log(POP8)) + s(x, y)

  Resid. Df Resid. Dev     Df Deviance P(>|Chi|)
1    277.00        353
2    269.29        337   7.71       17     0.029
```

Generalised Estimating Equations (GEE) are an alternative to the estimation of GLMs when we have correlated data. They are often used in the analysis of longitudinal data, when we have several observations for the same subject. In a spatial setting, the correlation arises between neighbouring areas. The treatment in Dormann et al. (2007) is promising for the restricted case of clusters of grid cells, but has not yet been extended to irregular point or polygon support.

Generalised linear mixed-effect models (GLMM) extend GLMs by allowing the incorporation of mixed effects into the linear predictor; see Waller and Gotway (2004, pp. 387–392) and Schabenberger and Gotway (2005, pp. 359–369). These random effects can account for correlation between observations. Here we use `glmmPQL` from **MASS**, described in Venables and Ripley (2002, pp. 292–298), and a Gaussian spatial correlation structure as above when applying linear mixed-effect models. The `glmmPQL` function calls `lme` internally, so we can use the values of the `random` and `correlation` arguments used above on p. 288. Dormann et al. (2007) suggest the use of a single group, because the spatial correlation structure is applied group-wise,[3] but admit that this is an 'abuse' of the procedure.

```
> library(MASS)
> attach(as(NY8, "data.frame"))
> nyGLMMp <- glmmPQL(Cases ~ PEXPOSURE + PCTAGE65P + PCTOWNHOME +
+     offset(log(POP8)), data = NY8, family = poisson,
+     random = ~1 | AREAKEY, correlation = scor)
> detach("as(NY8, \"data.frame\")")

> summary(nyGLMMp)

Linear mixed-effects model fit by maximum likelihood
 Data: NY8
  AIC BIC logLik
   NA  NA     NA

Random effects:
 Formula: ~1 | AREAKEY
        (Intercept) Residual
StdDev:   7.325e-05    1.121

Correlation Structure: Gaussian spatial correlation
 Formula: ~x + y | AREAKEY
 Parameter estimate(s):
    range
0.0005343
Variance function:
 Structure: fixed weights
 Formula: ~invwt

Fixed effects: Cases ~ PEXPOSURE + PCTAGE65P + PCTOWNHOME +
    offset(log(POP8))

             Value Std.Error  DF t-value p-value
(Intercept) -8.134    0.2062 277  -39.45  0.0000
PEXPOSURE    0.149    0.0352 277    4.23  0.0000
PCTAGE65P    3.998    0.6750 277    5.92  0.0000
PCTOWNHOME  -0.357    0.2148 277   -1.66  0.0976
 Correlation:
           (Intr) PEXPOS PCTAGE
PEXPOSURE  -0.472
PCTAGE65P  -0.634  0.030
PCTOWNHOME -0.768  0.134  0.230
```

[3] They report that results from PROC GLIMMIX in SAS can be reproduced using only a single group.

```
Standardized Within-Group Residuals:
       Min         Q1        Med         Q3        Max
-1.7839 -0.7476 -0.1731  0.6003  3.8928

Number of Observations: 281
Number of Groups: 281
```

The fitting functions require that the `Spatial*DataFrame` object be coerced to a `data.frame` object, and `attach` be used to make the variables visible in the global environment in this case. The outcome is very close to the GAM results, and again we find that closeness to the TCE sites is a significant covariate; again, the percentage owning their own homes is not significant. Since it is fitted by penalised quasi-likelihood, no log likelihood value is available, and the summary reports `NA` for AIC, BIC, and log likelihood.

10.5.2 Moran Eigenvectors

In the previous chapter, we touched on the use of eigenvalues in the Saddlepoint approximation and exact tests for Moran's I. The Moran eigenvector approach (Dray et al., 2006; Griffith and Peres-Neto, 2006) involved the spatial patterns represented by maps of eigenvectors; by choosing suitable orthogonal patterns and adding them to a linear or generalised linear model, the spatial dependence present in the residuals can be moved into the model.

It uses brute force to search the set of eigenvectors of the matrix $\mathbf{MWM}$, where

$$\mathbf{M} = \mathbf{I} - \mathbf{X}(\mathbf{X}^{\mathrm{T}}\mathbf{X})^{-1}\mathbf{X}^{\mathrm{T}}$$

is a symmetric and idempotent projection matrix and $\mathbf{W}$ are the spatial weights. In the spatial lag form of `SpatialFiltering` and in the GLM `ME` form below, $\mathbf{X}$ is an n-vector of ones, that is the intercept only.

In its general form, `SpatialFiltering` chooses the subset of the n eigenvectors that reduce the residual spatial autocorrelation in the error of the model with covariates. The lag form adds the covariates in assessment of which eigenvectors to choose, but does not use them in constructing the eigenvectors. `SpatialFiltering` was implemented and contributed by Yongwan Chun and Michael Tiefelsdorf, and is presented in Tiefelsdorf and Griffith (2007); `ME` is based on Matlab code by Pedro Peres-Neto and is discussed in Dray et al. (2006) and Griffith and Peres-Neto (2006).

```
> nySFE <- SpatialFiltering(Z ~ PEXPOSURE + PCTAGE65P +
+     PCTOWNHOME, data = NY8, nb = NY_nb, style = "W",
+     verbose = FALSE)
> nylmSFE <- lm(Z ~ PEXPOSURE + PCTAGE65P + PCTOWNHOME +
+     fitted(nySFE), data = NY8)
> summary(nylmSFE)
```

```
Call:
lm(formula = Z ~ PEXPOSURE + PCTAGE65P + PCTOWNHOME + fitted(nySFE),
    data = NY8)

Residuals:
    Min      1Q  Median      3Q     Max
-1.5184 -0.3523 -0.0105  0.3221  3.1964

Coefficients:
                    Estimate Std. Error t value Pr(>|t|)
(Intercept)          -0.5173     0.1461   -3.54  0.00047 ***
PEXPOSURE             0.0488     0.0323    1.51  0.13172
PCTAGE65P             3.9509     0.5578    7.08  1.2e-11 ***
PCTOWNHOME           -0.5600     0.1569   -3.57  0.00042 ***
fitted(nySFE)vec13  -2.0940     0.6053   -3.46  0.00063 ***
fitted(nySFE)vec44  -2.2400     0.6053   -3.70  0.00026 ***
fitted(nySFE)vec6    1.0298     0.6053    1.70  0.09007 .
fitted(nySFE)vec38   1.2928     0.6053    2.14  0.03361 *
fitted(nySFE)vec20   1.1006     0.6053    1.82  0.07015 .
fitted(nySFE)vec14  -1.0511     0.6053   -1.74  0.08366 .
fitted(nySFE)vec75   1.9060     0.6053    3.15  0.00183 **
fitted(nySFE)vec21  -1.0633     0.6053   -1.76  0.08014 .
fitted(nySFE)vec36  -1.1786     0.6053   -1.95  0.05258 .
fitted(nySFE)vec61  -1.0858     0.6053   -1.79  0.07399 .
---
Signif. codes:  0 '***' 0.001 '**' 0.01 '*' 0.05 '.' 0.1 ' ' 1

Residual standard error: 0.605 on 267 degrees of freedom
Multiple R-squared: 0.34,   Adjusted R-squared: 0.308
F-statistic: 10.6 on 13 and 267 DF,  p-value: <2e-16

> anova(nylm, nylmSFE)

Analysis of Variance Table

Model 1: Z ~ PEXPOSURE + PCTAGE65P + PCTOWNHOME
Model 2: Z ~ PEXPOSURE + PCTAGE65P + PCTOWNHOME + fitted(nySFE)
  Res.Df   RSS  Df Sum of Sq     F Pr(>F)
1    277 119.6
2    267  97.8  10      21.8 5.94  4e-08 ***
---
Signif. codes:  0 '***' 0.001 '**' 0.01 '*' 0.05 '.' 0.1 ' ' 1
```

Since the `SpatialFiltering` approach does not allow weights to be used, we see that the residual autocorrelation of the original linear model is absorbed, or 'whitened' by the inclusion of selected eigenvectors in the model, but that the covariate coefficients change little. The addition of these eigenvectors – each representing an independent spatial pattern – relieves the residual autocorrelation, but otherwise makes few changes in the substantive coefficient values.

The ME function also searches for eigenvectors from the spatial lag variant of the underlying model, but in a GLM framework. The criterion is a permutation bootstrap test on Moran's I for regression residuals, and in this case, because of the very limited remaining spatial autocorrelation, is set at $\alpha = 0.5$. Even with this very generous stopping rule, only two eigenvectors are chosen; their combined contribution just improves only the fit of the GLM model.

```
> nyME <- ME(Cases ~ PEXPOSURE + PCTAGE65P + PCTOWNHOME,
+     data = NY8, offset = log(POP8), family = "poisson",
+     listw = NYlistwW, alpha = 0.5)

> nyME

  Eigenvector ZI pr(ZI)
0          NA NA   0.26
1          24 NA   0.47
2         223 NA   0.52

> nyglmME <- glm(Cases ~ PEXPOSURE + PCTAGE65P + PCTOWNHOME +
+     offset(log(POP8)) + fitted(nyME), data = NY8, family = "poisson")

> summary(nyglmME)

Call:
glm(formula = Cases ~ PEXPOSURE + PCTAGE65P + PCTOWNHOME +
    offset(log(POP8)) + fitted(nyME), family = "poisson", data = NY8)

Deviance Residuals:
   Min       1Q   Median       3Q      Max
-2.569   -1.068   -0.212    0.610    3.166

Coefficients:
                     Estimate Std. Error z value Pr(>|z|)
(Intercept)           -8.1269     0.1834  -44.30  < 2e-16 ***
PEXPOSURE              0.1423     0.0314    4.53  5.8e-06 ***
PCTAGE65P              4.1105     0.5995    6.86  7.1e-12 ***
PCTOWNHOME            -0.3827     0.1924   -1.99    0.047 *
fitted(nyME)vec24      1.5266     0.7226    2.11    0.035 *
fitted(nyME)vec223     0.8142     0.7001    1.16    0.245
---
Signif. codes:  0 '***' 0.001 '**' 0.01 '*' 0.05 '.' 0.1 ' ' 1

(Dispersion parameter for poisson family taken to be 1)

    Null deviance: 428.25  on 280  degrees of freedom
Residual deviance: 347.34  on 275  degrees of freedom
AIC: Inf

Number of Fisher Scoring iterations: 5

> anova(nyGLMp, nyglmME, test = "Chisq")
```

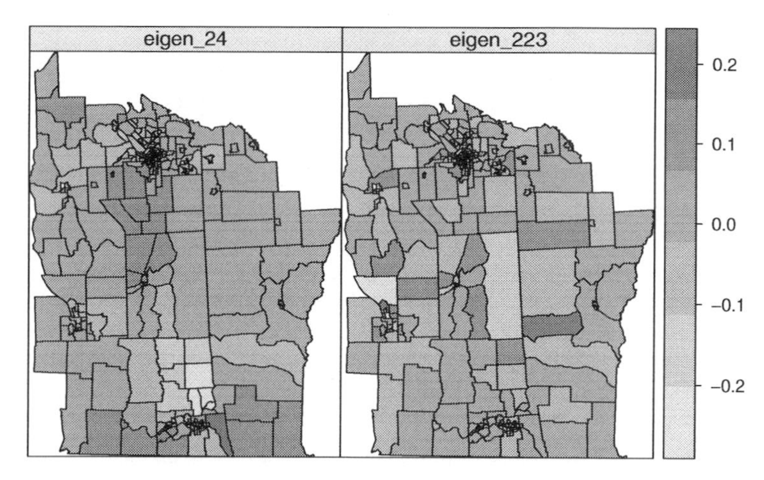

Fig. 10.6. Maps of the two eigenvalues selected for inclusion in the Poisson regression model

```
Analysis of Deviance Table

Model 1: Cases ~ PEXPOSURE + PCTAGE65P + PCTOWNHOME +
    offset(log(POP8))

Model 2: Cases ~ PEXPOSURE + PCTAGE65P + PCTOWNHOME +
    offset(log(POP8)) + fitted(nyME)

  Resid. Df Resid. Dev  Df Deviance P(>|Chi|)
1       277        353
2       275        347   2        6      0.05
```

Figure 10.6 shows the spatial patterns chosen to match the very small amount of spatial autocorrelation remaining in the model. As with the other Poisson regressions, the closeness to TCE sites is highly significant. Since, however, many TCE sites are also in or close to more densely populated urban areas with the possible presence of both point-source and non-point-source pollution, it would be premature to take such results simply at their face value. There is, however, a potentially useful contrast between the cities of Binghampton in the south of the study area with several sites in its vicinity, and Syracuse in the north without TCE sites in this data set.

10.5.3 Geographically Weighted Regression

Geographically weighted regression (GWR) is an exploratory technique mainly intended to indicate where non-stationarity is taking place on the map, that is where locally weighted regression coefficients move away from their global values. Its basis is the concern that the fitted coefficient values of a global model, fitted to all the data, may not represent detailed local variations in the data adequately – in this it follows other local regression implementations. It differs, however, in not looking for local variation in 'data' space, but by moving a weighted window over the data, estimating one set of coefficient values at every chosen 'fit' point. The fit points are very often the points at which observations were made, but do not have to be. If the local coefficients vary in space, it can be taken as an indication of non-stationarity.

The technique is fully described by Fotheringham et al. (2002) and involves first selecting a bandwidth for an isotropic spatial weights kernel, typically a Gaussian kernel with a fixed bandwidth chosen by leave-one-out cross-validation. Choice of the bandwidth can be very demanding, as n regressions must be fitted at each step. Alternative techniques are available, for example for adaptive bandwidths, but they may often be even more compute-intensive. GWR is discussed by Schabenberger and Gotway (2005, pp. 316–317) and Waller and Gotway (2004, p. 434), and presented with examples by Lloyd (2007, pp. 79–86).

```
> library(spgwr)

> bwG <- gwr.sel(Z ~ PEXPOSURE + PCTAGE65P + PCTOWNHOME,
+     data = NY8, gweight = gwr.Gauss, verbose = FALSE)
> gwrG <- gwr(Z ~ PEXPOSURE + PCTAGE65P + PCTOWNHOME, data = NY8,
+     bandwidth = bwG, gweight = gwr.Gauss, hatmatrix = TRUE)

> gwrG

Call:
gwr(formula = Z ~ PEXPOSURE + PCTAGE65P + PCTOWNHOME, data = NY8,
    bandwidth = bwG, gweight = gwr.Gauss, hatmatrix = TRUE)
Kernel function: gwr.Gauss
Fixed bandwidth: 179943
Summary of GWR coefficient estimates:
                 Min. 1st Qu.  Median 3rd Qu.    Max. Global
X.Intercept. -0.5220 -0.5210 -0.5200 -0.5140 -0.5110  -0.52
PEXPOSURE     0.0472  0.0480  0.0495  0.0497  0.0505   0.05
PCTAGE65P     3.9100  3.9300  3.9600  3.9600  3.9800   3.95
PCTOWNHOME   -0.5590 -0.5580 -0.5580 -0.5550 -0.5550  -0.56
Number of data points: 281
Effective number of parameters: 4.4
Effective degrees of freedom: 276.6
Sigma squared (ML): 0.4255
AICc (GWR p. 61, eq 2.33; p. 96, eq. 4.21): 568
AIC (GWR p. 96, eq. 4.22): 561.6
Residual sum of squares: 119.6
```

Once the bandwidth has been found, or chosen by hand, the `gwr` function may be used to fit the model with the chosen local kernel and bandwidth. If the `data` argument is passed a `SpatialPolygonsDataFrame` or a `SpatialPointsDataFrame` object, the output object will contain a component, which is an object of the same geometry populated with the local coefficient estimates.

If the input objects have polygon support, the centroids of the spatial entities are taken as the basis for analysis. The function also takes a `fit.points` argument, which permits local coefficients to be created by geographically weighted regression for other support than the data points.

The basic GWR results are uninteresting for this data set, with very little local variation in coefficient values; the bandwidth is almost 180 km. Neither `gwr` nor `gwr.sel` yet take a `weights` argument, as it is unclear how non-spatial and geographical weights should be combined. A further issue that has arisen is that it seems that local collinearity can be induced, or at least observed, in GWR applications. A discussion of the issues raised is given by Wheeler and Tiefelsdorf (2005).

As Fotheringham et al. (2002) describe, GWR can also be applied in a GLM framework, and a provisional implementation permitting this has been added to the `spgwr` package providing both cross-validation bandwidth selection and geographically weighted fitting of GLM models.

```
> gbwG <- ggwr.sel(Cases ~ PEXPOSURE + PCTAGE65P + PCTOWNHOME +
+     offset(log(POP8)), data = NY8, family = "poisson",
+     gweight = gwr.Gauss, verbose = FALSE)
> ggwrG <- ggwr(Cases ~ PEXPOSURE + PCTAGE65P + PCTOWNHOME +
+     offset(log(POP8)), data = NY8, family = "poisson",
+     bandwidth = gbwG, gweight = gwr.Gauss)

> ggwrG

Call:

ggwr(formula = Cases ~ PEXPOSURE + PCTAGE65P + PCTOWNHOME +
    offset(log(POP8)), data = NY8, bandwidth = gbwG, gweight =
    gwr.Gauss, family = "poisson")

Kernel function: gwr.Gauss
Fixed bandwidth: 179943
Summary of GWR coefficient estimates:
                Min. 1st Qu. Median 3rd Qu.   Max. Global
X.Intercept. -8.140  -8.140 -8.140  -8.130 -8.130  -8.13
PEXPOSURE     0.147   0.148  0.149   0.149  0.150   0.15
PCTAGE65P     3.980   3.980  3.980   4.010  4.020   4.00
PCTOWNHOME   -0.357  -0.355 -0.355  -0.349 -0.346  -0.36
```

The local coefficient variation seen in this fit is not large either, although from Fig. 10.7 it appears that slightly larger local coefficients for the closeness to TCE site covariate are found farther away from TCE sites than close to them. If, on the other hand, we consider this indication in the light of Fig. 10.8, it is clear that the forcing artefacts found by Wheeler and Tiefelsdorf (2005) in a different data set are replicated here.

Further ways of using R for applying different methods for modelling areal data are presented in Chap. 11. It is important to remember that the availability of implementations of methods does not mean that any of them are

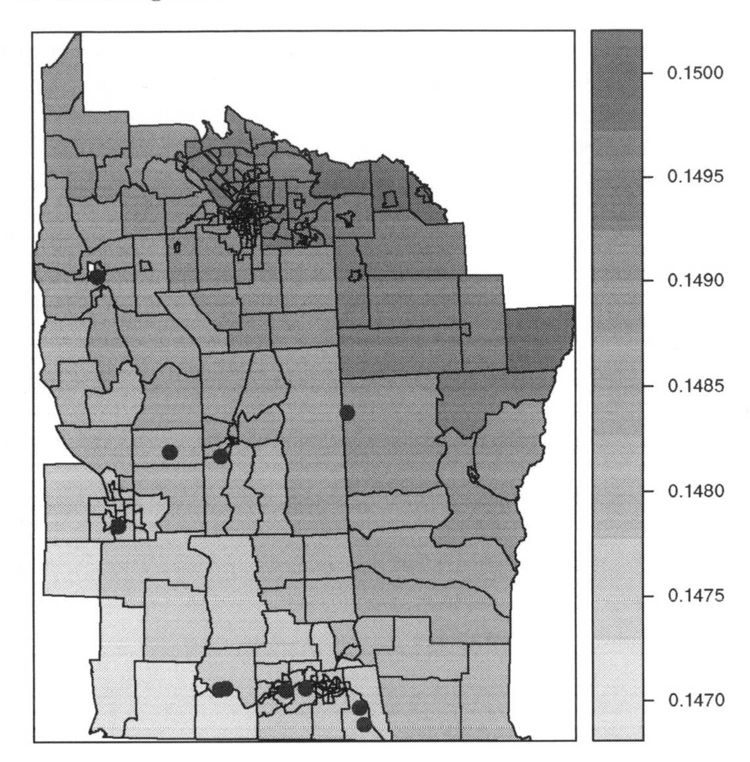

Fig. 10.7. GWR local coefficient estimates for the exposure to TCE site covariate

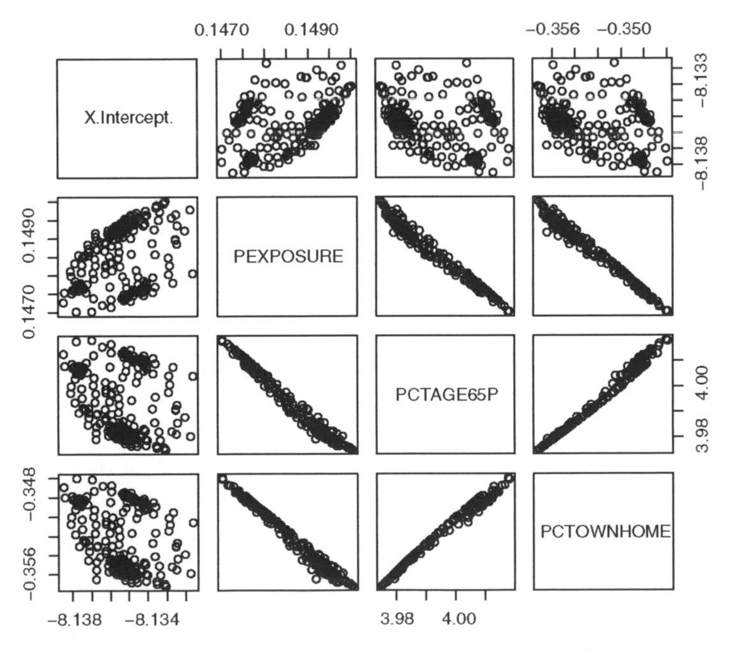

Fig. 10.8. Pairs plots of GWR local coefficient estimates showing the effects of GWR collinearity forcing

'best practice' as such. It is the analyst who has responsibility for choices of methods and implementations in relation to situation-specific requirements and available data. What the availability of a range of methods in R does make possible is that the analyst has choice and has tools for ensuring that the research outcomes are fully reproducible.

11

Disease Mapping

Spatial statistics have been widely applied in epidemiology for the study of the distribution of disease. As we have already shown in Chap. 7, displaying the spatial variation of the incidence of a disease can help us to detect areas where the disease is particularly prevalent, which may lead to the detection of previously unknown risk factors. As a result of the growing interest, Spatial Epidemiology (Elliott et al., 2000) has been established as a new multidisciplinary area of research in recent years.

The importance of this field has been reflected in the appearance of different books and special issues in some scientific journals. To mention a few, recent reviews on the subject can be found in Waller and Gotway (2004), while the special issues of the journal *Statistical Methods in Medical Research* (Lawson, A., 2005) and *Statistics in Medicine* (Lawson et al., 2006) also summarise novel developments in disease mapping and the detection of clusters of disease. Walter and Birnie (1991) compared many different atlases of disease and they compile the main issues to pay attention to when reporting disease maps. Banerjee et al. (2004, pp. 88–97, 158–174) also tackle the problem of disease mapping and develop examples that can be reproduced using `S-PLUS`™ SpatialStats (Kaluzny et al., 1998) and WinBUGS. In addition, some data sets and code with examples are available from the book website.[1] Haining (2003) considers different issues in disease mapping, including a Bayesian approach as well and provides examples, data, and code to reproduce the examples in the book. Schabenberger and Gotway (2005, pp. 394–399) briefly describe the smoothing of disease rates. Finally, Lawson et al. (2003) provide a practical approach to disease mapping, with a number of examples (with full data sets and WinBUGS code) that the reader should be able to reproduce after reading this chapter.

In this chapter we refer to the analysis of data which have been previously aggregated according to a set of administrative areas. The analysis of data available at individual level requires different techniques, which have

[1] `http://www.biostat.umn.edu/~brad/data2.html`

been described in Chap. 7. These kinds of aggregated data are continuously collected by Health Authorities and usually cover mortality and morbidity counts. Special registers have also been set up in several countries to record the incidence of selected diseases, such as cancer or congenital malformations. Spatial Epidemiology often requires the integration of large amounts of data, statistical methods, and geographic information. R offers a unique environment for the development of these types of analysis given its good connectivity to databases and the different statistical methods implemented.

Therefore, the aim of this chapter is not to provide a detailed and comprehensive description of all the methods currently employed in Spatial Epidemiology, but to show those which are widely used. A description as to how they can be computed with R and how to display the results will be provided. From this description, it will be straightforward for the user to adapt the code provided in this chapter to make use of other methods. Other analysis of health data, as well as contents on which this chapter is built, can be found in Chaps. 9 and 10.

The North Carolina SIDS data, which have already been displayed in Chap. 3 (Fig. 3.6), will be used throughout this chapter in the examples that accompany the statistical methodology described here. The SIDS data set records the number of sudden infant deaths in North Carolina for 1974–1978 and 1979–1984 and some other additional information. It is available as `nc.sids` in package **spdep** and further information is available in the associated manual page. Cressie and Read (1985) and Cressie and Chan (1989), for example, provide a description of the data and study whether there is any clustered pattern of the cases.

11.1 Introduction

The aim of disease mapping is to provide a representation of the spatial distribution of the risk of a disease in the study area, which we assume is divided into several non-overlapping smaller regions. The risk may reflect actual deaths due to the disease (mortality) or, if it is not fatal, the number of people who suffer from the disease (morbidity) in a certain period of time for the population at risk.

Hence, basic data must include the population at risk and the number of cases in each area. These data are usually split according to different variables in a number of groups or strata, which can be defined using sex, age, and other important variables. When available, a deprivation index (Carstairs, 2000) is usually employed in the creation of the strata. By considering data in different groups, the importance of each variable can be explored and potential confounding factors can be removed (Elliott and Wakefield, 2000) before doing any other analysis of the data. For example, if the age is divided into 13 groups and sex is also considered, this will lead to 26 strata in the population. Note that depending on the type of study the population at risk may be a reduced subset of the total population. For example, in our examples, it is reduced to the number of children born during the period of study.

Following this structure, we denote by P_{ij} and O_{ij} the population and observed number of cases in region i and stratum j. Summing over all strata j we can get the total population and number of cases per area, which we denote by P_i and O_i. Summing again over all the regions will give the totals, which will be denoted by P_+ and O_+.

Representing the observed number of cases alone gives no information about the risk of the disease given that the cases are mainly distributed according to the underlying population. To obtain an estimate of the risk, the observed number of cases must be compared to an *expected* number of cases.

If P_i and O_i are already available, which is the simplest case, the expected number of cases in region i can be calculated as $E_i = P_i r_+$, where r_+ is the overall incidence ratio equal to $\frac{O_+}{P_+}$. This is an example of the use of *indirect standardisation* (Waller and Gotway, 2004, pp. 12–15) to compute the expected number of cases for each area.

When data are grouped in strata, a similar procedure can be employed to take into account the distribution of the cases and population in the different strata. Instead of computing a global ratio $\frac{O_+}{P_+}$ for all regions, a different ratio is computed for each stratum as $r_j = \frac{\sum_i O_{ij}}{\sum_i P_{ij}}$. In other words, we could compute the ratio between the sum of all cases at stratum j over the population at stratum j. In this situation, the expected number of cases in region i is given by $E_i = \sum_j P_{ij} r_j$.

This standardisation is also called *internal standardisation* because we have used the same data to compute reference rates r_j. Sometimes they are known because another reference population has been used. For example, national data can be used to compute the reference rates to be used later in regional studies.

The following code, based on that available in the `nc.sids` manual page, will read the SIDS data, boundaries of North Carolina, and the adjacency structure of the North Carolina counties (as in Cressie and Read, 1985) in GAL format (see Chap. 9). By using the argument `region.id` we make sure that the order of the list of neighbours `ncCR85` is the same as the areas in the `SpatialPolygonDataFrame` `nc`.

```
> library(maptools)
> library(spdep)
> nc_file <- system.file("shapes/sids.shp", package = "maptools")[1]
> llCRS <- CRS("+proj=longlat +datum=NAD27")
> nc <- readShapePoly(nc_file, ID = "FIPSNO", proj4string = llCRS)
> rn <- sapply(slot(nc, "polygons"), function(x) slot(x,
+     "ID"))
> gal_file <- system.file("etc/weights/ncCR85.gal",
+     package = "spdep")[1]
> ncCR85 <- read.gal(gal_file, region.id = rn)
```

11.2 Statistical Models

A common statistical assumption to model the number of observed number of cases in region i and stratum j is that it is drawn from a Poisson distribution with mean $\theta_i E_{ij}$. Thus, a relative risk of 1 means that the risk is as the average in the reference region (from where the rates r_j are obtained) and it will be of interest in the location of the regions where the relative risk is significantly higher than 1. This basic model is described in Banerjee et al. (2004, pp. 158–159), Haining (2003, pp. 194–199), and Lawson et al. (2003, pp. 2–8).

Note that implicitly we are assuming that there is no interaction between the risk and the population strata, i.e. the relative risk θ_i depends only on the region.

At this point, a basic estimate of the risk in a given region can be computed as $\text{SMR}_i = O_i/E_i$, which is known as the *Standardised Mortality Ratio*. This is why the data involving the cases are often referred to as the *numerator* and the data of the population as the *denominator*, because they are used to compute a ratio that estimates the relative risk. Figure 11.1 shows the SMRs of the SIDS data for the period 1974–1978. Waller and Gotway (2004, pp. 11–18) describe in detail this and other types of standardisation, together with other risk ratios frequently used in practise.

```
> nc$Observed <- nc$SID74
> nc$Population <- nc$BIR74
> r <- sum(nc$Observed)/sum(nc$Population)
> nc$Expected <- nc$Population * r
> nc$SMR <- nc$Observed/nc$Expected
```

Using the fact that O_i is Poisson distributed, we can obtain a confidence interval for each SMR (using function `pois.exact` from package **epitools**). Figure 11.2 displays the 95% confidence interval of the SMR computed for

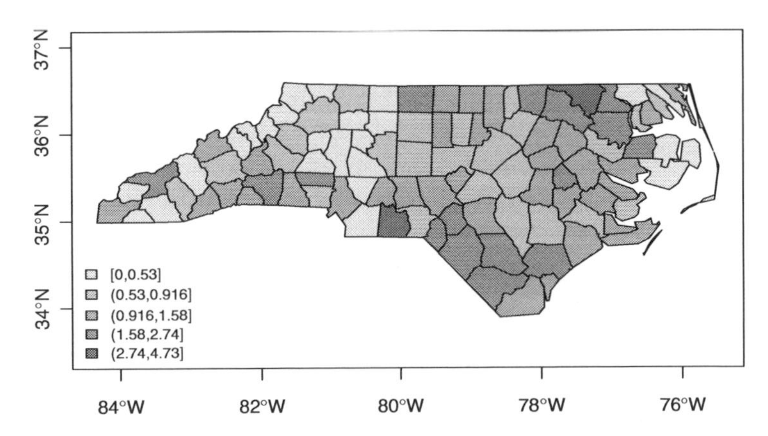

Fig. 11.1. Standardised mortality ratio of the North Carolina SIDS data in the period 1974–1978

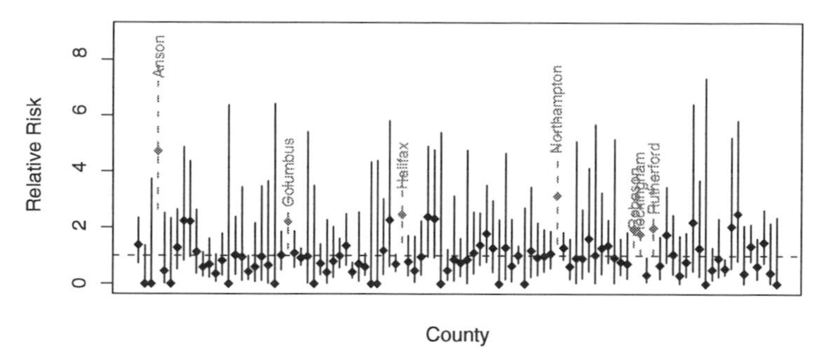

Fig. 11.2. Confidence intervals of the SMR obtained with an exact formula. The black dot represents the SMR of each area. The confidence intervals shown by dashed lines are significantly higher than 1

each area. Highly significant risks (i.e. those whose confidence interval is above one) have been drawn using a dashed line and the name of the county has been added as a label. Anson county, which has been pointed out as a clear extreme value in previous studies (Cressie and Chan, 1989), is the one with the highest confidence interval.

11.2.1 Poisson-Gamma Model

Unfortunately, using a Poisson distribution implies further assumptions that may not always hold. One key issue is that for this distribution the mean and the variance of O_i are supposed to be the same. It is often the case that data are 'over-dispersed', so that the variance of the data is higher than their mean and the statistical model needs to be expanded. A simple way to allow for a higher variance is to use a negative binomial distribution instead of the Poisson.

The negative binomial distribution can also be regarded as a mixed model in which a random effect following a Gamma distribution for each region is considered. This formulation is known as the Poisson-Gamma (PG) model, because it can be structured as the following two-level model:

$$\begin{aligned} O_i|\theta_i, E_i &\sim \mathrm{Po}(\theta_i E_i), \\ \theta_i &\sim \mathrm{Ga}(\nu, \alpha). \end{aligned}$$

In this model, we also consider the relative risk θ_i as a random variable, which is drawn from a Gamma distribution with mean ν/α and variance ν/α^2. Note that now the distribution of O_i is conditioned on the value of θ_i. The unconditioned distribution for each O_i is easy to derive and it is a negative binomial with size parameter ν and probability $\frac{\alpha}{\alpha+E_i}$.

In addition, the posterior distribution of θ_i, i.e. its distribution given the observed data $\{O_i\}_{i=1}^n$, can also be derived and it is a Gamma with parameters

$\nu + O_i$ and $\alpha + E_i$. In other words, the information provided by observing the data has *updated* our prior knowledge or assumptions on θ_i. The posterior expectation of θ_i is

$$E[\theta_i|O_i, E_i] = \frac{\nu + O_i}{\alpha + E_i},$$

which can also be expressed as a compromise between the prior mean of the relative risks and SMR_i, so that this is a *shrinkage* estimator:

$$E[\theta_i|O_i, E_i] = \frac{E_i}{\alpha + E_i}\text{SMR}_i + (1 - \frac{E_i}{\alpha + E_i})\frac{\nu}{\alpha}.$$

Two issues should be noted from this estimator. First of all, when E_i is small, as often happens in low populated areas, a small variation in O_i can produce dramatic changes in the value of SMR_i. For this reason, according to the previous expectation, the SMR_i will have a low weight, as compared to that of the prior mean. Secondly, information is borrowed from all the areas in order to construct the posterior estimates given that ν and α are the same for every region. This concept of *borrowing strength* can be modified and extended to take into account a different set of areas or neighbours.

```
> library(DCluster)
> eb <- empbaysmooth(nc$Observed, nc$Expected)
> nc$EBPG <- eb$smthrr
```

Given that ν and α are unknown, we need a procedure to estimate them. They can be easily estimated from the data using the method of moments, following formulae given by Clayton and Kaldor (1987) to produce Empirical Bayes (EB) estimates, implemented in package **DCluster**. In this example, the values are $\nu = 4.6307$ and $\alpha = 4.3956$, which gives a prior mean of the relative risks of 1.0535 (very close to 1).

Probability maps (Choynowski, 1959) are a convenient way of representing the significance of the observed values. These maps show the probability of a value being higher than the observed data according to the assumption we have made about the model. In other words, probability maps show the p-value of the observed number of cases under the current model. Figure 11.3 represents the probability maps for the Poisson and Poisson-Gamma models. The reason to compare both maps is to show how significance varies with the model. We noted that the Poisson-Gamma model was more appropriate in this case due to over-dispersion, and we should try to make inference based on this model. As expected, the p-values for the Poisson-Gamma model are higher because more variability is permitted. Nevertheless, there are still two zones of high risk to the northeast and south.

11.2.2 Log-Normal Model

Clayton and Kaldor (1987) proposed another risk estimator based on assumption that the logarithm of the relative risks ($\beta_i = \log(\theta_i)$) follows a multivariate

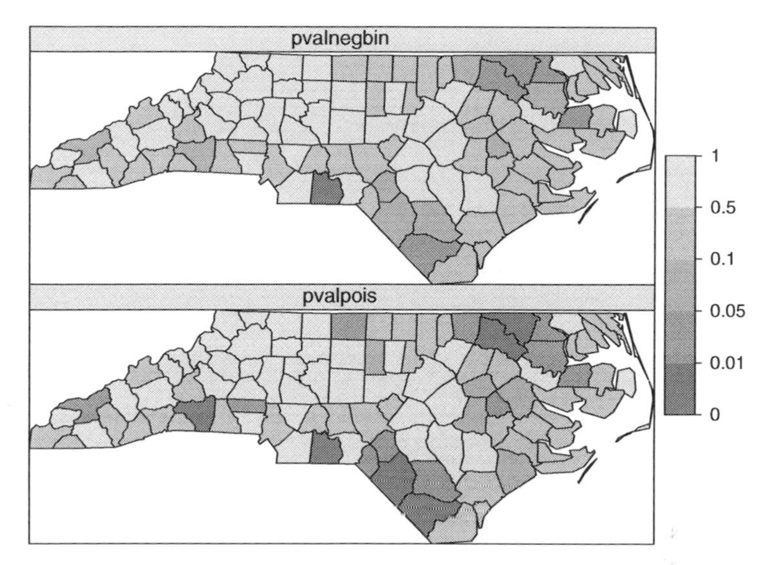

Fig. 11.3. Probability maps for the Poisson and negative binomial models

normal distribution with common mean ϕ and variance σ^2. In this case, the estimate of the log-relative risk is not taken as $\log(O_i/E_i)$ but $\log((O_i+1/2)/E_i)$, because the former is not defined if O_i is zero. The EM algorithm is used to obtain estimates of the mean and variance of the model, which can be plugged in to the following Empirical Bayes estimator of β_i:

$$\hat{\beta}_i = b_i = \frac{\hat{\phi} + (O_i + \frac{1}{2})\hat{\sigma}^2 \log[(O_i + \frac{1}{2})/E_i] - \hat{\sigma}^2/2}{1 + (O_i + \frac{1}{2})\hat{\sigma^2}},$$

where $\hat{\phi}$ and $\hat{\sigma}^2$ are the estimates of the prior mean and variance, respectively. These are given by

$$\hat{\phi} = \frac{1}{n}\sum_{i=1}^{n} b_i = \bar{b}$$

and

$$\hat{\sigma}^2 = \frac{1}{n}\left\{\hat{\sigma}^2 \sum_{i=1}^{n}[1 + \hat{\sigma}^2(O_i + 1/2)]^{-1} + \sum_{i=1}^{n}(b_i - \hat{\phi})^2\right\}.$$

Estimates b_i are updated successively using previous formulae until convergence. Hence, the estimator for θ_i is $\hat{\theta}_i = \exp\{\hat{\beta}_i\}$. Note that now the way information is borrowed is to estimate the common parameters ϕ and σ^2, and that the resulting estimates are a combination of the local estimate of the log relative risk and ϕ. Unfortunately, the current estimator is more complex than the previous one and it cannot be reduced to a shrinkage expression.

```
> ebln <- lognormalEB(nc$Observed, nc$Expected)
> nc$EBLN <- exp(ebln$smthrr)
```

11.2.3 Marshall's Global EB Estimator

Marshall (1991) developed a new EB estimator assuming that the relative risks θ_i have a common prior mean μ and variance σ^2, but without specifying any distribution. By using the method of moments, he is able to work a new estimator out employing a shrinkage estimator as follows:

$$\hat{\theta}_i = \hat{\mu} + C_i(\mathrm{SMR}_i - \hat{\mu}) = (1 - C_i)\hat{\mu} + C_i\mathrm{SMR}_i,$$

where

$$\hat{\mu} = \frac{\sum_{i=1}^n O_i}{\sum_{i=1}^n E_i}$$

and

$$C_i = \frac{s^2 - \hat{\mu}/\overline{E}}{s^2 - \hat{\mu}/\overline{E} + \hat{\mu}/E_i}.$$

$\overline{E}$ stands for the mean of the E_i's and s^2 is the usual unbiased estimate of the variance of the SMR_i's. Unfortunately, this estimator can produce negative estimates of the relative risks when $s^2 < \hat{\mu}/\overline{E}$, in which case $\hat{\theta}_i = \hat{\mu}$ is taken.

The shrinkage of this estimator highly depends (again) on the value of E_i. If it is high, which means that the SMR_i is a reliable estimate, C_i will be close to 1 and the estimator will give more weight to the SMR_i. On the other hand, if E_i is small, more weight is given to the estimate of the prior mean $\hat{\mu}$ because the SMR_i is less reliable and so it borrows more information from other areas.

```
> library(spdep)
> EBMarshall <- EBest(nc$Observed, nc$Expected)
> nc$EBMarshall <- EBMarshall[, 2]
```

Figure 11.4 represents the different estimates obtained by the different estimators described so far. All EB estimators seem to produce very similar estimates in all the areas. By comparing those maps to the map that shows the SMR, it is possible to see how very extreme values (either high or low) have been shifted towards the global mean. In other words, these values have been *smoothed* by taking into account global information in the computation of the estimate.

To compare the variability of the estimates produced by each method, we have created a boxplot of each set of values, which appear in Fig. 11.6. From the plot it is clear that the SMR is the most variable and that the other three have been shrunk towards the global mean, which is approximately 1. Hence, we might expect similar results when using any of the EB estimators. As pointed by Marshall (1991), the estimation procedure based on the Poisson-Gamma model proposed by Clayton and Kaldor (1987) may not converge in

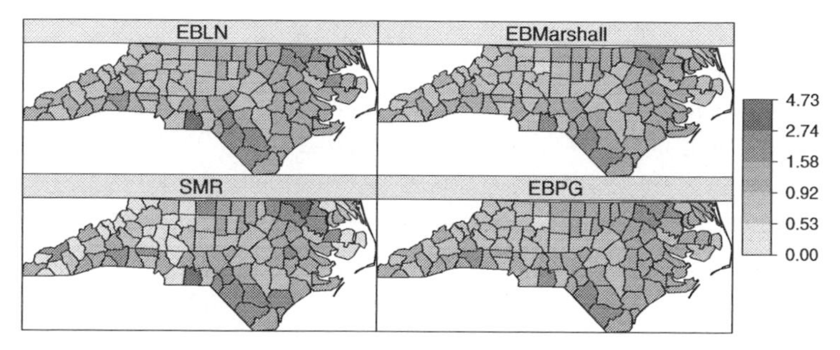

Fig. 11.4. Comparison of different risk estimators. SMR displays the standardised mortality ratio, whilst EBPG, EBLN, and EBMarshall show different empirical Bayes estimates using the Poisson-Gamma model, the log-normal model, and Marshall's global estimator

some circumstances and another estimator should be used. The EB proposed by Marshall (1991) can also be unfeasible in similar circumstances. Hence, the EB estimator based on the log-Normal model seems to be the most computationally stable and reliable.

All these EB estimators produce smoothed estimates of the risk rates *borrowing information* from the global area but, depending on the size and extension of the total area under study, it could be more reasonable to consider only a small set of areas that are close to each other. A common example is to use only the areas that share a boundary with the current region to compute its risk estimate. Unfortunately, this procedure involves the use of more complex models that require the use of additional software and will be discussed in the following sections.

11.3 Spatially Structured Statistical Models

Although borrowing strength globally can make sense in some cases, it is usually better to consider a reduced set of areas to borrow information from. A sensible choice is to take only neighbouring areas or areas which are within a certain distance from the current area.

Marshall (1991) proposed another estimator that requires only local information to be computed. For each region, a set of neighbours is defined and local means, variances, and shrinkage factors are defined in a similar way as in the global estimator, but considering only the areas in the neighbourhood. This produces a local shrinkage for each area, instead of the global shrinkage provided by the previous estimator.

```
> nc$EBMrshloc <- EBlocal(nc$Observed, nc$Expected, ncCR85)$est
```

The way this estimator is computed raises a new question about how areas are related to each other. In the previous models, no account for how areas were distributed in the study region was considered, so that the influence of a region did not depend on its location at all. That is, we would obtain the same estimates if the distribution of the regions were permutated at random. With the new estimator the exact location of the areas is crucial, and different locations of the regions will give different estimates as a result. The way regions are placed in a map can be described by means of its *topology*, which accounts for the neighbours of a given region. See Chap. 9 for more details on this and how to obtain it.

Although neighbours are usually defined as two regions that share a common boundary, Cressie and Chan (1989) define two regions as neighbours if the distance between their centroids is within 30 miles. This is not a trivial issue since different definitions of neighbourhood will produce different results.

The two estimators proposed by Marshall have been displayed in Fig. 11.5. The version that uses only local information produces smoothed estimates of the relative risks that are shrunk towards the local mean that turned out to be less shrunk towards the global mean. In addition, the shrinkage produced by the local estimator is in general lower than that for the global estimator.

The boxplot presented in Fig. 11.6 compares the different EB estimators discussed so far. Marshall's local estimator also shows a general shift towards the global mean, but it is less severe than for the others because only local information is employed. In general, EB smoothed estimators have been criticised because they fail to cope with the uncertainty of the parameters of the model (Bernardinelli and Montomoli, 1992) and to produce an overshrinkage since the parameters of the prior distributions are estimated from the data

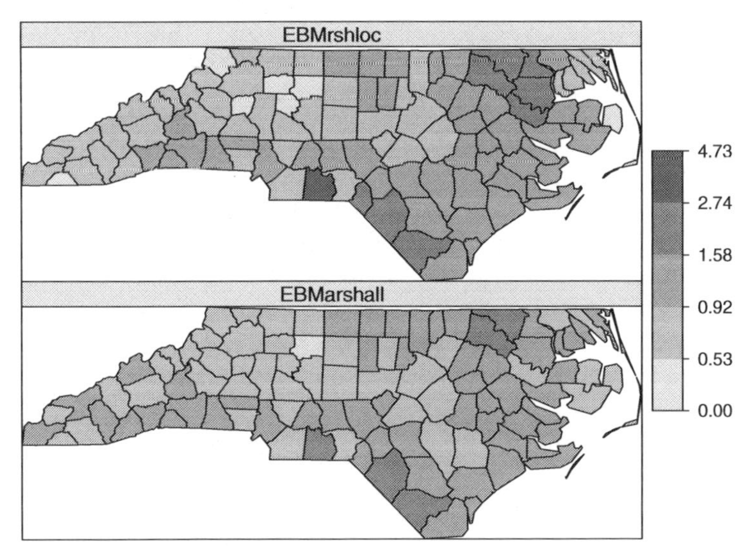

Fig. 11.5. Marshall's EB estimator using local (*top*) and global (*bottom*) information

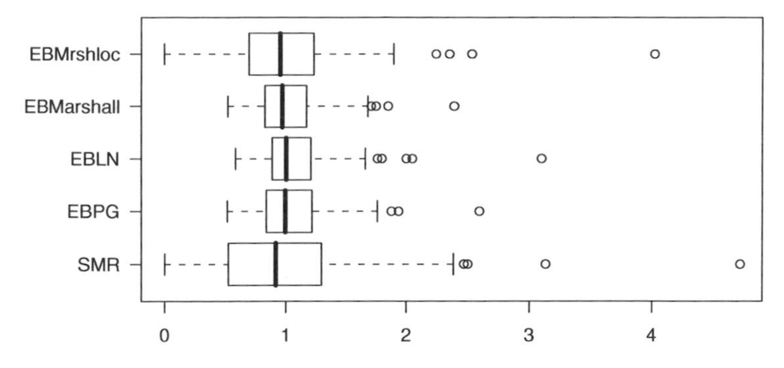

Fig. 11.6. Comparison of raw and EB estimators of the relative risk

and remain fixed. To solve this problem several *constrained* EB estimators have been proposed to force the posterior distribution of the smoothed estimates to resemble that of the raw data (Louis, 1984; Devine and Louis, 1994; Devine et al., 1994).

Full Bayes methods allow setting the prior distributions for these parameters and, hence, permit a greater variability and produce more suitable smoothed estimates. More standard smoothed risk estimators that borrow information locally can be developed by resorting to Spatial Autoregressive and Conditional Autoregressive specifications (Waller and Gotway, 2004). Basically, these models condition the relative risk in an area to be similar to the values of the neighbouring areas. More details are given in the next sections of this chapter and in Chap. 10 for non-Bayesian models.

11.4 Bayesian Hierarchical Models

Bayesian Hierarchical Models make an appropriate framework for the development of spatially structured models. The model is specified in different layers, so that each one accounts for different sources of variation. For example, they can cope with covariates at the same time as borrowing strength from neighbours to improve the quality of estimates. The use of these models in disease mapping is considered in Haining (2003, pp. 307–311, 367–376), Waller and Gotway (2004, pp. 409–429), Banerjee et al. (2004, pp. 159–169), and Schabenberger and Gotway (2005, pp. 394–399). Lawson et al. (2003) offer a specific volume on the subject, with reproducible examples.

Besag et al. (1991), BYM henceforth, introduced in their seminal paper a type of models that split the variability in a region as the sum of a spatially correlated variable (which depends on the values of its neighbours) plus an area-independent effect (which reflects local heterogeneity). Although direct estimates of the variables in the model can seldom be obtained when using Bayesian Hierarchical Models, their posterior distributions can be obtained by means of Markov Chain Monte Carlo (MCMC) techniques. Basically, MCMC methods generate simulations of the parameters of the model which, after a suitable burn-in period, become realisations of their posterior distributions. An introduction to MCMC and its main applications, including disease mapping, can be found in Gilks et al. (1996).

WinBUGS (Spiegelhalter et al., 2003) is software that uses MCMC methods (in particular, Gibbs Sampling; Gelman et al., 2003) to simulate from the posterior distributions of the parameters in the model. Starting from a set of initial values, one sample of each variable is simulated at the time using the full conditional distribution of the parameter given the other parameters. After a suitable burn-in period, the simulations generated correspond to the joint posterior distribution.

Although WinBUGS is the main software package, it was previously known as BUGS and currently it comes in different flavours. OpenBUGS, for example is the open source alternative to WinBUGS and it is actually a fork of the main WinBUGS software. Apart from the advantage of coming with the source code, OpenBUGS can be called from R using package **BRugs**. In addition, some specific plug-ins have been developed for WinBUGS to deal with certain applications. It is worth mentioning GeoBUGS, which provides a graphical interface to the management of maps and compute adjacency relationships within WinBUGS and OpenBUGS, and it can create maps with the results. Lawson et al. (2003) have described extensively how to do a disease mapping using Multilevel Models with WinBUGS (and MLwiN), and is a complete reference for those readers willing to go deeper in this subject.

Another package to use WinBUGS from R is **R2WinBUGS** (Sturtz et al., 2005). This package calls WinBUGS using its scripting facilities so that the resulting log file containing all the results can be loaded into R after the computations have finished. **R2WinBUGS** will be the package used in this book. The main reason is that at the time of writing **BRugs** only works on Windows (although the authors claim that it should also work on Linux with minor modifications) whilst **R2WinBUGS** can be used on several platforms with minor adjustments. Under Linux, for example it can be run using the Wine programme. Finally, it is worth noting that Gelman and Hill (2007) provide a good and accessible text on data analysis using Bayesian hierarchical models and describe the use of R and WinBUGS via **R2WinBUGS** in Chaps. 16 and 17.

11.4.1 The Poisson-Gamma Model Revisited

The following example shows a full Bayesian Poisson-Gamma formulation (i.e. assigning priors to the parameters ν and α) to produce smoothed estimates of the relative risks that can be run from R using **R2WinBUGS**. In this model, ν and α have been assigned vague gamma priors so that as little prior information as possible is introduced. The WinBUGS code needed to run the Poisson-Gamma model is shown in Fig. 11.7.

```
model
{

  for(i in 1:N)
  {
     observed[i]~dpois(mu[i])
     mu[i]<-theta[i]*expected[i]
     theta[i]~dgamma(nu, alpha)
  }

  nu~dgamma(.01, .01)
  alpha~dgamma(.01, .01)
}
```

Fig. 11.7. Code of the Poisson-Gamma model for WinBugs

The next chunk of code shows how to convert all the necessary data into the structure used by WinBUGS. In addition, we need to set up the initial values for some of the parameters of the model. Data and initial values must be saved into a separated file.

```
> library(R2WinBUGS)
> N <- length(nc$Observed)
> d <- list(N = N, observed = nc$Observed, expected = nc$Expected)

> pgmodelfile <- paste(getwd(), "/PG-model.txt", sep = "")
> wdir <- paste(getwd(), "/PG", sep = "")
> if (!file.exists(wdir)) {
+     dir.create(wdir)
+ }
> BugsDir <- "/home/asdar/.wine/dosdevices/c:/Program Files/WinBUGS14"
> MCMCres <- bugs(data = d, inits = list(list(nu = 1, alpha = 1)),
+     working.directory = wdir, parameters.to.save = c("theta",
+         "nu", "alpha"), n.chains = 1, n.iter = 20000,
+     n.burnin = 10000, n.thin = 10, model.file = pgmodelfile,
+     bugs.directory = BugsDir, WINEPATH = "/usr/bin/winepath")
```

Briefly explained, the `bugs` function will take data, initial values, model file, and other information required and it will create a script that will be run with WinBUGS.[2] `bugs` will create the necessary files (data, initial values, and script) that will be placed under `working.directory`. After running the model, the output will be stored here as well. The WinBUGS script will basically check the syntax of the model, load the data, and compile the model. The following step is to read (or generate from the priors) the initial values for the parameters of the model and 10,000 simulations of the Markov Chain

[2] Windows users must modify the paths in `working.directory`, `model.file`, and `bugs.directory` accordingly, and remove the argument `WINEPATH`, which is not needed.

are generated (keeping just 1 every 10). Note that, since we need a burn-in period, these are not saved. Then, we set that variables 'nu', 'alpha', and 'theta' will be saved and 10,000 more simulations are generated, of which only 1 of every 10 are saved to avoid autocorrelation and improve mixing and convergence. Finally, the summary statistics and plots are saved into the log files under the working directory. Two such files are created: an ODC file (WinBUGS format) with summary statistics and plots, and an ASCII file with the summary statistics. In addition, a summary of the output is stored as a series of lists in `MCMCres`. The posterior mean and median of the relative risks can be extracted as follows:

```
> nc$PGmean <- MCMCres$mean$theta
> nc$PGmedian <- MCMCres$median$theta
```

Although it will not be described here in detail, it is essential to check that the Markov Chain has converged so that the values that we are using have been drawn from the posterior distribution of the parameters. A example using package **coda** (Best et al., 1995) is shown later in a more complex model.

As we have obtained samples from the posterior distributions of ν and α, it is possible to compute pointwise estimates and probability intervals for both parameters. For the sake of simplicity and to be able to compare the values obtained with those from the EB approach, the pointwise estimates (posterior means) of these values were $\hat{\nu} = 6.253$ and $\hat{\alpha} = 5.967$, which are slightly higher than the ones obtained with the EB estimator. Similar estimates can be obtained for the relative risks, but note that now they are not based on single values of ν and α, but that the relative risk estimates are *averaged* over different values of those parameters.

Even though point estimates of the relative risks are usually very useful, for most applications it is better to give a credible interval, for it can be used to detect areas of significantly high risk, if the interval is over 1. Figure 11.8 summarises the 95% credible intervals for each region. The median has been

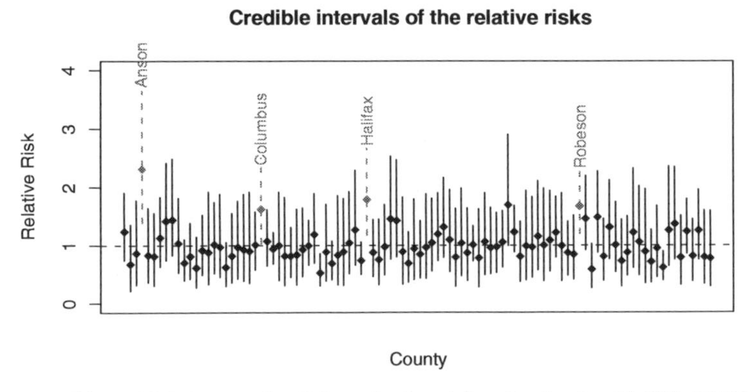

Fig. 11.8. 95% credible intervals of the relative risks obtained with WinBUGS using a full Bayes Poisson-Gamma model

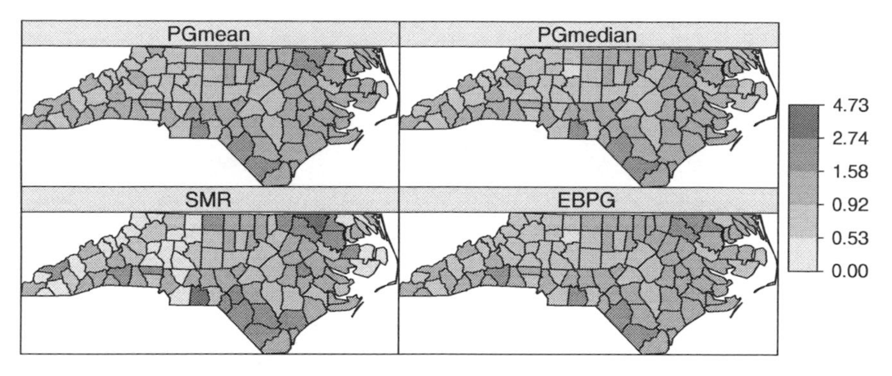

Fig. 11.9. Comparison of empirical Bayes and full Bayes estimates of the relative risks using a Poisson-Gamma model

included (black dot), and the areas whose credible intervals are above 1 have been highlighted using a dashed line and the county name displayed. As we mentioned before, Anson county is of special interest because it shows the highest risk.

In Fig. 11.9 we have compared the estimates of the relative risks provided by the Poisson-Gamma model using both Empirical Bayes and Full Bayes approaches. Both estimation procedures lead to very similar estimates and they only differ in a few areas. Note how they all provide smoothed estimates of the relative risks, as compared to the raw SMRs.

11.4.2 Spatial Models

Additional spatial structure can be included by considering a CAR model and covariates can be used to explain part of the variability of the relative risks. Cressie and Chan (1989) considered the proportion of non-white births as an important factor related to the incidence of SIDS. A full description of these models can be found in Banerjee et al. (2004, Chap. 5). In general, these models are far more complex than the Poisson-Gamma described before, and they should be used with extreme caution because of the high number of parameters and possible interactions between them.

As described in Sect. 10.2.1, the CAR specification for a set of random variables $\{v_i\}_{i=1}^n$ can be written as follows:

$$v_i|v_{-i} \sim N\left(\sum_{j\sim i} \frac{w_{ij}v_j}{\sum_j w_{ij}}, \sigma_v^2/\sum_j w_{ij}\right),$$

where w_{ij} is a weight that measures the strength of the relationship between (neighbour) regions i and j and σ_v^2 indicates the conditional variance of the CAR specification.

Although the conditional distributions are proper, it is not the case for the joint distribution. Nevertheless, this CAR specification is often used as a prior distribution of the spatial random effects and it can lead to a proper posterior under some constraints (Ghosh et al., 1998).

Given the structure of the CAR specification, it is necessary to know the neighbours of each region. They can be defined in different ways, depending on the type of relationship that exists between the areas. In our example, we use the same neighbourhood structure as in Cressie and Read (1985), which can be found in package `spdep`. In addition, it is necessary to assign a weight to each pair of neighbours, which measure the strength of the interaction. Following Besag et al. (1991), we set all the weights to 1 if regions are neighbours and 0 otherwise.

The flexibility of the Bayesian Hierarchical Models allows us to perform an Ecologic Regression (English, 1992) at the same time as we consider independent and spatial random effects. By including covariates in our model we aim to assess and remove the effect of potential confounders or risk factors. The assessment of the importance of a covariate is indicated by the estimated value of its coefficient and its associated probability interval. If, for example, the 95% credible interval does not contain the value 0, we may assume that the coefficient is significant and, if greater than zero, it will indicate a positive relationship between the risk and the variable.

The results of an Ecologic Regression can be potentially misleading if we try to make inference at the individual level, since the effects that operate at that level may not be the same as those reflected at the area level. In the extreme case, the effects might even be reversed. A solution to this is to combine the aggregated data with some individual data from a specific survey, which can be also used to improve the estimation of the effects of the covariates (Jackson et al., 2006).

In our example, we have the available number of non-white births in each county. The variable ethnicity is often used in the United States as a surrogate of the deprivation index (Krieger et al., 1997). Considering this variable in our model may help to explain part of the spatial variability of the risk of SIDS. To account for the ethnicity, we use the proportion of non-white births in the area. This also allow us to compare the values for different counties. Figure 11.10 shows the spatial variation of the proportion of non-white births. Notice how there exists a similar pattern to that shown by the spatial distribution of the SMR and the different EB estimates. Finally, the WinBUGS model used in this case can be found in Fig. 11.11. We have used the priors suggested in Best et al. (1999) to allow a better identifiability of the random effects u_i and v_i.

The chunk of code shown below converts the neighbours of each county as specified in Cressie and Read (1985) into the format required by WinBUGS. Note that these are already available in an R object and that they have been matched so that the list of neighbours is in the right order. When this is not the case, proper matching must be done. Function `nb2WB` can be used to

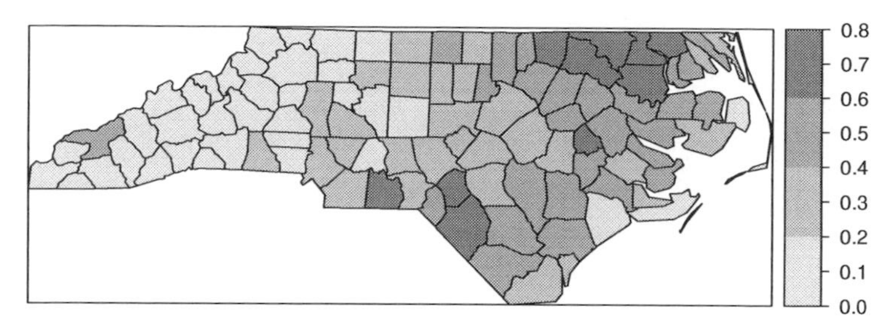

Fig. 11.10. Proportion of non-white births in Carolina, 1974–1978. Notice the similar pattern to the relative risk estimates

```
model
{

  for(i in 1:N)
  {
     observed[i] ~ dpois(mu[i])
     log(theta[i]) <-  alpha + beta*nonwhite[i] + u[i] + v[i]
     mu[i] <- expected[i]*theta[i]

     u[i] ~ dnorm(0, precu)
  }

  v[1:N] ~ car.normal(adj[], weights[], num[], precv)

  alpha ~ dflat()
  beta ~ dnorm(0,1.0E-5)
  precu ~ dgamma(0.001, 0.001)
  precv ~ dgamma(0.1, 0.1)

  sigmau<-1/precu
  sigmav<-1/precv
}
```

Fig. 11.11. Code of the Besag-York-Mollié model for WinBugs

convert an `nb` object into a list containing the three elements (`adj`, `weights`, and `num`) required to use a CAR specification in WinBUGS. Similarly, the function `listw2WB` can be used for a `listw` object. The main difference is that `nb2WB` sets all the weights to 1, whilst `listw2WB` keeps the values of the weights as in the `listw` object.

```
> nc.nb <- nb2WB(ncCR85)
```

The last step is to compute the proportion of non-white births in each county and create the R lists with the data and initial values.

```
> nc$nwprop <- nc$NWBIR74/nc$BIR74
> d <- list(N = N, observed = nc$Observed, expected = nc$Expected,
```

```
+     nonwhite = nc$nwprop, adj = nc.nb$adj, weights = nc.nb$weights,
+     num = nc.nb$num)
> dwoutcov <- list(N = N, observed = nc$Observed,
+     expected = nc$Expected, adj = nc.nb$adj, weights = nc.nb$weights,
+     num = nc.nb$num)
> inits <- list(u = rep(0, N), v = rep(0, N), alpha = 0,
+     beta = 0, precu = 0.001, precv = 0.001)
```

The procedure to run this model is very similar to the previous one. We only need to change the file names of the model, data, and initial values. Notice that not all initial values must be provided and that some can be generated randomly. In this model, we are going to keep the summary statistics for a wide range of variables. In addition to the relative risks θ_i, we want to summarise the values of the intercept (α), the coefficient of the covariate (β), and the values of the non-spatial (u_i) and spatial (v_i) random effects.

```
> bymmodelfile <- paste(getwd(), "/BYM-model.txt", sep = "")
> wdir <- paste(getwd(), "/BYM", sep = "")
> if (!file.exists(wdir)) {
+     dir.create(wdir)
+ }
> BugsDir <- "/home/asdar/.wine/dosdevices/c:/Program Files/WinBUGS14"

> MCMCres <- bugs(data = d, inits = list(inits),
+     working.directory = wdir, parameters.to.save = c("theta",
+         "alpha", "beta", "u", "v", "sigmau", "sigmav"),
+     n.chains = 1, n.iter = 30000, n.burnin = 20000,
+     n.thin = 10, model.file = bymmodelfile, bugs.directory = BugsDir,
+     WINEPATH = "/usr/bin/winepath")
```

After running the model, the summary statistics are added to the spatial object that contains all the information about the North Carolina SIDS data so that it can be displayed easily.

```
> nc$BYMmean <- MCMCres$mean$theta
> nc$BYMumean <- MCMCres$mean$u
> nc$BYMvmean <- MCMCres$mean$v
```

Convergence of the Markov Chain must be assessed before attempting any valid inference from the results. Cowles and Carlin (1996) provide a summary of several methods and a useful discussion. They state the difficulty to assess convergence in practise. Some of the criteria discussed in the paper are implemented in package **coda**. These criteria can be applied to the *deviance* of the model to monitor convergence of the joint posterior. Ideally, several chains (each one starting at a sufficiently different point) can be run in parallel so that the traces can be compared (Gelman and Rubin, 1992).

WinBUGS can produce the output in the format required by **coda**. Basically, it will produce an index file (`codaIndex.txt`) plus another file with the values of the variables (`coda1.txt`) that can be read using function `read.coda`.

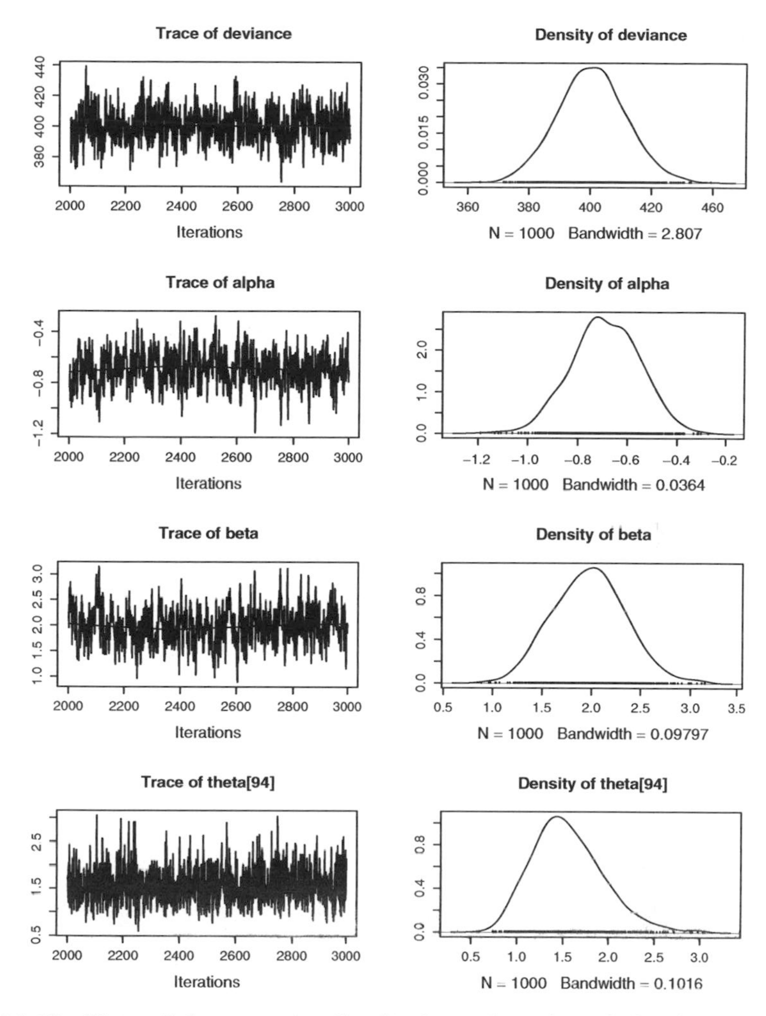

Fig. 11.12. Plots of the posterior distributions of α, β, and the deviance of the model

This will create an object of type `mcmc`, which contains the simulations from all the variables saved in WinBUGS. Figure 11.12 shows the trace and density of the posterior distribution of the deviance and the parameters α, β and the relative risk of Robeson county (area number 94 and cluster centre in Fig. 11.18).

For a single chain, Geweke's criterion (Geweke, 1992) can be computed to assess convergence. It is a score test based on comparing the means of the first and the last part of the Markov Chain (by default, the 10% initial values to the 50% last values). If the chain has converged, both means should be equal. Given that it is a score test, values of the test statistics between -1.96 and 1.96 indicate convergence, whilst more extreme value will denote lack of convergence. For the selected parameters, it seems that convergence has been reached:

```
> geweke.diag(ncoutput[, c("deviance", "alpha", "beta",
+     "theta[94]")])

Fraction in 1st window = 0.1
Fraction in 2nd window = 0.5

 deviance     alpha      beta theta[94]
   0.1216   -0.8550    0.4239   -1.0669
```

Figure 11.13 shows the SMR and the smoothed estimate of the relative risks obtained. When the posterior distribution is very skewed, the posterior median can be a better summary statistic, but it is not the case now.

According to the posterior density of β shown in Fig. 11.12, the coefficient of the covariate can be considered as significantly positive given that its posterior mean is greater than 0 and its 95% credible interval is likely not to contain the value 0. This means that there is an actual risk increase in those regions with a high proportion of non-white births. Point posterior estimates (mean) of the random effects u_i and v_i are shown in Figs. 11.14 and 11.15, respectively. They seem to have a very small variation, specially the former, but this is not so because they are in the log-scale.

It should be noted that if the spatial pattern is weak or appropriate covariates are included in the model, the random effects u_i and v_i may become unidentifiable. However, following Besag et al. (1995), valid inference could

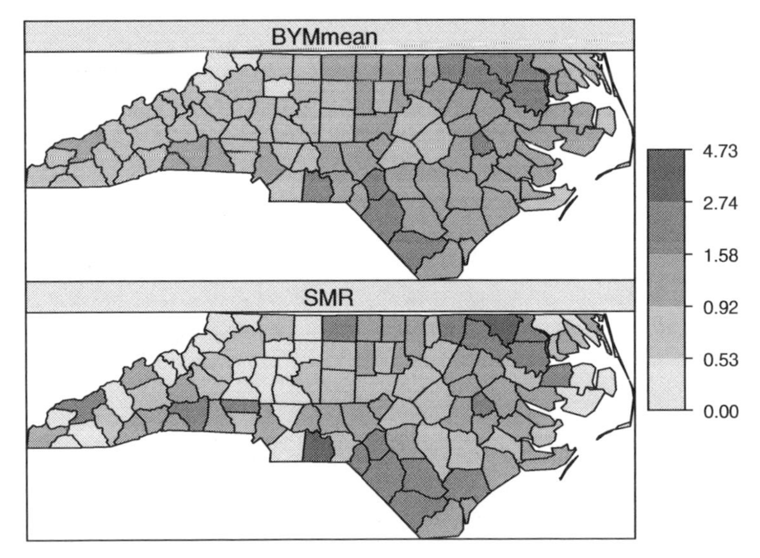

Fig. 11.13. Standardised Mortality Ratio and posterior means of the relative risks obtained with the BYM model

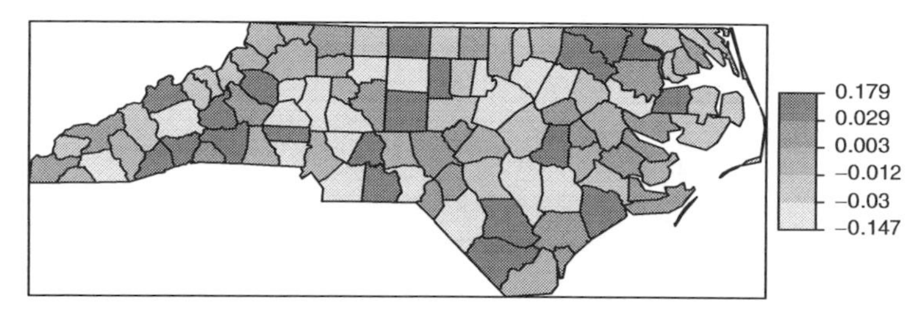

Fig. 11.14. Posterior means of the non-spatial random effects (u_i) estimated with the BYM model

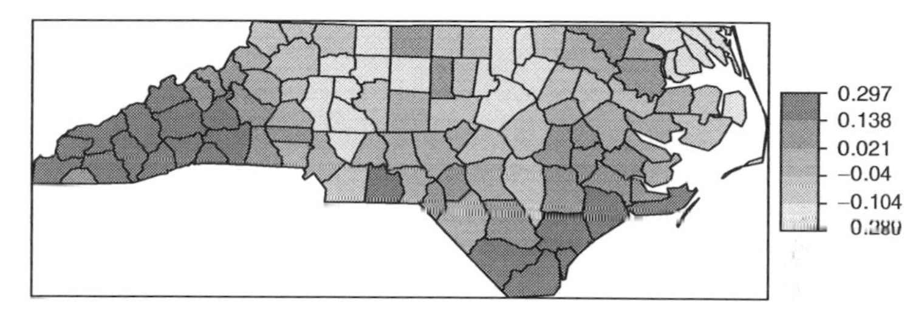

Fig. 11.15. Posterior means of the spatial random effects (v_i) estimated with the BYM model

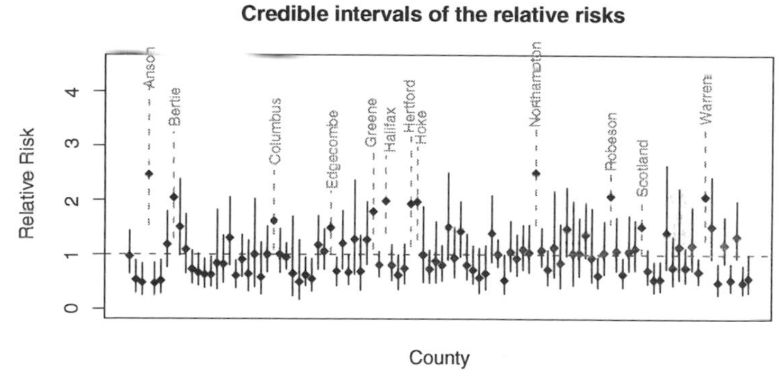

Fig. 11.16. 95% credible intervals of the relatives risks obtained with the BYM model

still be done for the relative risks, but care should be taken to avoid having an improper posterior. For this reason, we can monitor $u_i + v_i$ to assess that these values are stable and that they do not have an erratic behaviour that could have an impact on the posterior estimates of the relative risks and the coefficients of the covariates.

The credible intervals of the relative risks have been plotted in Fig. 11.16. The intervals in dashed line show the counties where the relative risk is significantly higher than one. All these regions are among the ones that appear in the two zones of high risk, plus Anson county.

A few words must be said as to how we have selected the intervals and colours to display the relative risks in the maps. The intervals have been chosen by taking the cut points equally spaced in the range of the relative risks in the log scale (N. Best, personal communication). As discussed in Chap. 3, the colours used to produce the maps can be based on the palettes developed by Brewer et al. (2003), which are available in package **RColorBrewer**. The research was initiated by Brewer et al. (1997) to produce an atlas of disease in the United States. Brewer and Pickle (2002) study how the variable intervals and colours affect how maps are perceived, and Olson and Brewer (1997) developed a useful set of palettes to be used in disease mapping and that are suitable for colour-blind people.

11.5 Detection of Clusters of Disease

Disease mapping provides a first insight to the spatial distribution of the disease, but it may be required to locate the presence of zones where the risk tends to be unusually higher than expected. Besag and Newell (1991) distinguish between methods for clustering and the assessment of risk around putative pollution sources. The former tackle the problem of assessing the presence of clusters, whilst the latter evaluate the risk around a pre-specified source. A third type of method is related to the location of the clusters themselves, which usually involve the examination of small portions of the whole study area each at a time.

Wakefield et al. (2000) provide a review of some classic methods for the detection of clusters of disease. Haining (2003, pp. 237–264) summarises a good number of well-known methods. Waller and Gotway (2004) also cover in detail most of the methods described in this chapter and many others, providing a discussion on the statistical performance of the tests (pp. 259–263). Lawson et al. (2003, Chap. 7) describe the use of Hierarchical Bayesian models for the analysis of risk around pollution sources.

Some of these methods have been implemented in package **DCluster** (Gómez-Rubio et al., 2005), which uses different models and bootstrap (Davison and Hinkley, 1997) to compute the significance of the observed values. This can be done in a general way by resampling the observed number of cases in each area and re-computing the value of the test statistic for each of the simulated data sets. Then, a p-value can be computed by ranking the observed value of the test statistic among the values obtained from the simulations.

Under the usual assumption that O_i is drawn from a Poisson with mean $\theta_i E_i$ and conditioning on the total number of cases, the distribution of $(O_1, \ldots, O_n)$ is Multinomial with probabilities $(E_1/E_+, \ldots, E_n/E_+)$. In addition to the multinomial model, **DCluster** offers the possibility of sampling using a non-parametric bootstrap, or from a Poisson (thus, not conditioning on O_+) or Negative Binomial distribution, to account for over-dispersion in the data. As discussed below, over-dispersion may affect the p-value of the test and when data are highly over-dispersed it may be worth re-running the test sampling from a Negative Binomial distribution.

11.5.1 Testing the Homogeneity of the Relative Risks

Before conducting any analysis of the presence of clusters, the heterogeneity of the relative risks must be assessed. In this way, we can test whether there are actual differences among the different relative risks. The reasons for this heterogeneity may be related to many different factors, such as the presence of a pollution source in the area, which may lead to an increase in the risk around it. Other times the heterogeneity is due to a spatially varying risk factor, and higher risks are related to a higher exposure to this risk factor.

Given that for each area we have computed its expected and observed number of cases, a chi-square test can be carried out to test for (global) significant differences between those two quantities. The statistic is defined by the following formula:

$$\chi^2 = \sum_{i=1}^{n} \frac{(O_i - \theta E_i)^2}{\theta E_i},$$

where θ is the global SMR $= \sum_i O_i / \sum_i E_i$ and, asymptotically, it follows a chi-square distribution with n degrees of freedom. If internal standardisation has been used to obtain E_i, then θ is equal to one and the number of degrees of freedom are reduced to $n-1$ because the additional constraint $\sum_{i=1}^n O_i = \sum_{i=1}^n E_i$ holds (Wakefield et al., 2000).

```
> chtest <- achisq.test(Observed ~ offset(log(Expected)),
+     as(nc, "data.frame"), "multinom", 999)
> chtest

Chi-square test for overdispersion

    Type of boots.: parametric
    Model used when sampling: Multinomial
    Number of simulations: 999
    Statistic:  225.5723
    p-value :  0.001
```

Note that in this case we know that the asymptotic distribution of the test statistic is a chi-square with $n-1$ degrees of freedom and that an exact test can be done instead of re-sampling (however, it may still be useful for small samples and recall that we may be interested in a Monte Carlo Test using a Negative Binomial).

```
> 1 - pchisq(chtest$t0, 100 - 1)

[1] 7.135514e-12
```

Potthoff and Whittinghill (1966) proposed another test of homogeneity of the means of different Poisson distributed variables, which can be used to test the homogeneity of the relative risks (Wakefield et al., 2000). The alternative hypothesis is that the relative risks are drawn from a gamma distribution with mean λ and variance σ^2:

$$H_0 : \theta_1 = \ldots = \theta_n = \lambda,$$
$$H_1 : \theta_i \sim Ga(\lambda^2/\sigma^2, \lambda/\sigma^2).$$

The test statistic is given by

$$\text{PW} = E_+ \sum \frac{O_i(O_i - 1)}{E_i}. \tag{11.1}$$

The alternative hypothesis of this test is that the O_i are distributed following a Negative Binomial distribution, as explained before and, therefore, this test can also be considered as a test of over-dispersion.

```
> pwtest <- pottwhitt.test(Observed ~ offset(log(Expected)),
+     as(nc, "data.frame"), "multinom", 999)
```

The asymptotic distribution of this statistic is Normal with mean $O_+(O_+ - 1)$ and variance $2nO_+(O_+ - 1)$, so a one-side test can be done as follows:

```
> Oplus <- sum(nc$Observed)
> 1 - pnorm(pwtest$t0, Oplus * (Oplus - 1), sqrt(2 * 100 *
+     Oplus * (Oplus - 1)))

[1] 0
```

Other tests for over-dispersion included in **DCluster** are the likelihood ratio test and some of the score tests proposed in Dean (1992). Although they are not described here, all these tests agree with the previous results obtained before and support the fact that the relative risks are not homogeneous and the observed cases are over-dispersed. Therefore, we have preferred to use a Negative Binomial to produce the simulations needed to assess the significance of some of the methods described in the remainder of this section. McMillen (2003) has addressed the importance of choosing the right model in a statistical analysis, and how autocorrelation can appear as a result of a wrong specification of the model.

In addition, Loh and Zhou (2007) discuss the effect of not accounting for extra-Poisson variation and sampling from the wrong distribution when detection of clusters of disease employs the spatial scan statistic (see Sect. 11.5.6). Loh and Zhou (2007) propose a correction based on estimating the distribution of the test statistics by sampling from a distribution that accounts for spatial correlation and other factors (for example, covariates). This approach produces more reliable p-values than the original test. Cressie and Read (1989) already mentioned that the Poisson model was not appropriate for the SIDS data due to the presence of over-dispersion and that other models that take it into account would be more appropriate.

In case of doubt, the reader is advised to assess the significance of a given test by using the Multinomial distribution. This is the standard procedure to assess the significance of the test statistic by Monte Carlo in this scenario. See Waller and Gotway (2004, pp. 202–203) for a discussion on this issue.

A first evaluation of the presence of clusters in the study region can be obtained by checking the spatial autocorrelation. Note that using the chi-square test, for example we can only detect that there are clear differences among the relative risks but not if there is any spatial structure in these differences. In other words, if neighbours tend to have similar (and higher) values. Note that a possible scenario is that of regions having significantly different (low and high) relative risks but with no spatial structure, in which the chi-square test will be significant but there will not be any spatial autocorrelation. This can happen if the scale of aggregation of the data is not taken properly into account or the scale of the risk factors does not exceed the scale of aggregation.

11.5.2 Moran's *I* Test of Spatial Autocorrelation

We have already discussed the use of Moran's I statistic to assess the presence of spatial autocorrelation. Here we apply Moran's I statistic to the SMR to account for the spatial distribution of the population. If we computed Moran's statistic for the O_i, we could find spatial autocorrelation only due to the spatial distribution of the underlying population, because it is well known that the higher the population, the higher the number of cases. Binary weights are used depending on whether two regions share a common boundary or not. Spatial autocorrelation is still found even after accounting for over-dispersion.

```
> col.W <- nb2listw(ncCR85, zero.policy = TRUE)
> moranI.test(Observed ~ offset(log(Expected)), as(nc,
+     "data.frame"), "negbin", 999, listw = col.W, n = length(ncCR85),
+     S0 = Szero(col.W))

Moran's I test of spatial autocorrelation

	Type of boots.: parametric
	Model used when sampling: Negative Binomial
	Number of simulations: 999
	Statistic:  0.2385172
	p-value :  0.001
```

11.5.3 Tango's Test of General Clustering

Tango (1995) proposed a similar test of global clustering by comparing the observed and expected number of cases in each region. He points out that different types of interactions between neighbouring regions can be considered and he proposes a measure of strength based on a decaying function of the distance between two regions.

Briefly, the statistic proposed by Tango is

$$T = (r - p)^{\mathrm{T}} A (r - p) \begin{cases} r^{\mathrm{T}} = [O_1/O_+, \ldots, O_n/O_+], \\ p^{\mathrm{T}} = [E_1/E_+, \ldots, E_n/E_+], \\ A = (a_{ij}) \text{ closeness matrix,} \end{cases} \quad (11.2)$$

where $a_{ij} = \exp\{-d_{ij}/\phi\}$ and d_{ij} is the distance between regions i and j, measured as the distance between their centroids. ϕ is a (positive) constant that reflects the strength of the dependence between areas and the scale at which the interaction occurs.

In our example, we construct the dependence matrix as suggested by Tango and, in addition, we take $\phi = 100$ to simulate a smooth decrease of the relationship between two areas as their relative distance increases. It is advisable to try different values of ϕ because this can have an important impact on the results and the significance of the test. Constructing this matrix in R is straightforward using some functions from package **spdep**, as shown in the following code below. In the computations the weights are globally re-scaled, but this does not affect the significance of the test since they all have simply been divided by the same constant. Furthermore, we have taken the approximate location of the county seats from `nc.sids` (columns `x` and `y`), which are in UTM (zone 18) projection. Note that using the centroids as the county seats – as computed by `coordinates(nc)` – may lead to slightly different coordinates and this may have an impact on the results of this and other tests.

```
> data(nc.sids)
> idx <- match(nc$NAME, rownames(nc.sids))
> nc$x <- nc.sids$x[idx]
> nc$y <- nc.sids$y[idx]
> coords <- cbind(nc$x, nc$y)
> dlist <- dnearneigh(coords, 0, Inf)
> dlist <- include.self(dlist)
> dlist.d <- nbdists(dlist, coords)
> phi <- 100
> col.W.tango <- nb2listw(dlist, glist = lapply(dlist.d,
+     function(x, phi) {
+         exp(-x/phi)
+     }, phi = phi), style = "C")
```

After computing the adjacency matrix we are ready to compute Tango's test of general clustering, which points out the presence of global clustering.

```
> tango.test(Observed ~ offset(log(Expected)), as(nc, "data.frame"),
+     "negbin", 999, listw = col.W.tango, zero.policy = TRUE)
```

```
Tango's test of global clustering
    Type of boots.: parametric
    Model used when sampling: Negative Binomial
    Number of simulations: 999
    Statistic:  0.000483898
    p-value :   0.049
```

11.5.4 Detection of the Location of a Cluster

So far we have considered methods that assess only the presence of heterogeneity of risks in the study area and give a general evaluation of the presence of clusters. To detect the actual location of the clusters present in the area a different approach must be followed. A useful family of methods that can help in this purpose are *scan statistics* (Hjalmars et al., 1996). These methods are based on a moving window that covers only a few areas each time and for which a test of clustering is carried out locally. By repeating this procedure throughout the study area, it will be possible to detect the locations of clusters of disease.

Scan methods usually differ in the way the window is defined, how it is moved over the area, and how the local test of clustering is carried. A recent review of these methods has appeared in *Statistics in Medicine* (Lawson et al., 2006). In this section we only refer to Openshaw's Geographical Analysis Machine (Openshaw et al., 1987) and Kulldorff's statistic (Kulldorff and Nagarwalla, 1995), because the latter is probably the first scan method proposed and the former is a widely established (and used) methodology.

11.5.5 Geographical Analysis Machine

Openshaw's Geographical Analysis Machine considers a regular grid of points $\{(x_i, y_i)\}_{k=1}^{p}$ over the study region at which a circular window is placed in turn. The test only considers the regions whose centroids are inside the window and it is based on comparing the total number of observed cases in the window (O_{k+}) to the total of expected cases in the window (E_{k+}) to assess if the latter is significantly high. Openshaw et al. (1987) define this test as the (one tailed) p-value of O_{k+}, assuming that it follows a Poisson distribution with mean E_{k+}. This procedure can be generalised and, if we have signs that the observed number of cases does not follow a Poisson distribution, the p-value can be obtained by simulation (Gómez-Rubio et al., 2005). Finally, if the current test is significant, the circle is plotted on the map. Alternatively, only the centre of each significant cluster can be plotted for the sake of simplicity and visualisation. Note also that we need to project the cluster centres back to longitude/latitude to be able to plot them on the map of North Carolina.

```
> sidsgam <- opgam(data = as(nc, "data.frame"), radius = 30,
+     step = 10, alpha = 0.002)
> gampoints <- SpatialPoints(sidsgam[, c("x", "y")] * 1000,
+     CRS("+proj=utm +zone=18 +datum=NAD27"))
```

Fig. 11.17. Results of Openshaw's GAM. The dots represent the centre of the clusters

```
> library(rgdal)
> ll <- CRS("+proj=longlat +datum=NAD27")
> gampoints <- spTransform(gampoints, ll)
> gam.layout <- list("sp.points", gampoints)
```

When the complete area has been screened, we will probably have found several places where many overlapping clusters have been found, as shown in Fig. 11.17, where the centres of the clusters found have been plotted. This is due to the fact that the tests performed are not independent and, hence, very similar clusters (i.e. most of their regions are the same) are tested. That is the reason why Openshaw's GAM has been highly criticised by the statistical community and why, in order to maintain global significance, the significance level of the local tests should be corrected. Despite this, the GAM is still helpful as an exploratory method and to generate epidemiological hypotheses (Cromley and McLafferty, 2002).

11.5.6 Kulldorff's Statistic

To overcome this and other problems, Kulldorff and Nagarwalla (1995) developed a new test for the detection of clusters based on a window of variable size that considers only the most likely cluster around a given region. Kulldorf's statistic works with the regions within a given circular window and the overall relative risk in the regions inside the window is compared to that of the regions outside the window. This scan method is available in the SatScan™ software (`http://www.satscan.org/`), which includes enhancements to handle covariates, detect space-time clusters, and some other functionalities.

The null hypothesis, of no clustering, is that the two relative risks are equal, while the alternative hypothesis (clustering) is that the relative risk inside the window is higher. This is resolved by means of a likelihood ratio test, which has two main advantages. First, the most likely cluster can be detected as the window with the highest value of the likelihood ratio and, second, there is no need to correct the p-value because the simulations for different centres are independent (Waller and Gotway, 2004, p. 220). For a Poisson model, the expression of the test statistic is as follows:

$$\max_{z \in Z_i} \left(\frac{O_z}{E_z}\right)^{O_z} \left(\frac{O_+ - O_z}{E_+ - E_z}\right)^{O_+ - O_z}, \tag{11.3}$$

where z is an element of Z_i, the set of all circles centred at region i. These circles are constructed so that only those that contain up to a fixed proportion of the total population are considered.

Note that, even though we select the most likely cluster around each region, it might not be significant. On the other hand, we may have more than one significant cluster, around two or more different regions, and that some clusters may overlap each other. When more than one cluster is found, we can consider the cluster with the lowest p-value as the *primary* or most prominent in the study region. *Secondary* clusters, that do not overlap with the former, may be considered too.

Loh and Zhou (2007) show that when data are over-dispersed, the *classical* spatial scan statistic will produce more false positives than the nominal significance level. To correct for this, they propose sampling from a different distribution that accounts for spatial correlation. The Negative Binomial can be used to account for the extra-variability, which may be caused by spatial autocorrelation coming from unmeasured covariates, and estimate the distribution of the test statistic under over-dispersion.

```
> mle <- calculate.mle(as(nc, "data.frame"), model = "negbin")
> thegrid <- as(nc, "data.frame")[, c("x", "y")]
> knresults <- opgam(data = as(nc, "data.frame"),
+     thegrid = thegrid, alpha = 0.05, iscluster = kn.iscluster,
+     fractpop = 0.15, R = 99, model = "negbin",
+     mle = mle)
```

The most likely cluster for the SIDS data set is shown in Fig. 11.18. The p-value is 0.04, which means that the cluster is significant.

The general procedure of application of this method includes testing each area as the centre of a possible cluster, although it can only be used on a single

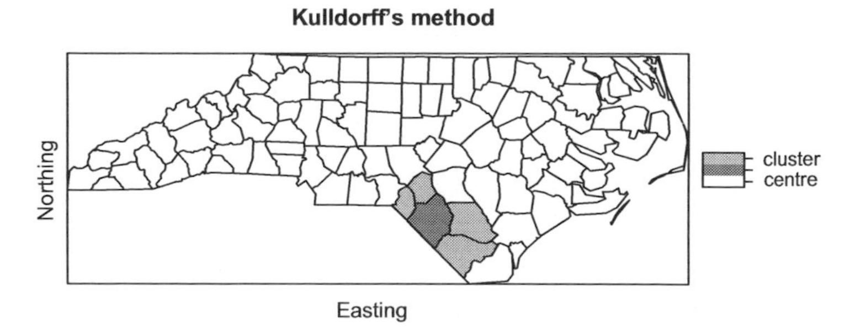

Fig. 11.18. Results of Kulldorff's test. The circles show the most likely cluster

point to test whether it is the centre of a cluster. This is specially helpful to assess the risk around putative pollution sources. Note that no assumption about the variation of the risk around the source is made. This is discussed in Sect. 11.5.7.

11.5.7 Stone's Test for Localised Clusters

As an alternative to the detection of clusters of disease, we may have already identified a putative pollution source and wish to investigate whether there is an increased risk around it. Stone (1988) developed a test that considers the alternative hypothesis of a descending trend around the pollution source. Basically, if we consider $\theta_{(1)}, \ldots, \theta_{(n)}$, the ordered relative risks of the regions according to their distances to the source, the test is as follows:

$$H_0 : \theta_{(1)} = \ldots = \theta_{(n)} = \lambda,$$
$$H_1 : \theta_{(1)} \geq \ldots \geq \theta_{(n)}.$$

λ is the overall relative risk, which may be one if internal standardisation has been used. The test statistic proposed by Stone is the maximum accumulated risk up to a certain region:

$$\max_i \frac{\sum_{j=1}^{i} O_j}{\sum_{j=1}^{i} E_j}.$$

A word of caution must be given here because, as already discussed by many authors (Hills and Alexander, 1989, for example), focused tests should be employed before checking the data, because a bias is introduced when we try to use these tests on regions where an actual increased risk has been observed. In those cases, it will be more likely to detect a cluster than usual.

As an example, we try to assess whether there is an increased risk around Anson county, which has been spotted as an area of high risk. A call to `stone.stat` will give us the value of the test statistic and the number of regions for which the maximum accumulated risk is achieved. Later, we can use `stone.test` to compute the significance of this value.

```
> stone.stat(as(nc, "data.frame"), region = which(nc$NAME ==
+     "Anson"))

            region
4.726392 1.000000

> st <- stone.test(Observed ~ offset(log(Expected)), as(nc,
+     "data.frame"), model = "negbin", 99, region = which(nc$NAME ==
+     "Anson"))
> st
```

```
Stone's Test for raised incidence around locations

    Type of boots.: parametric
    Model used when sampling: Negative Binomial
    Number of simulations: 99
    Statistic:  4.726392
    p-value :  0.01
```

As the results show, the size of the cluster is 1 (just Anson county), which turns out to be highly significant.

11.6 Other Topics in Disease Mapping

Although we have tried to cover a wide range of analyses in this chapter, we have not been able to include other important topics, such as the detection of non-circular clusters (see, for example Tango and Takahashi, 2005), spatio-temporal disease mapping (see, for example Martínez-Beneito et al., 2008, and the references therein), or the joint modelling of several diseases (Held et al., 2005). Other data sets and models could be used by making the corresponding modifications to the R and WinBUGS code shown here. Some examples are availabe in Lawson et al. (2003). Furthermore, Banerjee et al. (2004) describe a number of other possible Bayesian analyses of spatial data and provide data and WinBUGS code in the associated website, which the reader should be able to reproduce using the guidelines provided in this chapter.

Afterword

Both parts of this book have quite consciously tried not to give authoritative advice on choices of methods or techniques.[1] The handling and analysis of spatial data with R continues to evolve – this is implicit in open source software development. It is also an important component attempting to offer applied researchers access to accepted and innovative alternatives for data analysis, and applied statisticians with representations of spatial data that make it easier to test and develop new analytical tools.

A further goal has been to provide opportunities for bringing together the various camps and traditions analysing spatial data, to make it somewhat easier to see that their ways of conducting their work are not so different from one another in practise. It has always been worrying that fields like disease mapping or spatial econometrics, with very similar data scenarios, make different choices with regard to methods, and treatments of the assumptions underlying those methods, in their research practise. Research practise evolves, and learning from a broader spread of disciplines must offer the chance to avoid choices that others have found less satisfactory, to follow choices from which others have benefitted and to participate in innovation in methods.

This makes participation in the R community, posting questions or suggestions, reporting apparent bugs not only a practical activity, but also an affirmation that science is fostered more by openness than the unwarranted restriction of findings. In the context of this book, and as we said in the preface, we would be grateful for messages pointing out errors; errata will be posted on the book website (`http://www.asdar-book.org`).

[1] An illustration from an email exchange between the authors: "I think we are trying to enable people to do what they want, even if they shoot themselves in the feet (but in a reproducible way)!"

R and Package Versions Used

- R version 2.6.2 (2008-02-08), `i686-pc-linux-gnu`
- Base packages: base, datasets, graphics, grDevices, methods, stats, utils
- Other packages: adapt 1.0-4, boot 1.2-32, class 7.2-41, classInt 0.1-9, coda 0.13-1, DCluster 0.2, digest 0.3.1, e1071 1.5-18, epitools 0.4-9, foreign 0.8-24, gpclib 1.4-1, graph 1.16.1, gstat 0.9-44, lattice 0.17-6, lmtest 0.9-21, maps 2.0-39, maptools 0.7-7, Matrix 0.999375-9, mgcv 1.3-29, nlme 3.1-88, pgirmess 1.3.6, pkgDepTools 1.4.1, R2WinBUGS 2.1-8, RandomFields 1.3.30, RBGL 1.14.0, RColorBrewer 1.0-2, rgdal 0.5-24, Rgraphviz 1.16.0, sandwich 2.1-0, sp 0.9-24, spam 0.13-2, spatialkernel 0.4-8, spatstat 1.12-9, spdep 0.4-20, spgrass6 0.5-3, spgwr 0.5-1, splancs 2.01-23, tripack 1.2-11, xtable 1.5-2, zoo 1.5-0
- Loaded via a namespace (and not attached): cluster 1.11.10, grid 2.6.2, MASS 7.2-41, rcompgen 0.1-17, tools 2.6.2

Data Sets Used

- Auckland 90 m Shuttle Radar Topography Mission: downloaded on 26 September 2006 from the US Geological Survey, National Map Seamless Server `http://seamless.usgs.gov/`, GeoTiff file, 3 arcsec 'Finished' (90 m) data; file `70042108.zip` on book website.
- Auckland shoreline: downloaded on 7 November 2005 from the National Geophysical Data Center coastline extractor `http://www.ngdc.noaa.gov/mgg/shorelines/shorelines.html`; file `auckland_mapgen.dat` on book website.
- Biological cell centres: available as `data(cells)` from **spatstat**, documented in Ripley (1977).
- Broad Street cholera mortalities: original files provided by Jim Detwiler, who had collated them for David O'Sullivan for use on the cover of O'Sullivan and Unwin (2003), based on earlier work by Waldo Tobler and others; this version is available as a compressed archive of a GRASS location in file `snow_location.tgz`, and a collection of GeoTiff and shapefiles exported from this location in file `snow_files.zip` on the book website.
- California redwood trees: available as `data(redwoodfull)` from **spatstat**, documented in Strauss (1975).
- Cars: available as `data(cars)` from **datasets**.
- CRAN mirrors: locations of CRAN mirrors 1 October 2005; file on book website `CRAN051001a.txt`.
- Japan shoreline: available in the `'world'` database provided by **maps**.
- Japanese black pine saplings: available as `data(japanesepines)` from **spatstat**, documented in Numata (1961).
- Lansing Woods maple trees: available as `data(lansing)` from **spatstat**, documented in Gerard (1969).

- Loggerhead turtle: downloaded on 2 November 2005 with permission from SEAMAP, (Read et al., 2003), data set 105; data described in Nichols et al. (2000); file `seamap105_mod.csv` on book website.
- Manitoulin Island: created using `Rgshhs` in **maptools** from the GSHHS high resolution file `gshhs_h.b`, version 1.5, of 3 April 2007, downloaded from `ftp://ftp.soest.hawaii.edu/pwessel/gshhs`.
- Maunga Whau volcano: available as `data(volcano)` from **datasets**.
- Meuse bank: available as `data(meuse)` from **sp**, supplemented by `data(meuse.grid)` and `data(meuse.riv)`, and documented in Rikken and Van Rijn (1993) and Burrough and McDonnell (1998).
- New York leukemia: used and documented extensively in Waller and Gotway (2004) and with data made available in Chap. 9 of `http://www.sph.emory.edu/~lwaller/WGindex.htm`; the data import process is described in the help file of `NY_data` in **spdep**; geometries downloaded from the CIESIN server at `ftp.ciesin.columbia.edu`, file `/pub/census/usa/tiger/ny/bna_st/t8_36.zip`, and extensively edited; a zip archive `NY_data.zip` of shapefiles and a GAL format neighbours list is on the book website.
- North Carolina SIDS: shapefile `sids.shp` (based on geometries downloaded from `http://sal.agecon.uiuc.edu/datasets/sids.zip`) and GAL format neighbour lists `ncCC89.gal` and `ncCR85.gal` distributed with **spdep**, data from Cressie (1993), neighbour lists from Cressie and Chan (1989) and Cressie and Read (1985), documented in the `nc.sids` help page.
- North Derbyshire asthma study: the data has been studied by Diggle and Rowlingson (1994), Singleton et al. (1995), and Diggle (2003); the data are made available in anonymised form by permission from Peter Diggle as shapefiles in a zip archive `north_derby_asthma.zip` on the book website.
- Scottish lip cancer: Shapefile and data file downloaded from the book website of Waller and Gotway (2004), `http://www.sph.emory.edu/~lwaller/WGindex.htm`, Chaps. 2 and 9.
- Spearfish: downloaded as GRASS location from `http://grass.itc.it/sampledata/spearfish_grass60data-0.3.tar.gz`; this data set has been the standard GRASS location for tutorials and is documented in Neteler and Mitasova (2004).
- US 1999 SAT scores: state boundaries available in the `'state'` database provided by **maps**, original attribute data downloaded on 2 November 2005 from `http://www.biostat.umn.edu/~melanie/Data/` and supplemented with variable names and state names; the data set is also available from the website of Banerjee et al. (2004), `http://www.biostat.umn.edu/~brad/data/state-sat.dat`, and the modified version as file `state.sat.data_mod.txt` from the book website.
- US Census 1990 Counties: Three shapefiles for Virginia and North and South Carolina downloaded from the US Census Bureau cartographic boundary files site for 1990 county and county equivalent areas at `http://www.census.gov/geo/www/cob/co1990.html`; one text file by county defining

metropolitan area membership also from the US Census Bureau site http://blueprod.ssd.census.gov, file /population/estimates/metro-city/90mfips.txt available as file 90mfips.txt on the book website.
- World volcano locations: downloaded from the National Geophysical Data Center http://www.ngdc.noaa.gov/hazard/volcano.shtml, available as file data1964al.xy from book website.

References

Abrahamsen, P. and Benth, F. E. (2001). Kriging with inequality constraints. *Mathematical Geology*, 33:719–744. [229]

Akima, H. (1978). A method of bivariate interpolation and smooth surface fitting for irregularly distributed data points. *ACM Transactions on Mathematical Software*, 4:148–159. [233]

Andrade Neto, P. R. and Ribeiro Jr., P. J. (2005). A process and environment for embedding the R software into TerraLib. In *VII Brazilian Symposium on Geoinformatics, Campos do Jordão*. [109]

Anselin, L. (1988). *Spatial Econometrics: Methods and Models*. Kluwer, Dordrecht. [289, 290]

Anselin, L. (2002). Under the hood: Issues in the specification and interpretation of spatial regression models. *Agricultural Economics*, 27:247–267. [289, 290]

Anselin, L., Bera, A. K., Florax, R., and Yoon, M. J. (1996). Simple diagnostic tests for spatial dependence. *Regional Science and Urban Economics*, 26:77–104. [290]

Anselin, L., Syabri, I., and Kho, Y. (2006). GeoDa: An introduction to spatial data analysis. *Geographical Analysis*, 38:5–22. [256]

Assunção, R. and Reis, E. A. (1999). A new proposal to adjust Moran's I for population density. *Statistics in Medicine*, 18:2147–2162. [266]

Avis, D. and Horton, J. (1985). Remarks on the sphere of influence graph. In Goodman, J. E., editor, *Discrete Geometry and Convexity*. New York Academy of Sciences, New York, pp 323–327. [245]

Baddeley, A. and Turner, R. (2005). Spatstat: An R package for analyzing spatial point patterns. *Journal of Statistical Software*, 12(6):1–42. [156]

Baddeley, A., Möller, J., and Waagepetersen, R. (2000). Non- and semi-parametric estimation of interaction in inhomogeneous point patterns. *Statistica Neerlandica*, 54:329–350. [172, 186, 187]

Baddeley, A., Gregori, P., Mateu, J., Stoica, R., and Stoyan, D., editors (2005). *Case Studies in Spatial Point Process Modeling*. Lecture Notes in Statistics 185, Springer, Berlin. [190]

Baddeley, A. J. and Silverman, B. W. (1984). A cautionary example on the use of second-order methods for analysing point patterns. *Biometrics*, 40:1089–1093. [185]

Bailey, T. C. and Gatrell, A. C. (1995). *Interactive Spatial Data Analysis*. Longman, Harlow. [13]

Banerjee, S., Carlin, B. P., and Gelfand, A. E. (2004). *Hierarchical Modeling and Analysis for Spatial Data*. Chapman & Hall, London. [7, 13, 240, 259, 274, 296, 311, 314, 321, 325, 341, 345]

Bavaud, F. (1998). Models for spatial weights: A systematic look. *Geographical Analysis*, 30:153–171. [251]

Beale, C. M., Lennon, J. J., Elston, D. A., Brewer, M. J., and Yearsley, J. M. (2007). Red herrings remain in geographical ecology: A reply to Hawkins et al. (2007). *Ecography*, 30:845–847. [11]

Becker, R. A., Chambers, J. M., and Wilks, A. R. (1988). *The New S Language*. Chapman & Hall, London. [2, 38]

Berman, M. and Diggle, P. J. (1989). Estimating weighted integrals of the second-order intensity of a spatial point process. *Journal of the Royal Statistical Society B*, 51:81–92. [165, 166]

Bernardinelli, L. and Montomoli, C. (1992). Empirical Bayes versus fully Bayesian analysis of geographical variation in disease risk. *Statistics in Medicine*, 11:983–1007. [320]

Besag, J. and Newell, J. (1991). The detection of clusters in rare diseases. *Journal of the Royal Statistical Society A*, 154:143–155. [332]

Besag, J., York, J., and Mollie, A. (1991). Bayesian image restoration, with two applications in spatial statistics. *Annals of the Institute of Statistical Mathematics*, 43:1–59. [321, 326]

Besag, J., Green, P., Higdon, D., and Mengersen, K. (1995). Bayesian computation and stochastic systems. *Statistical Science*, 10:3–41. [330]

Best, N., Cowles, M. K., and Vines, K. (1995). CODA: Convergence diagnosis and output analysis software for Gibbs sampling output, Version 0.30. Technical report, MRC Biostatistics Unit, Cambridge. [324]

Best, N. G., Waller, L. A., Thomas, A., Conlon, E. M., and Arnold, R. A. (1999). Bayesian models for spatially correlated diseases and exposure data. In Bernardo, J., Berger, J. O., Dawid, A. P., and Smith, A. F. M., editors, *Bayesian Statistics 6*. Oxford University Press, Oxford, pp 131–156. [326]

Bivand, R. S. (2000). Using the R statistical data analysis language on GRASS 5.0 GIS data base files. *Computers and Geosciences*, 26:1043–1052. [99]

Bivand, R. S. (2002). Spatial econometrics functions in R: Classes and methods. *Journal of Geographical Systems*, 4:405–421. [151, 289]

Bivand, R. S. (2006). Implementing spatial data analysis software tools in R. *Geographical Analysis*, 38:23–40. [289]

Bivand, R. S. (2008). Implementing representations of space in economic geography. *Journal of Regional Science*, 48:1–27. [12, 259]

Bivand, R. S. and Portnov, B. A. (2004). Exploring spatial data analysis techniques using R: The case of observations with no neighbours. In Anselin, L.,

Florax, R. J. G. M., and Rey, S. J., editors, *Advances in Spatial Econometrics: Methodology, Tools, Applications.* Springer, Berlin, pp 121–142. [255]

Bivand, R. S. and Szymanski, S. (1997). Spatial dependence through local yardstick competition: Theory and testing. *Economics Letters*, 55:257–265. [12]

Bivand, R. S., Müller, W., and Reder, M. (2008). Power calculations for global and local Moran's *I*. Technical report, Department of Applied Statistics, Johannes Kepler University, Linz, Austria. [264]

Bordignon, M., Cerniglia, F., and Revelli, F. (2003). In search of yardstick competition: A spatial analysis of Italian municipality property tax setting. *Journal of Urban Economics*, 54:199–217. [12]

Braun, W. J. and Murdoch, D. J. (2007). *A First Course in Statistical Programming with R.* Cambridge University Press, Cambridge. [23, 127]

Brewer, C. A. and Pickle, L. (2002). Comparison of methods for classifying epidemiological data on choropleth maps in series. *Annals of the Association of American Geographers*, 92:662–681. [332]

Brewer, C. A., MacEachren, A. M., Pickle, L. W., and Herrmann, D. J. (1997). Mapping mortality: Evaluating color schemes for choropleth maps. *Annals of the Association of American Geographers*, 87:411–438. [332]

Brewer, C. A., Hatchard, G. W., and Harrower, M. A. (2003). Colorbrewer in print: A catalog of color schemes for maps. *Cartography and Geographic Information Science*, 30:5–32. [76, 332]

Brody, H., Rip, M. R., Vinten-Johansen, P., Paneth, N., and Rachman, S. (2000). Map-making and myth-making in Broad Street: The London cholera epidemic, 1854. *Lancet*, 356:64–68. [104, 105]

Burrough, P. A. and McDonnell, R. A. (1998). *Principles of Geographical Information Systems.* Oxford University Press, Oxford. [4, 6, 116, 191, 345]

Calenge, C. (2006). The package adehabitat for the R software: A tool for the analysis of space and habitat use by animals. *Ecological Modelling*, 197:516–519. [107]

Carstairs, V. (2000). Socio-economic factors at areal level and their relationship with health. In Elliot, P., Wakefield, J., Best, N., and Briggs, D., editors, *Spatial Epidemiology: Methods and Applications.* Oxford University Press, Oxford, pp 51–67. [312]

Chambers, J. M. (1998). *Programming with Data.* Springer, New York. [3, 27, 127]

Chambers, J. M. and Hastie, T. J. (1992). *Statistical Models in S.* Chapman & Hall, London. [24, 25, 26]

Chilès, J. and Delfiner, P. (1999). *Geostatistics: Modeling Spatial Uncertainty.* Wiley, New York. [191]

Choynowski, M. (1959). Map based on probabilities. *Journal of the American Statistical Society*, 54:385–388. [316]

Chrisman, N. (2002). *Exploring Geographic Information Systems.* Wiley, New York. [6, 8]

Christensen, R. (1991). *Linear Models for Multivariate, Time Series, and Spatial Data.* Springer, New York. [191]

Clark, A. B. and Lawson, A. B. (2004). An evaluation of non-parametric relative risk estimators for disease mapping. *Computational Statistics and Data Analysis*, 47:63–78. [166]

Clayton, D. and Kaldor, J. (1987). Empirical Bayes estimates of age-standardized relative risks for use in disease mapping. *Biometrics*, 43:671–681. [90, 316, 318]

Cleveland, W. S. (1993). *Visualizing Data.* Hobart Press, Summit, NJ. [57, 68, 192]

Cleveland, W. S. (1994). *The Elements of Graphing Data.* Hobart Press, Summit, NJ. [57, 68]

Cliff, A. D. and Ord, J. K. (1973). *Spatial Autocorrelation.* Pion, London. [257]

Cliff, A. D. and Ord, J. K. (1981). *Spatial Processes.* Pion, London. [12, 253]

Cowles, M. K. and Carlin, B. P. (1996). Markov Chain Monte Carlo convergence diagnostics: A comparative review. *Journal of the American Statistical Association*, 91:883–904. [328]

Cox, C. R. (1955). Some statistical methods connected with series of events (with discussion). *Journal of the Royal Statistical Society B*, 17:129–164. [187]

Crawley, M. J. (2005). *Statistics: An Introduction using R.* Wiley, Chichester. [25]

Crawley, M. J. (2007). *The R Book.* Wiley, Chichester. [25]

Cressie, N. (1985). Fitting variogram models by weighted least squares. *Mathematical Geology*, 17:563–586. [202]

Cressie, N. (1993). *Statistics for Spatial Data*, Revised Edition. Wiley, New York. [12, 15, 152, 191, 198, 240, 259, 274, 345]

Cressie, N. and Chan, N. H. (1989). Spatial modeling of regional variables. *Journal of the American Statistical Association*, 84:393–401. [312, 315, 320, 325, 345]

Cressie, N. and Read, T. R. C. (1985). Do sudden infant deaths come in clusters? *Statistics and Decisions*, 3:333–349. [312, 313, 326, 345]

Cressie, N. and Read, T. R. C. (1989). Spatial data analysis of regional counts. *Biometrical Journal*, 31:699–719. [334]

Cromley, E. K. and McLafferty, S. L. (2002). *GIS and Public Health.* Guilford Press, New York. [338]

Dalgaard, P. (2002). *Introductory Statistics with R.* Springer, New York. [25, 152]

Davison, A. C. and Hinkley, D. V. (1997). *Bootstrap Methods and Their Application.* Cambridge University Press, Cambridge. [332]

Dean, C. B. (1992). Testing for overdispersion in Poisson and Binomial regression models. *Journal of the American Statistical Association*, 87:451–457. [334]

Deutsch, C. and Journel, A. (1992). *GSLIB: Geostatistical Software Library and User's Guide.* Oxford University Press, New York. [191]

Devine, O. J. and Louis, T. A. (1994). A constrained empirical Bayes estimator for incidence rates in areas with small populations. *Statistics in Medicine*, 13:1119–1133. [321]

Devine, O. J., Louis, T. A., and Halloran, M. E. (1994). Empirical Bayes estimators for spatially correlated incidence rate. *Environmetrics*, 5:381–398. [321]

Diggle, P. J. (1985). A kernel method for smoothing point process data. *Applied Statistics*, 34:138–147. [165, 166]

Diggle, P. J. (1990). A point process modelling approach to raised incidence of a rare phenomenon in the vicinity of a prespecified point. *Journal of the Royal Statistical Society A*, 153:349–362. [173, 182]

Diggle, P. J. (2000). Overview of statistical methods for disease mapping and its relationship to cluster detection. In Elliott, P., Wakefield, J., Best, N., and Briggs, D., editors, *Spatial Epidemiology: Methods and Applications*. Oxford University Press, Oxford, pp 87–103. [173, 184]

Diggle, P. J. (2003). *Statistical Analysis of Spatial Point Patterns*. Arnold, London, second edition. [155, 156, 158, 161, 163, 164, 166, 168, 169, 170, 171, 172, 184, 190, 345]

Diggle, P. J. (2006). Spatio-temporal point processes: Methods and applications. In Finkenstadt, B., Held, L., and Isham, V., editors, *Statistical Methods for Spatio-Temporal Systems*. CRC, Boca Raton, pp 1–46. [190]

Diggle, P. J. and Chetwynd, A. (1991). Second-order analysis of spatial clustering for inhomogeneous populations. *Biometrics*, 47:1155–1163. [173, 184, 185]

Diggle, P. J. and Ribeiro Jr., P. J. (2007). *Model-Based Geostatistics*. Springer, New York. [235]

Diggle, P. J. and Rowlingson, B. (1994). A conditional approach to point process modelling of elevated risk. *Journal of the Royal Statistical Society A*, 157:433–440. [158, 159, 178, 182, 183, 184, 345]

Diggle, P. J., Elliott, P., Morris, S., and Shaddick, G. (1997). Regression modelling of disease risk in relation to point sources. *Journal of the Royal Statistical Society A*, 160:491–505. [184]

Diggle, P. J., Tawn, J. A., and Moyeed, R. A. (1998). Model-based geostatistics. *Applied Statistics*, 47:299–350. [230]

Diggle, P. J., Morris, S., and Wakefield, J. (2000). Point-source modelling using case-control data. *Biostatistics*, 1:89–105. [173]

Diggle, P. J., Gómez-Rubio, V., Brown, P. E., Chetwynd, A., and Gooding, S. (2007). Second-order analysis of inhomogeneous spatial point processes using case-control data. *Biometrics*, 63:550–557. [173, 175, 186, 187, 188]

Diniz-Filho, J. A., Bini, L. M., and Hawkins, B. A. (2003). Spatial autocorrelation and red herrings in geographical ecology. *Global Ecology and Biogeography*, 12:53–64. [11]

Diniz-Filho, J. A., Hawkins, B. A., Bini, L. M., De Marco Jr., P., and Blackburn, T. M. (2007). Are spatial regression methods a panacea or a Pandora's box? A reply to Beale et al. (2007). *Ecography*, 30:848–851. [11]

Dormann, C., McPherson, J., Araújo, M., Bivand, R., Bolliger, J., Carl, G., Davies, R., Hirzel, A., Jetz, W., Kissling, D., Kühn, I., Ohlemüller, R., Peres-Neto, P., Reineking, B., Schröder, B., Schurr, F., and Wilson, R. (2007). Methods to account for spatial autocorrelation in the analysis of species distributional data: A review. *Ecography*, 30:609–628. [274, 296, 300, 301]

Dray, S., Legendre, P., and Peres-Neto, P. R. (2006). Spatial modeling: A comprehensive framework for principle coordinate analysis of neighbor matrices (PCNM). *Ecological Modelling*, 196:483–493. [302]

Elliott, P. and Wakefield, J. C. (2000). Bias and confounding in spatial epidemiology. In Elliott, P., Wakefield, J., Best, N., and Briggs, D., editors, *Spatial Epidemiology: Methods and Applications.* Oxford University Press, Oxford, pp 68–84. [312]

Elliott, P., Wakefield, J., Best, N., and Briggs, D., editors (2000). *Spatial Epidemiology. Methods and Applications.* Oxford University Press, Oxford. [173, 311]

English, D. (1992). Geographical epidemiology and ecological studies. In Elliott, P., Cuzick, J., English, D., and Stern, R., editors, *Geographical and Environmental Epidemiology. Methods for Small-Area Studies.* Oxford University Press, Oxford, pp 3–13. [326]

Erle, S., Gibson, R., and Walsh, J. (2005). *Mapping Hacks.* O'Reilly, Sebastopol, CA. [7]

Faraway, J. J. (2004). *Linear Models with R.* Chapman & Hall, Boca Raton. [152]

Faraway, J. J. (2006). *Extending Linear Models with R: Generalized Linear, Mixed Effects and Nonparametric Regression Models.* Chapman & Hall, Boca Raton. [152]

Fortin, M.-J. and Dale, M. (2005). *Spatial Analysis: A Guide for Ecologists.* Cambridge University Press, Cambridge. [13, 240, 259, 268, 274]

Fotheringham, A. S., Brunsdon, C., and Charlton, M. E. (2002). *Geographically Weighted Regression: The Analysis of Spatially Varying Relationships.* Wiley, Chichester. [306, 307]

Fox, J. (2002). *An R and S-Plus Companion to Applied Regression.* Sage Publications, Thousand Oaks, CA. [152]

Gatrell, A. C., Bailey, T. C., Diggle, P. J., and Rowlingson, B. S. (1996). Spatial point pattern analysis and its application in geographical epidemiology. *Transactions of the Institute of British Geographers*, 21:256–274. [172]

Gelman, A. and Hill, J. (2007). *Data Analysis Using Regression and Multilevel/Hierarchical Models.* Cambridge University Press, Cambridge. [322]

Gelman, A. and Rubin, D. B. (1992). Inference from iterative simulation using multiple sequences (with discussion). *Statistical Science*, 7:457–472. [328]

Gelman, A., Carlin, J. B., Stern, H. S., and Rubin, D. B. (2003). *Bayesian Data Analysis.* CRC, Boca Raton. [322]

Gerard, D. J. (1969). Competition quotient: A new measure of the competition affecting individual forest trees. Research Bulletin 20, Agricultural Experiment Station, Michigan State University. [169, 344]

Geweke, J. (1992). Evaluating the accuracy of sampling-based approaches to calculating posterior moments. In Bernado, J. M., Berger, J. O., Dawid, A. P., and Smith, A. F. M., editors, *Bayesian Statistics 4*. Oxford University Press, Oxford, pp 169–194. [329]

Ghosh, M., Natarajan, K., Stroud, T. W. F., and Carlin, B. P. (1998). Generalized linear models for small-area estimation. *Journal of the American Statistical Association*, 93:273–282. [326]

Gilks, W. R., Richardson, S., and Spiegelhalter, D. J., editors (1996). *Markov Chain Monte Carlo in Practice*. Chapman & Hall, London. [322]

Gómez-Rubio, V. and López-Quílez, A. (2005). RArcInfo: Using GIS data with R. *Computers and Geosciences*, 31:1000–1006. [88, 93]

Gómez-Rubio, V., Ferrándiz-Ferragud, J., and López-Quílez, A. (2005). Detecting clusters of disease with R. *Journal of Geographical Systems*, 7:189–206. [332, 337]

Goovaerts, P. (1997). *Geostatistics for Natural Resources Evaluation*. Oxford University Press, Oxford. [191, 219, 227]

Gotway, C. A. and Young, L. J. (2002). Combining incompatible spatial data. *Journal of the American Statistical Association*, 97:632–648. [114]

Griffith, D. A. (1995). Some guidelines for specifying the geographic weights matrix contained in spatial statistical models. In Arlinghaus, S. L. and Griffith, D. A., editors, *Practical Handbook of Spatial Statistics*. CRC, Boca Raton, pp 65–82. [251]

Griffith, D. A. and Peres-Neto, P. R. (2006). Spatial modeling in ecology: The flexibility of eigenfunction spatial analyses. *Ecology*, 87:2603–2613. [302]

Guttorp, P. (2003). Environmental statistics – a personal view. *International Statistical Review*, 71:169–180. [114]

Haining, R. P. (2003). *Spatial Data Analysis: Theory and Practice*. Cambridge University Press, Cambridge. [13, 151, 311, 314, 321, 332]

Härdle, W., Müller, M., Sperlich, S., and Werwatz, A. (2004). *Nonparametric and Semiparametric Models*. Springer-Verlag, Berlin. [168]

Hastie, T. and Tibshirani, R. (1990). *Generalised Additive Models*. Chapman & Hall, London. [297]

Hawkins, B. A., Diniz-Filho, J. A., Bini, L. M., De Marco Jr., P., and Blackburn, T. M. (2007). Red herrings revisited: Spatial autocorrelation and parameter estimation in geographical ecology. *Ecography*, 30:375–384. [11]

Held, L., Natário, I., Fento, S. E., Rue, H., and Becke, N. (2005). Towards joint disease mapping. *Statistical Methods in Medical Research*, 14:61–82. [341]

Hepple, L. W. (1998). Exact testing for spatial autocorrelation among regression residuals. *Environment and Planning A*, 30:85–108. [264]

Heuvelink, G. B. M. (1998). *Error Propagation in Environmental Models with GIS.* Taylor & Francis, London. [115]

Heywood, I., Cornelius, S., and Carver, S. (2006). *An Introduction to Geographical Information Systems.* Pearson Education, Harlow, England. [6]

Hills, M. and Alexander, F. (1989). Statistical methods used in assessing the risk of disease near a source of possible environmental pollution: A review. *Journal of the Royal Statistical Society A*, 152:353–363. [340]

Hjalmars, U., Kulldorff, M., Gustafsson, G., and Nagarwalla, N. (1996). Childhood leukaemia in Sweden: Using GIS and a spatial scan statistic for cluster detection. *Statistics in Medicine*, 15:707–715. [337]

Hjaltason, G. and Samet, H. (1995). Ranking in spatial databases. In Egenhofer, M. J. and Herring, J. R., editors, *Advances in Spatial Databases – 4th Symposium, SSD'95*, Number 951 in Lecture Notes in Computer Science. Springer-Verlag, Berlin, pp 83–95. [215]

Hoef, J. M. V. and Cressie, N. A. C. (1993). Multivariable spatial prediction. *Mathematical Geology*, 25:219–240. [210]

Isaaks, E. and Srivastava, R. (1989). *An Introduction to Applied Geostatistics.* Oxford University Press, Oxford. [191]

Jackson, C., Best, N., and Richardson, S. (2006). Improving ecological inference using individual-level data. *Statistics in Medicine*, 25(12):2136–2159. [326]

Jacqmin-Gadda, H., Comenges, C., Nejjari, C., and Dartigues, J. (1997). Testing of geographical correlation with adjustment for explanatory variables: An application to dyspnoea in the elderly. *Statistics in Medicine*, 21:359–370. [298]

Jarner, M. F., Diggle, P., and Chetwynd, A. G. (2002). Estimation of spatial variation in risk using matched case–control data. *Biometrical Journal*, 44:936–945. [173]

Johnston, J. and DiNardo, J. (1997). *Econometric Methods.* McGraw Hill, New York. [290]

Journel, A. G. and Huijbregts, C. J. (1978). *Mining Geostatistics.* Academic Press, London. [191, 215]

Kaluzny, S. P., Vega, S. C., Cardoso, T. P., and Shelly, A. A. (1998). *S+SpatialStats, User Manual for Windows and UNIX.* Springer-Verlag, Berlin. [13, 311]

Kelejian, H. H. and Prucha, I. R. (1999). A generalized moments estimator for the autoregressive parameter in a spatial model. *International Economic Review*, 40:509–533. [295]

Kelsall, J. E. and Diggle, P. J. (1995a). Kernel estimation of relative risk. *Bernoulli*, 1:3–16. [166, 173, 174, 176]

Kelsall, J. E. and Diggle, P. J. (1995b). Non-parametric estimation of spatial variation in relative risk. *Statistics in Medicine*, 14:559–573. [166, 173, 174, 176]

Kelsall, J. E. and Diggle, P. J. (1998). Spatial variation in risk: A non-parametric binary regression approach. *Applied Statistics*, 47:559–573. [166, 173, 178, 179, 180]

Kirkwood, R., Lynch, M., Gales, N., Dann, P., and Sumner, M. (2006). At-sea movements and habitat use of adult male Australian fur seals (Arctocephalus pusillus doriferus). *Canadian Journal of Zoology*, 84:1781–1788. [130]

Kopczewska, K. (2006). *Ekonometria i Statystyka Przestrzenna.* CeDeWu, Warszawa. [VIII]

Krieger, N., Williams, D. R., and Moss, N. E. (1997). Measuring social class in US Public Health research: Concepts, methodologies, and guidelines. *Annual Review of Public Health*, 18:341–378. [326]

Kulldorff, M. and Nagarwalla, N. (1995). Spatial disease clusters: Detection and inference. *Statistics in Medicine*, 14:799–810. [337, 338]

Lawson, A., editor (2005). SMMR special issue on disease mapping. *Statistical Methods in Medical Research*, 14(1). [311]

Lawson, A., Gangnon, R. E., and Wartenburg, D., editors (2006). Special issue: Developments in disease cluster detection. *Statistics in Medicine*, 25(5). [311, 337]

Lawson, A. B., Browne, W. J., and Rodeiro, C. L. V. (2003). *Disease Mapping with WinBUGS and MLwiN.* Wiley, Chichester. [311, 314, 321, 322, 332, 341]

Leisch, F. (2002). Sweave: Dynamic generation of statistical reports using literate data analysis. In Härdle, W. and Rönz, B., editors, *Compstat 2002 – Proceedings in Computational Statistics.* Physica, Heidelberg, Verlag, pp 575–580. [VII]

Leisch, F. and Rossini, A. J. (2003). Reproducible statistical research. *Chance*, 16(2):46–50. [VII]

Lennon, J. J. (2000). Red-shifts and red herrings in geographical ecology. *Ecography*, 23:101–113. [11]

Leung, Y., Ma, J.-H., and Goodchild, M. F. (2004). A general framework for error analysis in measurement-based GIS Part 1: The basic measurement-error model and related concepts. *Journal of Geographical Systems*, 6:325–354. [115]

Lin, G. and Zhang, T. (2007). Loglinear residual tests of Moran's I autocorrelation and their applications to Kentucky breast cancer data. *Geographical Analysis*, 3:293–310. [298]

Lloyd, C. D. (2007). *Local Models for Spatial Analysis.* CRC, Boca Raton. [268, 306]

Loh, J. M. and Zhou, Z. (2007). Accounting for spatial correlation in the scan statistic. *The Annals of Applied Statistics*, 1:560–584. [334, 339]

Longley, P. A., Goodchild, M. F., Maguire, D. J., and Rhind, D. W. (2005). *Geographic Information Systems and Science.* Wiley, Chichester. [6]

Louis, T. A. (1984). Estimating a population of parameter values using Bayes and empirical Bayes methods. *Journal of the American Statistical Society*, 79:393–398. [321]

Marshall, R. J. (1991). Mapping disease and mortality rates using Empirical Bayes estimators. *Applied Statistics*, 40:283–294. [318, 319]

Martínez-Beneito, M. A., López-Quílez, A., and Botella-Rocamora, P. (2008). An autoregressive approach to spatio-temporal disease mapping. *Statistics in Medicine*, 27:2874-2889. [341]

Matula, D. W. and Sokal, R. R. (1980). Properties of Gabriel graphs relevant to geographic variation research and the clustering of points in the plane. *Geographic Analysis*, 12:205–222. [245]

McCulloch, C. and Searle, S. (2001). *Generalized, Linear, and Mixed Models*. Wiley, New York. [287]

McMillen, D. P. (2003). Spatial autocorrelation or model misspecification? *International Regional Science Review*, 26:208–217. [334]

Mitchell, T. (2005). *Web Mapping Illustrated: Using Open Source GIS Toolkits*. O'Reilly, Sebastopol, CA. [7, 81, 110]

Möller, J. and Waagepetersen, R. (2003). *Statistical Inference and Simulation for Spatial Point Processes*. CRC, Boca Raton. [155, 163, 164, 171, 190]

Murrell, P. (2006). *R Graphics*. CRC, Boca Raton. [38, 57]

Neteler, M. and Mitasova, H. (2004). *Open Source GIS: A GRASS GIS Approach*. Kluwer, Boston, Second Edition. [99, 345]

Neteler, M. and Mitasova, H. (2008). *Open Source GIS: A GRASS GIS Approach*. Springer, New York, Third Edition. [6, 99]

Nichols, W., Resendiz, A., J.A.Seminoff, and Resendiz, B. (2000). Transpacific migration of a loggerhead turtle monitored by satellite telemetry. *Bulletin of Marine Science*, 67:937–947. [37, 345]

Numata, M. (1961). Forest vegetation in the vicinity of Choshi. Coastal flora and vegetation at Choshi, Chiba Prefecture IV. *Bulletin of Choshi Marine Laboratory, Chiba University*, 3:28–48 [in Japanese]. [156, 157, 344]

Olson, J. M. and Brewer, C. A. (1997). An evaluation of color selections to accommodate map users with color-vision impairments. *Annals of the Association of American Geographers*, 87:103–134. [332]

Openshaw, S., Charlton, M., Wymer, C., and Craft, A. W. (1987). A Mark I geographical analysis machine for the automated analysis of point data sets. *International Journal of Geographical Information Systems*, 1:335–358. [337]

Ord, J. K. (1975). Estimation methods for models of spatial interaction. *Journal of the American Statistical Association*, 70:120–126. [285]

O'Sullivan, D. and Unwin, D. J. (2003). *Geographical Information Analysis*. Wiley, Hoboken, NJ. [13, 104, 116, 155, 160, 173, 240, 249, 253, 259, 268, 344]

Page, B., McKenzie, J., Sumner, M., Coyne, M., and Goldsworthy, S. (2006). Spatial separation of foraging habitats among New Zealand fur seals. *Marine Ecology Progress Series*, 323:263–279. [130]

Pebesma, E. J. (2004). Multivariable geostatistics in S: The gstat package. *Computers and Geosciences*, 30:683–691. [210]

Pebesma, E. J. and Bivand, R. S. (2005). Classes and methods for spatial data in R. *R News*, 5(2):9–13. [3]

Pinheiro, J. C. and Bates, D. M. (2000). *Mixed-Effects Models in S and S-Plus*. Springer, New York. [287]

Potthoff, R. F. and Whittinghill, M. (1966). Testing for homogeneity: II. The Poisson distribution. *Biometrika*, 53:183–190. [334]

Prince, M. I., Chetwynd, A., Diggle, P. J., Jarner, M., Metcalf, J. V., and James, O. F. (2001). The geographical distribution of primary biliary cirrhosis in a well-defined cohort. *Hepatology*, 34:1083–1088. [173]

R Development Core Team (2008). *R: A Language and Environment for Statistical Computing*. R Foundation for Statistical Computing, Vienna, Austria. [VII, 2]

Read, A. J., Halpin, P. N., Crowder, L. B., Hyrenbach, K. D., Best, B. D., and Freeman, S. A. (2003). *OBIS-SEAMAP: Mapping marine mammals, birds and turtles*. Duke University. World Wide Web electronic publication. `http://seamap.env.duke.edu`, Accessed on April 01, 2008. [37, 345]

Revelli, F. (2003). Reaction or interaction? Spatial process identification in multi-tiered government structures. *Journal of Urban Economics*, 53:29–53. [12]

Revelli, F. and Tovmo, P. (2007). Revealed yardstick competition: Local government efficiency patterns in Norway. *Journal of Urban Economics*, 62:121–134. [12]

Rikken, M. G. J. and Van Rijn, R. P. G. (1993). Soil pollution with heavy metals – an inquiry into spatial variation, cost of mapping and the risk evaluation of copper, cadmium, lead and zinc in the floodplains of the meuse west of stein. Technical Report, Department of Physical Geography, Utrecht University. [345]

Ripley, B. D. (1976). The second order analysis of stationary point processes. *Journal of Applied Probability*, 13:255–266. [171]

Ripley, B. D. (1977). Modelling spatial patterns (with discussion). *Journal of the Royal Statistical Society B*, 39:172–212. [156, 157, 171, 344]

Ripley, B. D. (1981). *Spatial Statistics*. Wiley, New York. [12, 118]

Ripley, B. D. (1988). *Statistical Inference for Spatial Processes*. Cambridge University Press, Cambridge. [12]

Ripley, B. D. (2001). Spatial statistics in R. *R News*, 1(2):14–15. [13]

Rowlingson, B. and Diggle, P. J. (1993). Splancs: Spatial point pattern analysis code in S-PLUS™. *Computers and Geosciences*, 19:627–655. [156]

Sarkar, D. (2008). *Lattice: Multivariate Data Visualization with R*. Springer, New York. [57, 68]

Schabenberger, O. and Gotway, C. A. (2005). *Statistical Methods for Spatial Data Analysis*. Chapman & Hall, London. [12, 114, 155, 160, 163, 164, 171, 173, 190, 240, 259, 260, 268, 274, 282, 287, 296, 300, 306, 311, 321]

Shekar, S. and Xiong, H., editors (2008). *Encyclopedia of GIS*. Springer, New York. [6]

Sibson, R. (1981). A brief description of natural neighbor interpolation. In Barnett, V., editor, *Interpreting Multivariate Data*. Wiley, Chichester, pp 21–36. [233]

Silverman, B. W. (1986). *Density Estimation for Statistics and Data Analysis*. Chapman & Hall, London. [165, 166]

Singleton, C. D., Gatrell, A. C., and Briggs, J. (1995). Prevalence of asthma and related factors in primary school children in an industrial part of England. *Journal of Epidemiology and Community Health*, 49:326–327. [158, 345]

Slocum, T. A., McMaster, R. B., Kessler, F. C., and Howard, H. H. (2005). *Thematic Cartography and Geographical Visualization*. Pearson Prentice Hall, Upper Saddle River, NJ. [57, 77]

Spiegelhalter, D., Thomas, A., Best, N., and Lunn, D. (2003). *WinBUGS Version 1.4 User's Manual*. MRC Biostatistics Unit, Cambridge. `http://www.mrc-bsu.cam.ac.uk/bugs`. [322]

Stein, M. (1999). *Interpolation of Spatial Data: Some Theory for Kriging*. Springer, New York. [197]

Stineman, R. (1980). A consistently well behaved method of interpolation. *Creative Computing*, 6:54–57. [233]

Stone, R. A. (1988). Investigating of excess environmental risks around putative sources: Statistical problems and a proposed test. *Statistics in Medicine*, 7:649–660. [340]

Strauss, D. J. (1975). A model for clustering. *Biometrika*, 62:467–475. [156, 157, 344]

Sturtz, S., Ligges, U., and Gelman, A. (2005). R2WinBUGS: A package for running WinBUGS from R. *Journal of Statistical Software*, 12(3):1–16. [322]

Tait, N., Durr, P. A., and Zheng, P. (2004). Linking R and ArcGIS: Developing a spatial statistical toolkit for epidemiologists. In *Proceedings of GISVET'04*, Guelph, Canada. [110]

Tango, T. (1995). A class of tests for detecting general and focused clustering of rare diseases. *Statistics in Medicine*, 14:2323–2334. [335]

Tango, T. and Takahashi, K. (2005). A flexibly shaped spatial scan statistic for detecting clusters. *International Journal of Health Geographics*, 4:1–15. [341]

Tiefelsdorf, M. (1998). Some practical applications of Moran's I's exact conditional distribution. *Papers in Regional Science*, 77:101–129. [264]

Tiefelsdorf, M. (2000). *Modelling Spatial Processes: The Identification and Analysis of Spatial Relationships in Regression Residuals by Means of Moran's I*. Springer, Berlin. [264]

Tiefelsdorf, M. (2002). The saddlepoint approximation of Moran's I and local Moran's I_i reference distributions and their numerical evaluation. *Geographical Analysis*, 34:187–206. [264]

Tiefelsdorf, M. and Griffith, D. A. (2007). Semiparametric filtering of spatial autocorrelation: The eigenvector approach. *Environment and Planning A*, 39:1193–1221. [302]

Tiefelsdorf, M., Griffith, D. A., and Boots, B. (1999). A variance-stabilizing coding scheme for spatial link matrices. *Environment and Planning A*, 31:165–180. [251, 253]

Toussaint, G. T. (1980). The relative neighborhood graph of a finite planar set. *Pattern Recognition*, 12:261–268. [245]

Tukey, J. W. (1977). *Exploratory Data Analysis*. Addison-Wesley, Reading, MA. [151]

Unwin, D. J. (1996). Integration through overlay analysis. In Fischer, M. M., Scholten, H. J., and Unwin, D., editors, *Spatial Analytical Perspectives on GIS*. Taylor & Francis, London, pp 129–138. [116, 117]

Venables, W. N. and Dichmont, C. M. (2004). A generalised linear model for catch allocation: An example from Australia's northern prawn fishery. *Fisheries Research*, 70:409–426. [152]

Venables, W. N. and Ripley, B. D. (2000). *S Programming*. Springer, New York. [27, 127]

Venables, W. N. and Ripley, B. D. (2002). *Modern Applied Statistics with S*. Springer, New York, Fourth Edition. [11, 152, 156, 233, 300]

Venables, W. N., Smith, D. M., and the R Development Core Team (2008). *An Introduction to R*. R Foundation for Statistical Computing, Vienna, Austria. [23, 25, 26]

Wakefield, J. C., Kelsall, J. E., and Morris, S. E. (2000). Clustering, cluster detection and spatial variation in risk. In Elliott, P., Wakefield, J., Best, N., and Briggs, D., editors, *Spatial Epidemiology: Methods and Applications*. Oxford University Press, Oxford, pp 128–152. [332, 333, 334]

Wall, M. M. (2004). A close look at the spatial structure implied by the CAR and SAR models. *Journal of Statistical Planning and Inference*, 121:311–324. [46, 274]

Waller, L. A. and Gotway, C. A. (2004). *Applied Spatial Statistics for Public Health Data*. Wiley, Hoboken, NJ. [7, 13, 57, 82, 90, 91, 114, 155, 160, 163, 164, 171, 173, 174, 237, 239, 240, 241, 243, 259, 262, 265, 266, 268, 270, 271, 272, 274, 278, 283, 296, 300, 306, 311, 313, 314, 321, 332, 335, 339, 345]

Walter, S. D. and Birnie, S. E. (1991). Mapping mortality and morbidity patterns: An international comparison. *International Journal of Epidemiology*, 20:678–689. [311]

Wang, S. Q. and Unwin, D. J. (1992). Modelling landslide distribution on loess soils in China: An investigation. *International Journal of Geographical Information Systems*, 6:391–405. [117]

Wheeler, D. and Tiefelsdorf, M. (2005). Multicollinearity and correlation among local regression coefficients in geographically weighted regression. *Journal of Geographical Systems*, 7:161–187. [307]

Wikle, C. K. (2003). Hierarchical models in environmental science. *International Statistical Review*, 71:181–200. [114]

Wise, S. (2002). *GIS Basics.* Taylor & Francis, London. [6]

Wood, S. (2006). *Generalized Additive Models: An Introduction with R.* CRC, Boca Raton. [180, 233, 297]

Worboys, M. F. and Duckham, M. (2004). *GIS: A Computing Perspective.* CRC, Boca Raton, Second Edition. [6, 115]

Yao, T. and Journel, A. G. (1998). Automatic modeling of (cross) correlogram tables using fast Fourier transform. *Mathematical Geology*, 30:589–615. [201]

Zeileis, A. (2004). Econometric computing with HC and HAC covariance matrix estimators. *Journal of Statistical Software*, 11(10):1–17. [290]

Subject Index

Functions Index

Made in the USA
Lexington, KY
18 October 2011